Designing Knowledge Organizations

Designing Knowledge Organizations

A Pathway to Innovation Leadership

Joseph Morabito, Ira Sack, and Anilkumar Bhate

WILEY

This edition first published 2018

Registered Office
John Wiley & Sons, Inc., 111 River Street, Hoboken, NJ 07030, USA

Editorial Office
111 River Street, Hoboken, NJ 07030, USA

For details of our global editorial offices, customer services, and more information about Wiley products visit us at www.wiley.com.

Wiley also publishes its books in a variety of electronic formats and by print-on-demand. Some content that appears in standard print versions of this book may not be available in other formats.

Library of Congress Cataloging-in-Publication Data

Names: Morabito, Joseph, 1946– author. | Sack, Ira, 1942– author. | Bhate, Anilkumar, 1947– author.
Title: Designing knowledge organizations : a pathway to innovation leadership / Joseph Morabito, Ira Sack, Anilkumar Bhate.
Description: Hoboken, NJ : John Wiley & Sons, 2018. | Includes bibliographical references and index.
Identifiers: LCCN 2017009402 (print) | LCCN 2017011361 (ebook) | ISBN 9781118905845 (cloth : alk. paper) | ISBN 9781119078777 (pdf) | ISBN 9781119078784 (epub)
Subjects: LCSH: Knowledge management. | Organizational effectiveness.
Classification: LCC HD30.2 .M655 2018 (print) | LCC HD30.2 (ebook) | DDC 658.4/038–dc23
LC record available at https://lccn.loc.gov/2017009402

Cover image: © John Lund/Gettyimages
Cover design: Wiley

Set in 10/12pt Warnock by SPi Global, Pondicherry, India

Printed in the United States of America.

10 9 8 7 6 5 4 3 2 1

Contents

Acknowledgments

We want to thank our students for their participation in our graduate course on knowledge and organizations. Through their questions, participation in class exercises, and experiences in their respective organizations, we have been able to test ideas, create unique problems, and shape thought-provoking diagrams and text.

We deeply appreciate several faculty colleagues for their feedback on selected chapters that correspond to their respective areas of expertise.

We also express our sincere gratitude to our research assistants. In particular, we thank Rishikesh Bhosale, Priyam Mishra, Prannoy Pal, Shamanth Umesh and Anisha Vappala.

Introduction to Knowledge Systems

In his last book, *Management Challenges for the 21st Century*, management guru Peter Drucker asserts that the most important issue for 21st century organizations is the improvement of knowledge worker productivity. Drucker (1999) goes on to claim that knowledge management (KM) is not merely another management idea or tool but the harbinger of a new economy that will fundamentally change society and organizations. We only need to look around us to see how knowledge and technology are transforming the world.

More than a 100 years ago, Frederick Winslow Taylor, an alumnus of Stevens Institute of Technology, transformed the world with Scientific Management: the scientific *analysis* of work and the *separation* of work from worker—all elements of mechanical design. In fact, Scientific Management is largely responsible for the flourishing of industrial economies. However, the 21st century requires that we solve a new problem: the *design* and *management of organizational knowledge* and *knowledge work*. What is needed, in effect, is an approach that does for knowledge work what Scientific Management did for industrial work.

I.1 Machine Versus Art Metaphor

In his lectures at Harvard University in 1932, John Dewey, widely acknowledged as the foremost American philosopher for cultural criticism and aesthetics, compared a machine with a work of art. The machine is an outcome of imagination—there are many imagined possibilities, but one is selected to produce a final result. The machine operates in the physical realm, arranged by human contrivance. In contrast, "the work of art ... is not only the outcome of imagination, but operates imaginatively rather than in the realm of physical existences. What it does is to concentrate and enlarge an immediate experience." The machine–art metaphor accurately contrasts data and information with knowledge as well as industrial and knowledge societies. As with a

Designing Knowledge Organizations: A Pathway to Innovation Leadership, First Edition.
Joseph Morabito, Ira Sack, and Anilkumar Bhate.

machine, data and information are of the *world*; as with art, knowledge is of the *mind*. Whereas the machine is a tangible thing that gets diminished with use, knowledge is imaginative and enlarged with application. Hence, in contrast to the machinery of the industrial society, the knowledge society is characterized by intangible assets and resources. Finally, as with the machine, the industrial society is *of* the imagination; as with art, the knowledge society is *in* the imagination.

Continuing with the machine–art metaphor, Dewey (1934) states: "The formed matter of esthetic experience (art) directly *expresses* ... the meanings that are imaginatively evoked; it does not, like the material brought into new relations in a machine, merely provide *means* by which purposes ... may be executed. And yet the meanings imaginatively summoned, assembled, and integrated are embodied in material existence..." Similarly, unlike an industrial or service organization, the design of the knowledge organization is tied to the human imagination and the character of knowledge itself—this makes it substantially different than designing a manufacturing or information processing organization. Designing the knowledge organization requires that we formulate the *elements* to summon the imagination and create knowledge, assemble and integrate the new knowledge with other knowledge threads, and embody the outcome in a physical world. Such elements are the "stuff" of design.

The machine–art metaphor encapsulates our research into knowledge and the organization. The knowledge economy is handicapped by its industrial history—our education, management, and legal systems carry the legacy of the industrial era into the modern era. For example, a performance review is an artifact of industrial work where performance is compared to a predetermined specification or set of goals. The performance review still exists today—complete with forced distribution quotas and rankings. How is this useful in an organization which makes its *living by thinking*?

(We borrow from the title of Davenport's (2005) book *Thinking for a Living*—a good metaphor for knowledge work, which we elevate to include the knowledge organization as well as knowledge work.)

The knowledge organization concentrates, enlarges, and embodies its imagination into physical products and not-so-physical services. Knowledging, as with art making, is an imaginative endeavor that *operates imaginatively*, concentrating and expanding itself with use. In the knowledge organization, everything we design and manage should promote imagination and thinking, concentration and enlargement, integration and application.

I.2 Design and the Ordering of Ideas

How, we may ask, do we design a knowledge organization? Is it really different than designing traditional organizations? As we will see in subsequent chapters, one anecdotal solution is to "do the opposite." If the factory system requires control, the knowledge system (KS) demands autonomy; if the factory system requires the formal implementation of strategy and knowledge (i.e., top-down, early knowledge binding), the KS demands that we first gather or create knowledge and then implement through lateral relations (i.e., bottom-up, late knowledge binding); and so on. Of course, as interesting as "doing the opposite" may be, a formal approach to design is required.

In this regard, it is instructive to continue with the art metaphor. In 1908, Henri Matisse described the actual process of painting: "If, on a clean canvas, I put at intervals patches of blue, green and red, with every touch that I put on, each of those previously laid on loses in importance ... It is necessary that the different tones I use be balanced in such a way that they do not destroy one another. To secure that, I have to put my ideas in order; the relationship between tones must be instituted in such a way that they are built up instead of being knocked down. A new *combination* of colors will succeed to the first one and will give the wholeness of my conception."

How do we select and order our ideas so that each reinforces rather than undermines the other? In our previous book, *Organization Modeling*, we introduced several artifacts for the formal selection and ordering of organizational design elements (Morabito *et al.* 1999). Our approach to organization design included the idea of "organization molecules" for the domains of organizations—process, information, culture, and other domain-specific molecules. Organization molecules give structure and order to the domains and may be considered organizational building blocks. In Chapter 4, we introduce a "knowledge molecule" for the domain of organizational knowledge work.

While we finished *Organization Modeling* with a short discussion of knowledge (K), in this book it is appropriate that we concentrate fully on K. Despite the recommendation of our students, we never developed a "knowledge molecule." This is because designing KM carries with it the weight of knowledge itself—it is inherently different from, say, designing a business process. As with Matisse painting a picture, the colors and tones must be selected and balanced to bring out a larger whole, even when neither the goals nor the colors are fully known in advance. In fact, as we will subsequently see, the colors of knowledge are everywhere in our society. Furthermore, the colors and tones are in the hands and mind of an artist—they are not mechanical pieces readied for assembly according to a blueprint. Hence, we think *of K design more as an artistic rather than as an engineering endeavor,* operating dynamically in the imagination, much as Matisse painting a picture.

I.3 Organization of the Book

Despres and Chauvel (2000) have observed that KM is "intuitively important but intellectually elusive"; hence, from a management perspective KM is "everything and nothing." This is a critically important observation because, indeed, KM *is* everywhere and thus cries out for the structure—that is, the ordering of ideas—necessary to bring coherence to the discipline. The world is surrounded by knowledge—the impact of KM is everywhere (see Figure I.1a).

Whereas most approaches to KM concentrate on the organization, we have extended the discussion to the very small and the very large (see Figure I.1b). Again, KM is everywhere and demands an integrated line of thinking.

This book is based on our graduate course delivered at Stevens Institute of Technology, "Design of the Knowledge Organization." In this course, we discuss the discipline of KM from a variety of perspectives. Accordingly, the organization of this book is as follows:

1) *A targeted literature review.* This book draws a linkage between KM and organizational design. Since the book is based on a graduate course, our aim is not to provide a complete summarization of the literature but to present a foundation for our approach to design and KM.
2) *Enriching principles.* We do not undo existing ideas on organizational design or KM but enhance and elevate each through assimilation with

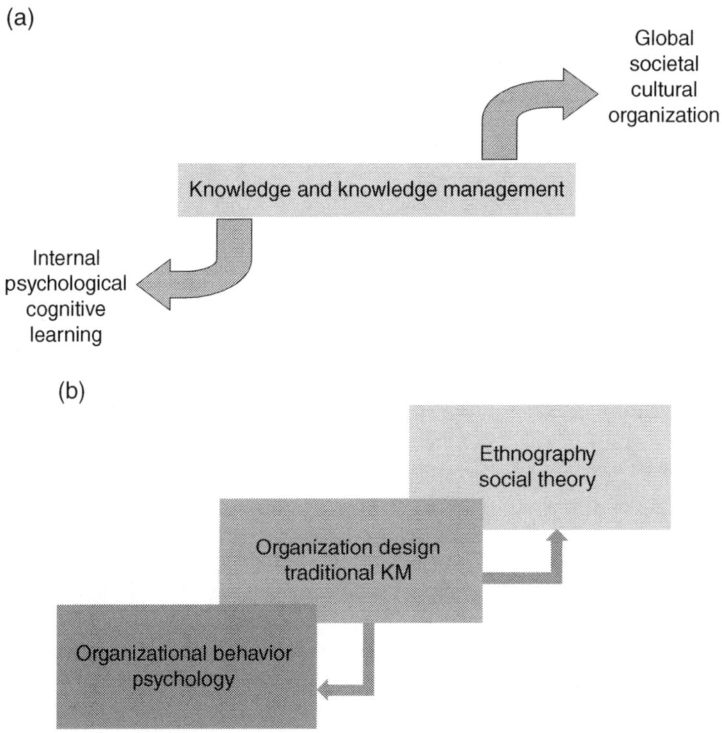

Figure I.1 (a) Impact of Knowledge Management. (b) Extending the Discipline of Knowledge Management.

concepts from other disciplines. Our discussion of philosophy, for example, is not merely a historical review, but an argument for shifting our view of K from a technical perspective to one based on moral behavior and social action, ethics and justice, and conscientiousness and altruism. From philosophy, we develop a model of a *K practice* as a mediated interaction between theoretical K and social K that is foundational to K work in and out of organizations.

3) *Knowledge systems.* KS exist everywhere in the world. Every human being is a KS, as is everything he or she creates and builds. In this book, *everything we discuss is a KS of some sort.* Naturally, we are selective, but by intertwining several disciplines, we are better positioned to accommodate new K waves or KS, such as social networking and big data. Accordingly, our approach is to elucidate the construction of knowledge competencies throughout the organization, from the small to the large—this is illustrated in Figure I.1a and b. We believe that an *integral design* approach represents the natural evolution of KM into both a more rigorous and a practical phase.
4) This book consolidates several years of research, teaching, and experience with graduate students from a variety of industries. In fact, we can argue that the *seeds* of this book lie with our students.
5) The *pedagogy* we employ is to follow a course-like structure with graphics and accompanying text. This is not the common course-book, but one organized around a classroom setting with extensive discussion, boardwork, examples, our questions and student responses, and visual representations. Rather than using standard cases, we use student exercises that represent real-world applications of KM.
6) Chapters 1–7 begin with *food for thought* questions that are suitable for student reflection and classroom discussion. Also, each chapter ends with *class exercises* that take the form of questions and projects that require a certain level of analysis. Several of both the food for thought and class exercise questions are suitable for research. We conclude our last chapter (Chapter 8) with a *final exam* that includes both new and existing food for thought questions that have been selected to allow the student to pursue his or her future knowledge work.
7) We started this book with an outline of 12 or more chapters. As we progressed in our writing, we made an important observation: we could not untangle the ideas we were discussing. Several concepts had an affinity for each other, though they were part of different topics. In effect, they were *integral in forming a larger whole.* For example, we discovered that knowledge work and technology were better understood in light of each other; social networks, office design, and culture belonged together; and so on. Moreover, these topics peek into each other, so that a discussion of one (in a given chapter) may appear with a discussion of another (in some other chapter).

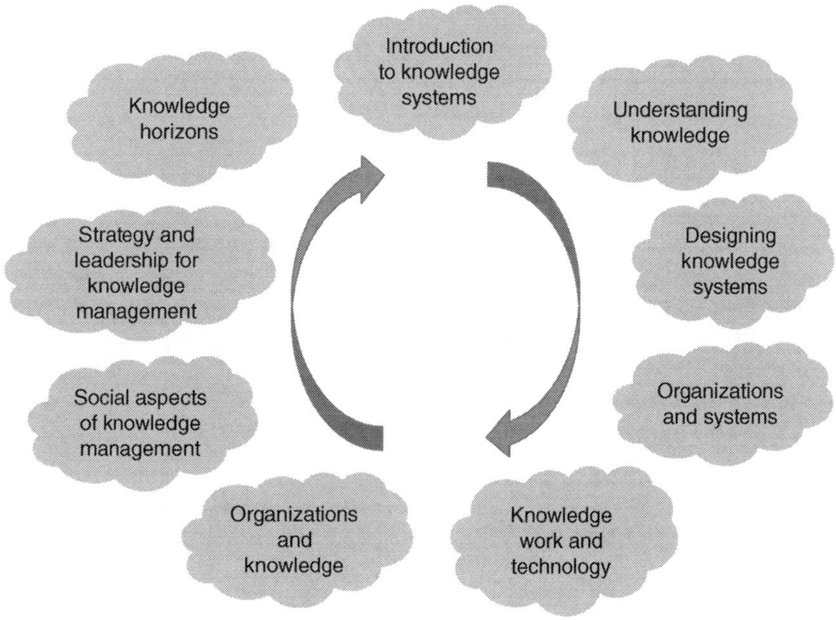

Figure I.2 Themes of Designing Knowledge Organizations

Knowledge is an integral construct and a discussion of K is better understood as such. This is in contrast to modular thinking, which characterizes mechanical design. Indeed, knowledge design *is* more like painting a picture than designing a machine—writing this book illustrated this point to the authors.

Our solution has been to combine topics into a handful of themes, each of which is somewhat dense and can be expanded into a separate book. We believe our organization will promote its use by students, faculty, researchers, and practicing managers. This is illustrated in Figure I.2.

I.4 How to Read This Book

The chapters in this book are organized around "themes." Themes have an architectural flavor and remain comparatively constant; only their content changes. In the new and dynamic discipline of KM, we believe it advantageous to establish stable themes. Each of these "territories" of KM contains a variety of ever-changing topics. Furthermore, since knowledge is like a spider web where everything is connected to everything else (in addition to being recursive), several topics require discussion in more than one context (i.e., territory). An obvious example is *knowledge creation*, a topic with a stand-alone discussion in KM literature but also one that manifests itself in every context—business process and strategy, for example. Accordingly, we recommend that you read each theme or chapter in several sittings and cross-reference other chapters when necessary:

- You may read this book as you would any other book—in sequence, starting at the beginning.
- You may reference a specific topic or territory and use the embedded backward and forward references to establish a fuller context.
- You may combine the discipline of reading a traditional book with a wiki-like reference experience.

I.5 A Journey Through KS

Culture, strategy, work, conversations—virtually every human activity or endeavor, even the layout of your office or school, is a K system. This book is a journey through such systems from the perspective of the organization. Again, we do not cover every possible KS, but we invite the reader to join our journey as we explore several of the endless possibilities. Also, we ask that you consider for yourself the KS with which you have experience—even your personal culture—and how it fits into the new world of knowledge organizations.

References

Davenport, T.H. (2005), *Thinking for a Living: How to Get Better Performance and Results from Knowledge Workers*. Harvard Business School Press.

Despres, C., and Chauvel, D. (2000), "A Thematic Analysis of the Thinking in Knowledge Management". In *Knowledge Horizons: The Present and the Promise of Knowledge Management*. Eds. Despres, C., and Chauvel, D. Butterworth Heinemann.

Dewey, J. (1934), *Art as Experience*. A Perigee Book, The Berkley Publishing Group.

Drucker, P.F. (1999), *Management Challenges for the 21st Century*. HarperBusiness.

Morabito, J., Sack, I., and Bhate, A. (1999), *Organization Modeling: Innovative Architectures for the 21st Century*. Prentice Hall.

1

Understanding Knowledge

Chapter Preview

Introduction

In this chapter, we discuss the new world of people and organizations. There is a revolution—based on new knowledge and technical artifacts—that is turning the world upside-down. Knowledge and technology are substantially changing the way people think and behave. The young are breaking with old ways and bringing new expectations to their work. They will interact with organizations differently than did their parents.

Organizations are changing too. As with technology and people, the organization finds itself an integral part of a global mind, what we refer to as the new Pangaea. In this chapter, we discuss the economic and social stresses—for people and organizations—the new Pangaea brings. We call attention to the role of culture in the development of knowledge. We ask, can we really separate knowledge from moral behavior or social responsibility? We conclude with a discussion of competence and organization-level models of intellectual capital (IC). No one really knows where all this will take us, but hopefully we are on a pathway to a better place!

Food for Thought

1) Can knowledge arise solely from the faculty of the mind?
2) Do you believe there are cultural or social tendencies in our thinking patterns that may lead to different perspectives on the meaning of knowledge?
3) Imagine a fully automated world. What would the social order look like, and what would you do about it?
4) What are some of the positive and negative impacts of globalization and technology on your organization or school?
5) How do you know what knowledge is most important in your organization or school?

Designing Knowledge Organizations: A Pathway to Innovation Leadership, First Edition.
Joseph Morabito, Ira Sack, and Anilkumar Bhate.

Topic Layout

The figure below illustrates the topic layout of Chapter 1. The left section informs the organization on the characteristics of the knowledge economy and its social implications. We discuss the cultural perspectives of knowledge through the prism of Indian, Chinese, and Greek/Western traditions. We also introduce the K-triangle as a metaphor to represent a cultural synthesis. The right side of the figure shows the artifacts associated with data, information, and knowledge (D-I-K). We discuss the information continuum as an artifact that characterizes levels of richness between data, information, and knowledge, as well as their interactions. The center of the figure represents the conceptions of knowledge in practice. We consider popular models, including Plato, Polanyi, and Nonaka, and Takeuchi. We discuss tacit and explicit knowledge. We further characterize how organizations may represent their K assets—intangibles and IC. We conclude with a student exercise on the development of an IC model for their organization.

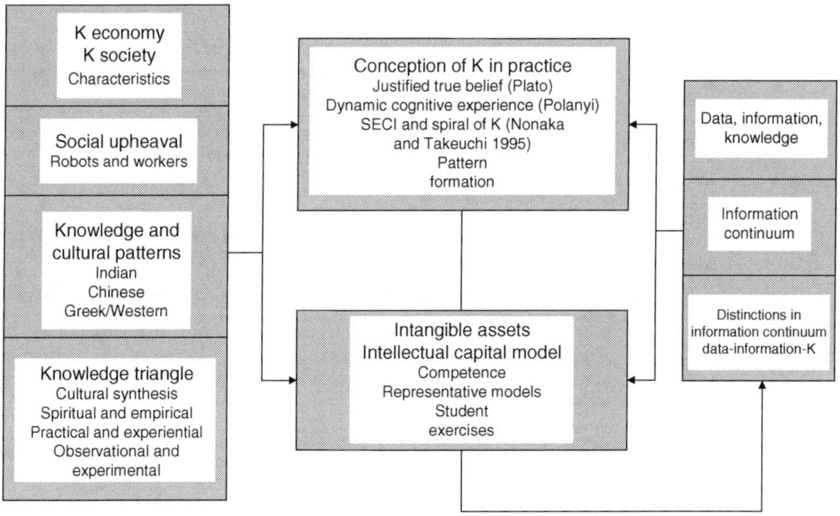

Concept roadmap to Chapter 1.

1.1 The New Pangaea

Once upon a time, say 300 million years ago, there was a single global supercontinent known as Pangaea that was home to all land life. This single land mass promoted a certain mixing and combining of life constrained, of course, by geological formations and weather conditions. Information flow was relatively easy.

In time, earth forces broke apart Pangaea, thereby creating the continents—and this movement continues today. With this separation of land, life too became isolated, and "cultures" soon developed. As an example, Charles Darwin observed the many species of finches in the Galapagos Islands—species that arose from isolation caused by the creation of islands in close proximity to one another. Information flow, initially inhibited by geographic isolation, was further inhibited by cultural isolation.

Periods of isolation characterize the history of life and humankind on earth. The many races and languages as well as beliefs and experiences demonstrate what happens when humans are separated from each other. Communication and information flow were also difficult, usually requiring travel by foot or occasionally by water. Conflict followed separation and culturation—and this too continues today. However, something different is happening now. It is the figurative formation of a new type of Pangaea. In this new Pangaea, there is once again a mixing and combining, not in a biological sense, but of human culture, information, and knowledge. Thomas Friedman (2005) calls this movement a "flat world"; others simply call it "globalization."

The ancient landmass of Pangaea serves as a useful starting point for our discussion of knowledge and knowledge management (KM). The history of knowledge is the history of human beings. First, isolation and, now, interaction are the creative forces that give rise to the fabric of human knowledge.

1.2 Characterizing the Knowledge Economy

How do we characterize KM? Despres and Chauvel (2000) have described two "divides" associated with KM. The first asserts that KM is a progression from managing by objectives to managing by (organizational) development, to total quality management, to business process reengineering, and finally to KM. In this perspective, KM is an evolutionary process creating yet another collection of organizational technologies, matrices, and techniques. The subtheme is that managing knowledge is not very different from managing data or information. In effect, the organization's management practices continue those of the industrial age with certain structural addendums, such as teams.

The second divide, advocated by Peter Drucker (1999) and supported by the authors, is that KM is a harbinger of a new age in both management and economics. The subtheme here is that KM is a global phenomenon, what we characterize as the new Pangaea, with dimensions that extend well beyond the organization into virtually every human endeavor. Its impact will be largely cultural and social, with profound effects on the design and management of organizations. And this takes us back to Drucker's assertion that the *single most important issue for 21st century organizations is the improvement of knowledge worker productivity.*

In this new way of thinking, Despres and Chauvel (2000) illustrate both the importance and dilemma that KM presents:

- KM is "intuitively important but intellectually elusive."
- "With rare exceptions, the productivity of a modern corporation or nation lies more in its intellectual and systems capabilities than in its hard assets."
- "To define knowledge in a non-abstract and non-sweeping way seems to be very difficult. Knowledge easily becomes everything and nothing."

There is a certain theme to these quotes: KM is of overwhelming importance to society yet difficult to grasp. This presents an organizational problem—if knowledge *is* everything, then in practice, it *becomes* nothing. And this is our central design challenge: if, indeed, knowledge is *everywhere,* its design must be based on new and creative concepts so as to avoid becoming *nothing.*

1.3 A Glimpse into the Knowledge Society

Lester Thurow (2009) has identified several attributes of our knowledge society. Of course, the list is endless and constantly changing, but Thurow (2009) makes two interesting observations. In the first, he asks from where will new ideas come. He answers, "corporate democracy." This approach is contrary to the way organizations typically create knowledge. Social networking, for example, extends knowledge transfer beyond the organization's boundaries into the global arena. As importantly, what of the inner workings of the organization itself? We believe that the organization must shift its corporate governance (and design) from one where, say, the organization chart exists to *implement* the knowledge of those at the top (or domain experts) to one where it exists to *gather* the knowledge of the organization's members. This gets to the very process of knowledge origination and construction, what we have termed *knowledge binding*: *early K-binding* (top-down approach) and *late K-binding* (bottom-up approach). It is late K-binding that comes closest to the corporate democracy described by Thurow (2009). We will discuss K-binding in Chapters 4 and 5.

The second observation is "winner-takes-all markets." The world's best opera singers, for example, may be seen directly on television or on the Internet—it is no longer necessary to patronize local opera houses to see lesser opera singers. But this raises an important KM question: where will the new opera singers come from? How will great opera singers become great in the first place? Indeed, the best talent may capture most of the markets, but our society, schools, and organizations still need a *farm system* in which to develop that talent.

What happens when we lose a farm system? Eliza Gray (2013) noted at the Institute for Advanced Study, "Demands for quick results are everywhere, from corporations focused on quarterly reports to universities increasingly obsessed with private enterprise partnerships that can spawn start-ups and burnish their image with students and donors." Further noting, "Pursuing questions for which the value of the answers isn't obvious may be a luxury that America can no longer afford—or at least appreciate the importance of...." "There will be less of these game-changing discoveries—at least in this country (i.e., USA)," commented Marc Kastner, Dean of MIT's School of Science.

1.4 Industrial Revolutions

Thurow (2009) goes on to describe three industrial revolutions—"industrial" being used broadly for the organization of work. The first industrial revolution, around 1800, was focused on transportation. It was the invention of the steam engine that propelled Great Britain to world power.

The second industrial revolution, around 1900, was distinguished by two breakthroughs. The first was the German development of systematic industrial research and development. German leadership in the chemical engineering and pharmaceutical industries followed this innovation. The second was electrification, which, in addition to the initial function of providing light, created opportunities for other industries to develop (much like the Internet does today).

The third industrial revolution, around 2000 (i.e., today), is characterized by six transformative technologies: biotechnology, machine tools and robotics, material sciences, microelectronics, telecommunications, and computers. These technologies, individually and in the aggregate, are driving innovation to new frontiers and transforming society.

As we can see, most of the innovations through the centuries have been technological. The one exception is the development of R&D, which may be described as an organizational innovation. We can add other prior organizational innovations, such as the division of labor, the factory system, and, of course, scientific management. *According to Peter Drucker, the next great organizational innovation will be the elucidation and refinement of knowledge work and with that the improvement in the productivity of knowledge workers.*

1.5 The Social Challenge of the Knowledge Economy

In discussing the transformation of America's economy, Robert Reich (Labor Secretary under President Clinton) asserts on *Moyers & Company* (PBS, 2013) and in his film *Inequality for All* (2013) that starting in the 1970s, two driving forces—*globalization* and *technology*—displaced middle-class workers and transformed the economy. Many blue-collar jobs were automated or sent overseas, permanently reducing wages, wealth, and dreams.

Reich went on to discuss "tipping points" which are political and social responses to perceived inequities and poverty. In 1901, for example, the rise of the progressive movement led to a progressive tax system and the breakup of trusts, among other reforms. And now we move to the 21st century: knowledge may be seen as a tipping point giving rise to a political and social revolution.

We can extend this revolution into the idea of knowledge work (K work) and the modern economy. We can confidently assert that (primarily) those with knowledge and knowledge-oriented jobs will flourish in the 21st century. A global renewal will come through the promotion of knowledge and knowledging (K&K), in all its forms, at all levels of organizations and society. The answer to the challenge of KM comes in many forms: conceptual and practical; individual and organizational; and political and social.

1.5.1 The Challenge of Robots

In discussing the robot economy, Von Drehle (2013) states, "If your job involves a set of logical rules to task after task—from grilling a hamburger to completing a tax return—you are ripe for replacement by a robot." What will millions of truck drivers do when robot trucks replace them? This shift in the nature of work extends to office as well as industrial workers, creating modern tipping points. Blue-collar jobs have been replaced by white-collar jobs; white-collar jobs now give way to gold-collar jobs. But there is a major economic issue: office work largely (but not completely) replaced industrial work, easing in part the stress in the transformation of work. However, gold-collar work cannot replace white-collar work in terms of employment numbers. K workers create knowledge intensity and, in a macroeconomic sense, are transforming our society, but far fewer such workers are needed. For example, social networking companies have capitalizations well above the very largest industrial and many service organizations but employ a fraction of the people.

Not long ago, our ancestors left the *farm* for the *factory* and then the factory for the *office*. Where will they now go? We're not sure, but the answer may be *home* or at least a *network of smaller organizations*. Our society and the entire world are in the throes of reconfiguring themselves to accommodate the requirements of K development. We are rapidly moving to a society centered on K workers as individuals or as participants in collaborative groups. And this brings us to the unanswered social question of the 21st century: *what happens to those who are not part of the K economy?*

1.6 A Macro Perspective of Knowledge Management

Illustrated in Figure 1.1 is our representation of KM. First, let us consider our use of the terms *knowledge* and *knowledging*. The former is the noun and the latter the verb, somewhat analogous to Earl's (1998) state of knowledge and

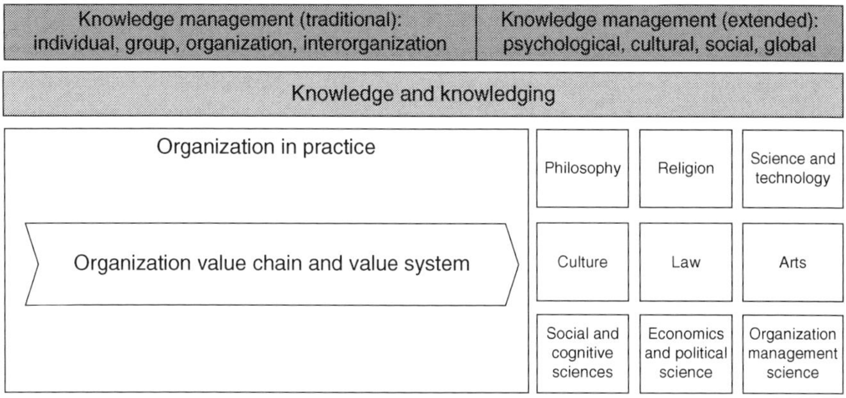

Figure 1.1 A macro perspective of knowledge management.

state of knowing, respectively. That is to say, knowledge represents a mental state, whereas knowledging is the process of knowledge creation or application. They are inextricably intertwined, and though we sometimes separate the two for analytical purposes, human experience embodies both simultaneously.

We consider K&K to be a human endeavor, and therefore they exist in all human activities and disciplines. This is illustrated on the right side of Figure 1.1 where we identify nine representative disciplines, each of which practices KM in its own distinctive way. That is to say, K&K is synonymous with human existence and is practiced (KM) according to the traditions and practices of a given discipline. On the left side of the diagram, we show the organization in practice, meaning knowledge developed in the field, outside of the laboratory, sometimes known as experiential or management knowledge of context. This is experiential or management knowledge of context referred to by James March (2006) (see Introduction). *Note too that we use the term KM to mean K&K within a given context or tradition.* This extends KM beyond a handful of techniques and management practices—and takes us into the second KM divide of a new age in organizations and societies.

In keeping with March's description of scholarship as the combination of academic and experiential knowledge, we consider K&K to be an abstraction of human discipline and practice (the layer above the level of disciplines in Figure 1.1). The term "knowledge management," as popularly used today, refers to the practice of K&K in organizations (top, left layer). Note too that we refer to this as traditional KM and include the individual, group, and organizational sublevels of organizations. Occasionally, an interorganization sublevel may be added. However, in our view, KM in the modern organization extends outside organizational life (not just outside the organization) into both the psychological and social realms, embracing all cultural traditions; in turn, this brings us full circle to the disciplines associated with human existence. This is shown in the upper right of Figure 1.1.

1.7 Architecture of the Organization

Illustrated in Figure 1.2 are two related arguments—an architectural perspective of organizations and the drivers of design. An organization may be represented with three architectural views: structural, process, and cognitive. The structural architecture refers to the layout of things—everything from the organization chart to the network architecture. This is the traditional industrial era view of organizations where organizational elements are designed from a structural perspective. For example, in this view, it is not uncommon to see the organization of work force fitted into the organization chart.

The process view is relatively new in organizational design and may be seen in the emphasis on systems development and business process design and reengineering. This perspective embraces the first divide to which Despres and Chauvel (2000) make reference; that is, traditional organizational scientists believe (unconsciously) KM to be an extension of the process or activity architecture.

The third perspective refers to the essential or cognitive architecture, what we refer to as the "thinking of the organization" (somewhat analogous to the term "thinking for a living" as used by Davenport 2005). This perspective is necessary if we are to embrace the second divide of Despres and Chauvel (2000) (and Drucker 1988, 1999) where K&K and KM are revolutionary ideas. This perspective requires that we illuminate *knowledge as such*.

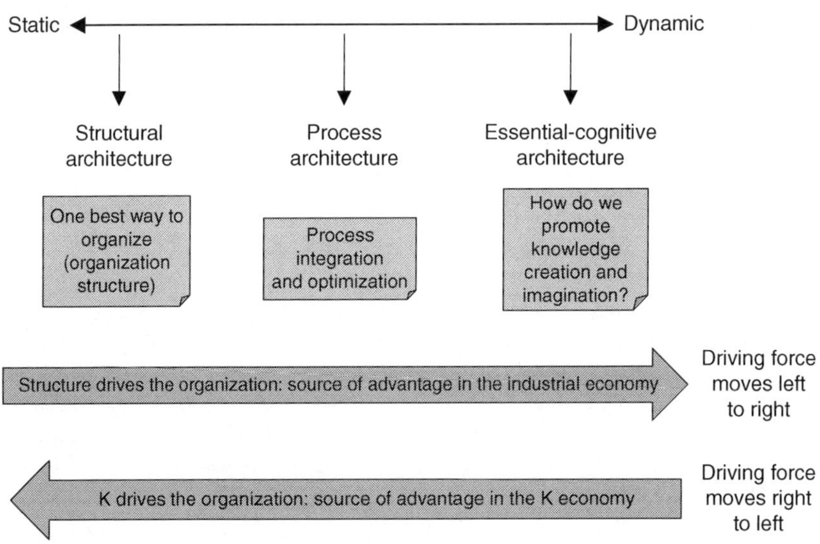

Figure 1.2 Architecture of the organization.

Finally, we show two driving forces. The left-to-right driver illustrates the architectural importance of structure and is characteristic of the industrial economy. In contrast, the right-to-left driver is characteristic of the knowledge economy. In this latter perspective, advantage comes from the development of the cognitive architecture of the organization, with the process and structural elements designed to conform to the requirements of the cognitive, not the other way around.

1.8 Data, Information, and Knowledge

We now come to our discussion of D-I-K. We do not intend to restate the endless discussions of D-I-K but to elucidate on each from *a unique architectural perspective.*

In our view, D-I-K lay on an *information continuum* that is illustrated in Figure 1.3. In its most general form, the information continuum moves from the less rich to the more rich. Increasing richness requires increasing cognitive and perhaps social activity. Each successive representation in Figure 1.3 is an instantiation of the continuum discussed.

The following is a brief description of the elements shown in the lowest representation. This is supplied to us from a young colleague who was describing the continuum to his students:

> Data can be thought of as variables, observations, numbers, as the building blocks of the information continuum. There is no richness in raw data.
>
> A database is an organized structure of a given dataset. This can be a simple text file, with one entry per line, or it could be a relational database with tables, constraints, etc. A database is used to bring order to a dataset, store it, and help us interact with it.
>
> Routine knowledge includes standards and policies known to everyone. It is procedural knowledge with very little richness.
>
> Explicit knowledge is objective knowledge that can be easily expressed and communicated. This includes well-known or obvious facts.
>
> Communal knowledge is a rich type of knowledge, shared only by the members of a population. This could be a team, a company, or an organization.
>
> Finally, Individual knowledge is the richest of the six. By using his own personal experience and expertise, a person can mine this type of knowledge from the data.

As seen in Figure 1.3, it is reasonable to observe that data is information with the least richness, whereas knowledge is information with the most richness. Though not entirely wrong, this is an engineering perspective and does not

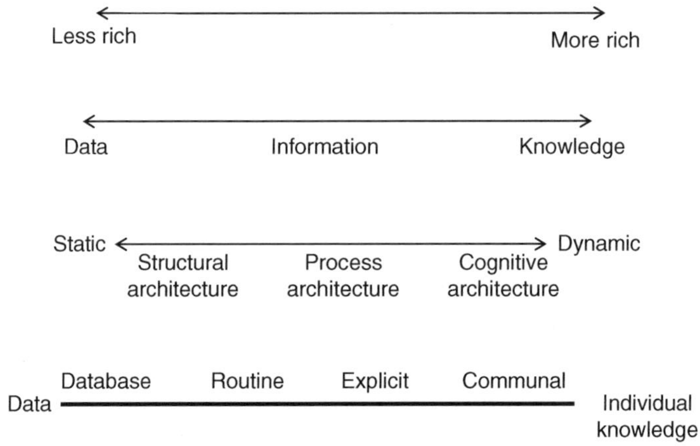

Figure 1.3 Information continuum.

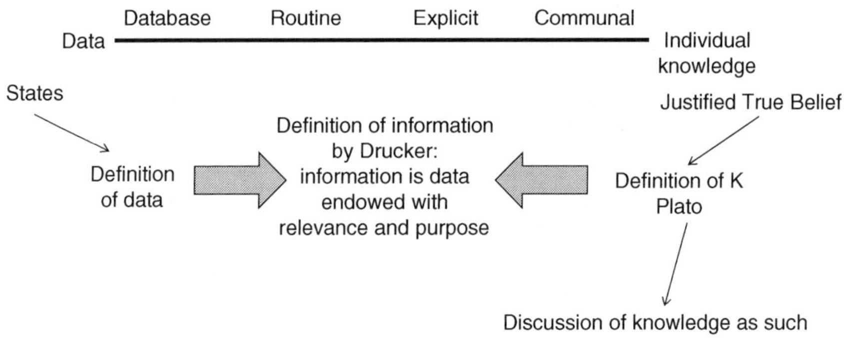

Figure 1.4 Illustration of information.

fully account for the human and dynamic nature of knowledge (but one cannot put everything into a single representation—see Figure 1.4 for another perspective, discussed later).

Data is a *state*, a reference to a business or an environmental event, such as a sales event. We record these events by collecting corresponding data in a database. Information, in turn, has many definitions or characterizations (e.g., "message"—Davenport and Prusak 1998); the one we prefer is Peter Drucker's definition that *information is data endowed with relevance and purpose.* (However, we need to elaborate a bit here: the data to which relevance and purpose are applied must be of sufficient quality in the first place; quality will be determined, in part, by context.) But where do relevance and purpose come from? From human beings and their organizations. This is shown in Figure 1.4

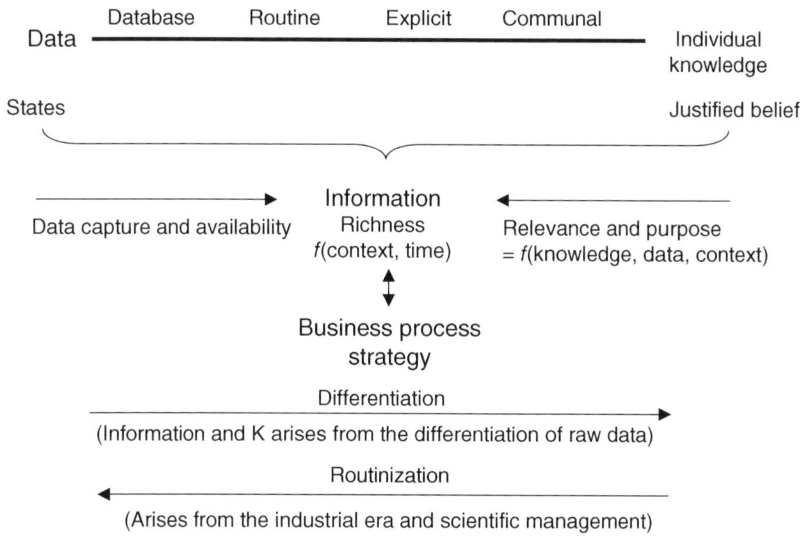

Figure 1.5 Information continuum in an organizational context.

as a collision between data and knowledge. The right side of Figure 1.4 is, of course, knowledge itself and is largely the subject of this book. Here we use Plato's definition of knowledge as *justified true belief*. It is elaborated further in "Knowledge As Such."

Finally, human beings often change their minds about what is relevant: Drucker has often stated, what we all intuitively know, that today's high knowledge is tomorrow's common knowledge or garbage. This change of mind suggests that information, like knowledge, is a function of context and time, indicated by the notation *f*(context, time) in Figure 1.5.

Similarly, relevance and purpose is shown as *f*(knowledge, data, context). Typical contexts include business process and strategy. Furthermore, observe that differentiation and routinization are opposite processes. This again is expressed by our colleague in his explanation of the continuum:

> Going back to the information continuum, we know that, in order to get from raw data to individual knowledge, we need to specialize, differentiate, look for patterns and trends, and generally interpret the data. On the other hand, the reverse process, which we refer to as routinization, is ALSO very important. Imagine how useful it would be to take this individual tacit knowledge and put it into writing. How useful it would be to distill the wisdom in the mind of a specialist, document it, and subsequently make it available to "everyone" in the company.

1.9 Distinctions in the Information Continuum

In our earlier book, we discussed three macro organizational configurations: *automating, informating,* and *knowledging.* Figure 1.6 is an adaptation from Michael Earl's (1998) distinctions in D-I-K and may, in part, be used as the basis for our organizational configurations. As we can see in the left column, data is distinguished by business events and transactional systems. The key information task is to adequately represent the data, as in, for example, a data model. There is no special human activity, other than observation of the transaction; on the other hand, the organization intent is to automate and modularize the transaction.

The middle column describes informating. This corresponds to traditional management science and includes business intelligence (BI) and analytics. The whole point is the reduction of uncertainty (defined as insufficient information) and the support of managerial decision making.

This central column distinguishes the observable work of knowledge workers in the field—this is what K workers and managers actually do. Therefore, we may ask, what does the right column represent? Knowledging! But how does this work? The K worker, living, as it were, in the middle column, interacts with the environment in a decision-making context. The *internal effect* of doing work (middle column) on the human being is what the right column in Figure 1.6 represents. This internal effect—at both the individual and organization levels—is the essence of knowledging. Note too we illustrate the boundary between information and knowledge with a heavy line representing a fundamental shift between the external and the internal, the visible and the not

	Data	Information	K
Content	Events	Trends	Expertise
Form	Transactions	Patterns	Learnings
Information task	Representation	Manipulation	Codification
Human element	Observation	Judgment	Experience
Organization intent	Automation	Decision making	Action
Value test	Building block	Uncertainty reduction	New understanding

External and visible ⟷ Internal and not visible

Figure 1.6 Distinguishing aspects of knowledging—Adapted from Earl (1998).

visible. The key idea is that the effect on the K worker is internal and concerns learning, competence building, and ultimately action.

There is an important qualification not shown in the diagram: knowledging requires *receptivity*; that is, the worker must be *receptive* to the new learning and understanding. For example, many experienced professionals—for example, physicians, professors—sometimes go "through the motions" associated with informating but do not necessarily learn anything from their respective work!

1.10 Revisiting the Information Continuum

In Figure 1.7, we illustrate additional details associated with D-I-K along the information continuum. Starting with raw data and moving left to right, data blends into information and may be shown with increasingly rich data models. The lowest set of models represents knowledge management systems (KMS) and organizational knowledge cycles and will be discussed in subsequent chapters.

Observe too that we show two barriers. By barrier, we mean a disjoint where the modes of knowledge are different on either side; consequently, flow across the barrier is impeded. The first barrier is between the visible and the not visible and is the equivalent of the heavy vertical line in Figure 1.6. This barrier is between two modes of knowledge: explicit knowledge on the left and tacit knowledge on the right. As we continue to move to the right, we encounter a second barrier between two forms of tacit knowledge: communal (or group) and individual tacit knowledge. An example of tacit–communal is the shared

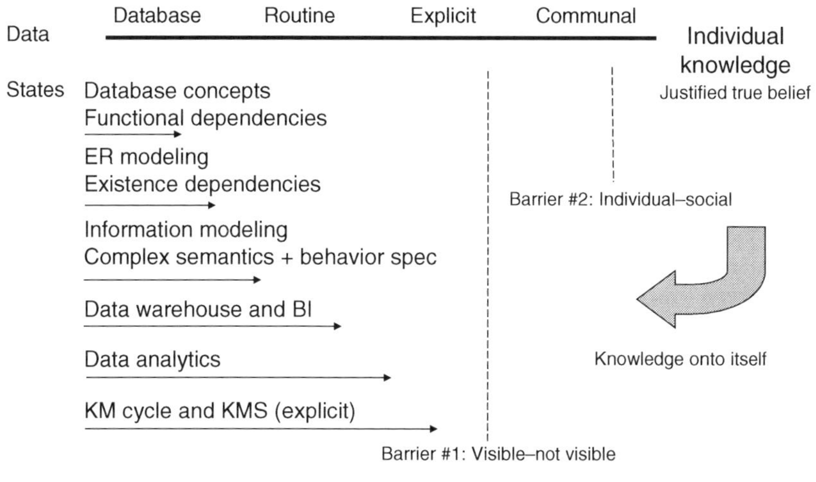

Figure 1.7 Details and barriers in D-I-K and the information continuum.

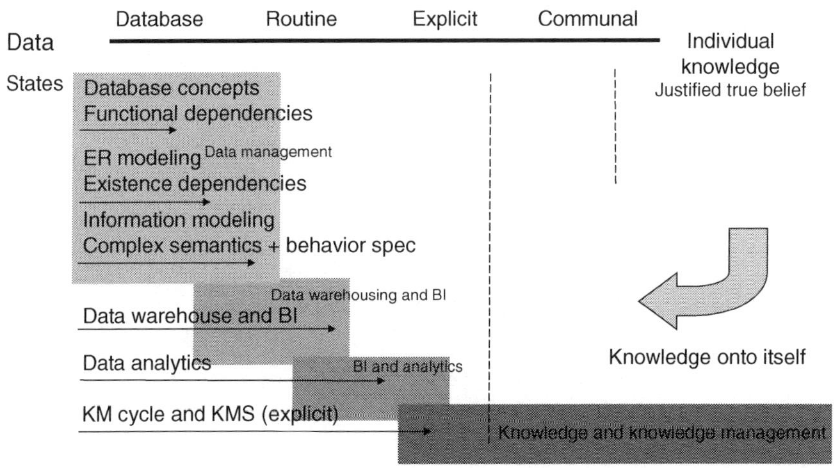

Figure 1.8 Data-oriented KM disciplines and the information continuum.

assumptions that embody culture. (We will discuss culture subsequently in the Chapters 6 and 7.)

Finally, Figure 1.8 shows the various data-oriented academic disciplines or courses associated with the information continuum: data management, data warehousing and BI, BI and data analytics, KM, and KMS. Note too that we refer to these courses as "data-oriented" only because (as we will see later in this book) KM is much more than an academic extension of data and information. It embraces many disciplines such as fine art, philosophy, sociology, and so on.

1.11 Knowledge As Such

In our discussion of knowledge, it is necessary that we include a brief historical perspective on the early conceptualizations of knowledge. Here we review selected arguments from around the world; we acknowledge that we are selective and a bit philosophical, with a strong historical thread running through our discussion. Further, it serves as a good foundation to our understanding of knowledge, and as a valuable insight into how we came to be where we are today.

We provide three broad traditions that serve as a foundation for our understanding—a *K-triangle* that serves as a metaphor for understanding and linking all of them together. In discussions of this type, names of individuals, ideas, themes, books, and movements (political, social, religious) become blurred. In Indian thought, names of books and other writings are highlighted; in Chinese thought, we orient our arguments toward themes and movements; in Greek/Western thought, we focus on the philosophical pronouncements of renowned people.

Though we make reference to several sources, we concentrate on certain noteworthy scholars in their respective fields. Dr. Sarvapalli Radhakrishnan (1929) is highly respected because of his scholarship and erudition on Vedas and other Hindu Scriptures. His two-volume book on Indian philosophy (basically Hindu philosophy) is considered a classic not only in India but all over the world. Interestingly, Radhakrishnan (1929) became the first Vice-President of India under India's first Prime Minister, Mr. Jawaharlal Nehru. Wing-tsit Chan (1967a, b, c) is a particularly important source—he is considered a leading scholar on Chinese philosophy and tradition. He has unwound the complexity of Chinese philosophy for Western readers, with an emphasis on its historical development and the essential humanism of Chinese thought. Finally, we use the work of Richard Tarnas (1991) to provide a historical backdrop to discuss Western thought and fill in the details with additional sources (e.g., Tarnas (1991), Scruton (1982), Kant).

1.11.1 Indian Thought

1.11.1.1 Vedic Origins and the Knowledge Network

A singular feature of Indian civilization has been the theory of knowledge that emanated from the Vedic tradition and wisdom. The fundamental principles belonging to the multiple threads of social, political, and economic life were intertwined into a single fabric consisting of a multidimensional, yet comprehensive theory that led to a coherent spirituality. However, although existing in a fabric, each thread still retained its own originality, leading to the formation of what may be identified as a "knowledge network." (This idea of a network or fabric will be detailed in the Polanyi model of knowledge.)

In time, the fabric structure of this Indian K-network led to the formation of a cosmopolitan culture. This culture consists of multiple aspects, each aspect having its own particular nuances that present a disjointed picture composed of multiple gods, deities, multiple castes, and multiple ways of worship. In time, a distributed social, political, economic, and religious fabric was woven out of a variety of all kinds of multiplicity. Here we highlight several of these features.

- Multiplicity of scriptures. The knowledge pertaining to the ancient Indian wisdom is not confined to a single holy book, such as the Bible. Rather, it needed a whole "library of holy books" for the proper and meaningful exposition of the knowledge content of the wisdom enclosed within the ancient Indian culture.
- Pluralism of knowledge. The original, or the fundamental and "foundational" source of this *knowledge complex*, consisting not of a singular knowledge, but a multitude of "pluralistic knowledges," has been the collection of four Vedas, with a large body of associated holy, spiritual, philosophical, and religious scriptures. The collection of not one but four Vedas clearly indicates the

collective complexity of the Vedic wisdom, starting at the source, with multiple threads emanating like rays in all directions. (The plural in the word *knowledges* is intentional, since ancient Indian philosophy looks upward on knowledge as not one but many.)
- Inheritance and polymorphism (variation) within scriptures. The variety of main books, consisting only of the prominent ones out of the large collection of ancient scriptures, is in the form of an *inverted tree* shown in Figure 1.9. The actual number of scriptures is so large that many associated scriptures have been omitted, with only the major scriptures shown in the illustration.

1.11.1.2 Inverted Tree of Indian Knowledge

This idea of the inverted tree has been an important theme within the collective complex of pluralistic knowledges contained in ancient Indian philosophy. The tree is inverted because its roots emanate from heaven (where the Supreme Being dwells) and its branches keep coming downward, toward the direction of this world, which is below, with its leaves finally reaching the earth.

There are four levels of branches shown, indicating the polymorphism that exists in the form of a variety of knowledges (hence the plural), within the collective complex of the K-network, described within the totality of scriptures.

Also indicated is the fact that each level below inherits one or more knowledges from the level directly above it. It also indicates that every level below helps to elaborate the knowledges coming down from the corresponding level above it.

- Culture of knowledge (or, knowledge is culture). Although the given "knowledge complex" is identified as Indian, in fact, the term does not have any geographical connotation. Instead of geography, knowledge is more appropriately represented by the culture within which it is embedded and is the most essential aspect of it. In other words, *knowledge itself is the culture and the culture is knowledge.*

It follows that "learning" in ancient India has always been the acquisition of knowledge for its own sake. Ancient Indian education is also to be understood as being ultimately the outcome of the Indian theory of knowledge as part of the corresponding scheme of life and values. This gives a particular angle of vision, a sense of perspective and proportion, in which the material and the moral, the physical and spiritual, the perishable and permanent interests and values of life are clearly defined and strictly differentiated. In this sense, then, learning and what we have called knowledging become inseparable.

1.11.1.3 Systems of Philosophy

Traditionally, the network of ancient Indian knowledge mentioned earlier is distributed into the following six systems of philosophy: (1) Saamkhya, an atheistic and strongly dualistic theoretical exposition of consciousness and matter; (2) Yoga, a school emphasizing meditation, contemplation, and

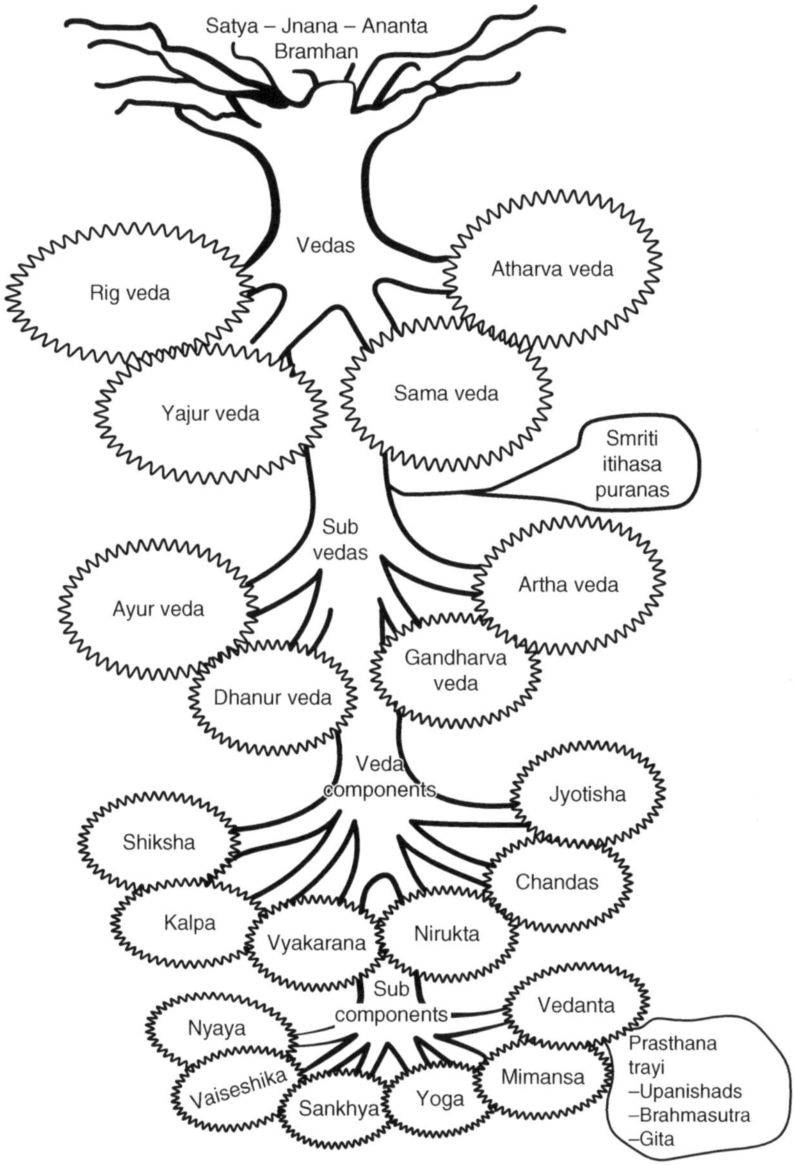

Figure 1.9 The inverted tree of ancient Indian wisdom.

liberation; (3) Nyaaya or logic, explores sources of knowledge; (4) Vaisheshika, an empiricist school of atomism; (5) Mimamsaa, an anti-ascetic and anti-mysticism school; (6) Vedanta, the ultimate repository of all knowledge pertaining to the ultimate reality.

Needless to say, philosophy in ancient India was essentially *spiritual*, with spirituality dominating every walk of life. It takes its origin in life, and enters back into life in a circular fashion, after passing through myriads of different schools of philosophy. Right from its origin, its founders have been striving for a continuous and persistent *socio-spiritual* reformation of society.

The Hindu religion has no dogma. Instead of being dogmatic, it presents a rational synthesis, which goes on gathering new conceptions into itself as the philosophy progresses. It is experimental and provisional in its nature, constantly attempting to keep pace with the progress of thought. In ancient India, therefore, religion was never a dogma, but a working hypothesis of human conduct, adapted to different stages of spiritual development as well as different conditions in life.

Without an unchanging dogma, we observe that when traditionally accepted beliefs become inadequate, then the insight of a new teacher or a new philosopher, such as Buddha or Shamkaraa, supervenes. All this is strong evidence of the intellectuality of the ancient Indian mind, which seeks to know the inner truth on one hand and the laws governing human behavior on the other. This intellectual impulse is not confined to philosophy and theology, but it extends into logic and grammar, rhetoric and language, medicine and astronomy, in fact, all arts and sciences. Everything either useful to life or interesting to the mind becomes an object of inquiry and criticism. This is depicted in the form of various branches in the theoretical model shown in Figure 1.9.

The traditional Hindu way of life in ancient India consisted of four aspects of ethical praxis: (1) Dharma, consisting of laws of life, laws of personal behavior, and universal laws of nature; (2) Artha, consisting of laws of wealth, materialistic possession, and power; (3) Kaama, laws controlling desire, ambition, and their satisfaction; (4) Moksha, spiritual salvation. Clearly, the four aspects of the ideal life indicate that ancient Indian philosophy is not "up in the air" or merely hypothetical, but is very down-to-earth and directly concerned with the practicalities of life.

Though the world has changed considerably in its outward material aspect, means of communication, scientific inventions, and so on, there has not been any great change in its inner spiritual side. *Material progress has never superseded philosophy, and hence, the ideas of great thinkers are never obsolete.* However, to set forth the development and growth of the principles of ancient Indian philosophy from the dawn of history is an undertaking of the most formidable kind.

1.11.1.4 Nature of Reality

The philosophic attempt to determine the nature of reality may start with the thinking of either self or the objects of thought. In the West, reality is always interpreted in terms of the objects of thought. Where the vision is turned outward, the rush of fleeting events engages the mind. However, in ancient India, the vision was always turned inward, and the interest of philosophy was in the "self" of personhood.

Ancient Indian psychology realized the value of concentration and meditation and looked upon it as the means of perceiving the truth. It believed that there were no ranges of life that were out-of-bound for the mind and which could not be reached by systematic and methodical training of the mind, including both will and knowledge.

Indian philosophy also recognized the close connection between mind and body. The psychic experiences, such as telepathy and clairvoyance, were considered neither abnormal nor miraculous. They were not supposed to be the products of diseased minds or inspiration of gods, but rather powers that the human mind can exhibit under carefully ascertained conditions. The Yoga system of philosophy (shown in the fourth level of branches at the bottom of Figure 1.9) can ultimately lead to a superior kind of knowledge (at the higher level of many knowledges illustrated earlier). The metaphysical schemes present therein are based on the data of psychological science.

The dominance of the subjective pursuit in ancient India, however, does not mean that the objective sciences were overlooked. Ancient Indians laid the foundation of mathematical science with the invention of the concept behind the number zero. They measured the land, divided the year, mapped out the heavens (Jyotisha, in level three of the branches in Figure 1.9), traced the course of the sun and the planets through the zodiacal belt, analyzed the constitution of matter, developed an advanced system of medicine (AyurVeda at the level of branches in Figure 1.9), and studied the nature of birds and beasts, plants and seeds (parts of AtharvaVeda shown at the first level of branches in Figure 1.9), as well as a systematic study of language and linguistics (Vyaakrana shown at the third level of branches in Figure 1.9).

The speculative mind is more synthetic, whereas the scientific one is more analytic. The former tends to create cosmic philosophies. These embrace in a singular vision the origin of all things as well as the history of ages and the decay and final dissolution of the world. The latter is inclined to linger over the particulars of the world and miss a sense of oneness and wholeness. Ancient Indian thought was of the former kind, attempting vast, impersonal views of human existence, through its idealistic and contemplative vision. It is this synthetic vision that prevailed within ancient India that has been the foundational source of the K-network mentioned at the beginning of our discussion on ancient Indian philosophy.

1.11.2 Chinese Thought

Let us move to a different tradition. Professor Wing-tsit Chan (1967a, b, c), one of the foremost scholars on Chinese philosophy and culture, has described three great currents in Chinese philosophy. The earliest tide, between the 6th and 2nd century B.C., established the intellectual tradition of Chinese thought, including Confucianism, Taoism, and Moism, with many other lesser schools, such as the Legalists and Logicians. The second movement was an intermingling of this thought with the introduction of Buddhism, which provided a foreign antithesis, and together these threads constituted medieval Chinese thought. The third wave, from the 11th century to the present, was the resurrection of Confucianism into what has become known as Neo-Confucianism. This last phase is characterized by a turning inward to solidify uniquely Chinese thought. (We will revisit this last movement later in the book when we discuss social network theory and the effect of isolation on knowledge creation.) Given here is our summary, largely taken from professors Chan (1967a, b, c), Chen (1964), Hughes (1967), and Mei (1967) (each a scholar in Chinese philosophy), of the highlights in the Chinese line of thinking. Also note that where appropriate we use Romanization for corresponding Chinese characters in order to be faithful to the original texts.

1.11.2.1 Confucianism

Confucianism is the most powerful intellectual system in China and is overwhelmingly humanistic in character, so much so that the essence of Chinese culture is humanistic and social in nature. In general, Chinese life is concerned with "this world" as opposed to the next. Confucius stated that there was one thread that ran through his doctrines—"conscientiousness and altruism." The foundation of the Confucian system, then, lies in the *moral realm* and *human experience.*

It follows that there developed the core Confucian doctrine of *harmony*, known as the *Golden Mean. The Doctrine of the Mean* became the metaphysical foundation in Confucianism. Confucianism held the Mean to be the highest ideal where anything too one-sided or extreme should be rejected. This has led Chinese culture to be regarded as a *both–and* as opposed to the *either–or* culture that distinguishes Western thought.

1.11.2.2 Taoism

In the West, Taoism is considered to be the opposite of Confucianism, but this is not so. Early Taoism was very much consonant with Confucianism and actually in opposition to Hinduism and Buddhism. Also, contrary to popular belief, Lao Tzu (the first leader and driving force of Taoism) did not renounce life and society. The main interest of both Confucianism and Taoism was life in this world, the chief difference being that in Confucianism the fulfillment of life comes with the full development of man (including his social order), whereas in Taoism the preservation of life comes from following nature.

The central idea in Taoism is simplicity, harmony, inaction, and spontaneity. Simplicity here means a life where profit and cleverness are discarded and selfishness and desire reduced. Similarly, inaction does not mean inactivity but a natural way of behavior; spontaneity, in turn, is contrasted with the artificial, cleverness, and superficial morality.

Lao Tzu strongly condemned knowledge, but "Enlightenment" in Taoism is contrasted with cleverness and deceit. True knowledge is "great knowledge," whereas great knowledge is "all-embracing and extensive." That is to say, true knowledge does not correspond with individual judgment but with a common all-inclusive principle: unity and synthesis are a virtue; in contrast, the particular, the concrete, and the specific are defects.

1.11.2.3 The Medieval Period and Buddhism

The Medieval period in Chinese philosophy was characterized by the development of Buddhism and several important schools of thought, chief among them the Yin–Yang school. Every school of Indian Buddhism was introduced into China, but only those that could be made consonant with Chinese temperament remained, such as Ch'an, which later became Zen in Japan. The Ch'an (i.e., Meditation) mentioned earlier and the Ching-tu (i.e., Pure Land) schools are essentially Chinese creations even though their origins are in India. Key elements in Ch'an are the ideas of "direct intuition" and "sudden enlightenment." Ch'an discouraged the study of texts and use of images and rituals; in essence, Ch'an resisted all forms of intellectual pursuits. In contrast, Ch'an advocated intuitiveness and quickness of mind, wit, and lightening decisions.

In his paper on epistemology in Chinese philosophy, E.R. Hughes (1967) observed the following: "The Chinese Ch'an movement started with a simple unlearned monk who revolted against all the deliberate refinements and elaboration of thought, ritual, yoga, and the like which he saw in his fellow monks. He discovered that enlightenment was the one thing he needed, and that it was the one thing he could not get for himself: the more he tried to achieve it, the more sophisticated he became, and the less possible it was for him to be enlightened. Therefore, he went on living as a man on the simplest possible basis of living. There the Taoist and to a certain extent the Confucianist spoke in him." We clearly see the simultaneous rejection of the complex philosophy of Indian thought and the embrace of Taoism and to a lesser extent Confucianism. *This is a typical Chinese tactic: the capture of foreign ideas, which are either rejected or transformed to fit (and into) Chinese thought and culture.*

The principles associated with the Yin–Yang school originated with both the Confucian classic *The Book of Changes* (*I-Ching*) and the original Taoist classic *Tao-te ching*. Yin and Yang represent two modes of understanding, which provided common ground for the intermingling of divergent philosophical schools, and largely conditioned the Chinese outlook toward reality. Yin and Yang provided a mechanism for structuring reality through correspondence of complementary, universal principles. The school flourished in the medieval period.

1.11.2.4 Neo-Confucianism

The last period of thought followed the Tang dynasty (618–907 A.D.), generally considered to be the zenith of Chinese culture and power. China was the most highly developed nation in terms of its inventiveness in commerce, science, technology, and philosophy. Then, under the Sung dynasty (960–1279 A.D.), a reaction in the form of Neo-Confucianism arose. Buddhism had been a major contributor to the culture of China in all its forms; however, though highly respected, it was still a foreign religion. This provided an opportunity for a synthesis of Buddhism and Taoism into Confucianism, known as Neo-Confucianism. Chinese scholars were never consonant with Buddhism: Chinese tradition emphasized "this-worldliness" and social responsibility, whereas Buddhism emphasized "otherworldliness" and the illusoriness of reality. The two viewpoints could not be more different.

A major contribution of Chinese culture to the world was the development of the civil service examination system, first introduced during the Han dynasty (206 B.C.–220 A.D.). The examination system imbued the Chinese nation and culture with an opportunity for success and prestige. The examination system focused on Confucianism and hence played a central role in its reinforcement under Neo-Confucianism.

Modern Neo-Confucianists—meaning the last 300 years—have called for a renewed harmony (i.e., synthesis) of principle and daily life. In particular, there has been a movement away from the speculative to the empirical, from the universal to the particular, from abstract metaphysics to the social and political interests of Confucius.

As we look to the future, let us quote Professor Chan: "There can be no doubt that Chinese philosophy will be baptized by Western science, logic, and epistemology." On the other hand, also from Chan, all those wonderful and thought-provoking ideas—the Golden Mean, the cordial relationship between man and Nature, the both–and attitude, humanism, the preservation of one's life, the full realization of one's nature, mental tranquility, spontaneous creation, the interaction of the active and passive universal principles, the harmony of the one and the many, the balance between the individual and society, and the essential goodness of human nature—all are essentially Chinese. The current Communist government, as is Chinese society wherever it is found, is in the process of making this accommodation.

There are several interesting common elements in Chinese thought through the ages. The Doctrine of the Mean, Yin–Yang, filial piety, and so on represent a synthesis of extremes; in fact, *synthesis* is a characteristic of Chinese thought and practice. Here are additional highlights of our discussion on Chinese thought.

1.11.2.5 Metaphysics

- Chinese metaphysics is simple and unsystematic, the elements of which have been debated through the centuries as primarily theoretical foundations for ethics. This is in sharp contrast to Indian philosophy.

- Western philosophy developed from metaphysics to social and moral philosophy, whereas Chinese philosophy developed the other way around.
- Whereas we may say that Confucianism is humanistic, Taoism is naturalistic, and Buddhism otherworldly, it is also true that Taoism and even Buddhism (i.e., in China) have strong humanistic strains.
- Reality in Chinese thought is dynamic and relational, and not static or absolute.
- Change may be regarded as expressions of the interaction of two forces, Yin and Yang. Change takes place in the form of supplementation and alternation, usually in the form of cycles or spirals, but never extremes.
- Neo-Confucianism has been the dominant line of thought for the last millennium. Future development will extend this synthesis and embrace scientific inquiry.

1.11.2.6 Ethics

- There runs through life and the universe one all-pervading principle, rational and ethical in nature.
- Man's duty is to follow this principle, which brings him into harmony with society and the universe.
- Every mortal has the capacity to become a sage, one who "assists the transforming and nourishing powers of Heaven and Earth."
- A balance is necessary between the individual and society. Confucius said, "The man of *jen* (i.e., humanity) wishing to establish his own character, also tries to help others to succeed." This expresses the central Chinese idea that a good man must also be a good member of society. (Aside: The Chinese character for *jen* is a compound word, consisting of two parts: one meaning "man," the other "many.")

1.11.2.7 Chinese Conception of Knowledge

- Developing one's nature requires education; education, in turn, involves both conduct and knowledge. *Knowledge and conduct are identical in Chinese thought.*
- Buddhism distinguished between the higher truth and the lower truth; Taoism distinguished the great knowledge and the small knowledge; Neo-Confucianism, borrowing and synthesizing Buddhism and Taoism, created "knowledge through information" and "knowledge through one's moral nature." Unlike Buddhism and Taoism which, respectively, considered lower truth and small knowledge to be untrustworthy, Neo-Confucianism considered "knowledge through information" to be an acceptable form of inquiry. However, "knowledge through one's moral nature" leads to the true understanding of *li* and the fulfillment of one's nature. (Aside: True understanding implies *li,* which is the universal principle and essence underlying all things; its complement *ch'i* is the particularizing principle that expresses the form and operation of *li.*)

- Knowledge requires both intellectual activity and actual practice. On the one hand, this may mean "extending the mind to things," and, on the other hand, the declaration that, for example, to play a musical instrument requires actually playing the instrument, not reading the musical score. There eventually developed a synthesis that knowledge combines the intuitive, the rational, and the empirical.
- In contrast to Buddhism, the Neo-Confucian perspective on *intuition* is that it is rational; there is no need for meditation. Today, we would describe the Chinese view of intuition as "informed judgment." (Aside: Neo-Confucianists practiced meditation, but only as an aid to mental hygiene, not as a way to knowledge.)
- The individual must obtain knowledge—there can be no extension of the mind through an artifact.
- Goodness is innate: both the knowledge and practice of the good are innate.
- Second to the question of human nature, the relationship between knowledge and conduct is the most persistent in Chinese philosophy. The two primary arguments have been "which came first" and "which is the more difficult to achieve." Generally, it is concluded that the two cannot be separated: "the full value of knowledge cannot be realized without action, and unless coupled with true knowledge, no action can be intelligent or correct."
- The Chinese terms *t'i-yen* and *t'i-jen* roughly mean "personally testing" and "personally understanding," respectively. The word *t'i* means "body," emphasizing *active personal experience.* (We will revisit this concept later in the Polanyi model of knowledge.) Active personal experience requires

1) Identification with the object of knowledge
2) Moral preparation and social action
3) Use of both the rational and intuitive methods.

This uniquely Chinese perspective (e.g., moral preparation) creates a harmony between metaphysics, epistemology, and ethics. Many contemporary scholars on Chinese philosophy believe this to be a major contribution that China has made to the world. Also, this perspective brings us very close to the notion of *wisdom.*

1.11.3 Greek and Western Thought

Our last argument centers on Greek/Western thought. In our discussion, we will rely on a handful of key thinkers, acknowledging that in our very selection we are advocating a particular line of thinking. We draw on several scholars, most particularly Professor Tarnas' (1991) classic on Western intellectual development, but we also refer to Nonaka and Scruton, among others.

Western intellectual tradition has its origins in Greek thought. In fact, much of what we call Western knowledge is Greek in origin. The birth of Greek philosophy begins with Thales of Miletus, in approximately 600 B.C. Thales and his followers made the assumption that there existed a rational unity and order in all that surrounded them, and that this was ascertainable by man. They started to observe the natural world about them and became, in essence, proto-scientists. Without necessarily giving up their beliefs, they started the break with mythology, and developed impersonal and conceptual explanations of the world. Plato, and for that matter most of Western philosophy and Western philosophers are very individualistic. We can visualize a continuum of differing approaches to knowledge: from the Greek and Western emphasis on the individual through the Indian and finally to the Chinese where social responsibility is the most fully developed.

1.11.3.1 Plato

A central doctrine in Platonism is the archetypal Ideas or Forms (in Greek, *idea* and *eidos*, respectively, are used interchangeably by Plato). The Ideas exist autonomously in a higher plane, with the objects of reality their imperfect derivatives. An important conceptualization associated with the Ideas is that they are not categorical abstractions concocted by humans from the particulars of the world; rather, they contain qualities of being and reality itself, but in a superior plane. That is, they stand apart and structure reality as we understand it—they are the hidden essences underlying things and processes. (Chinese philosophy also postulated abstract categories, but these were abstractions of reality, not supertypes responsible for structuring the world. Also, the German philosopher Immanuel Kant made reference to categories, in the sense that they are forms of thought.) In either case, such universals were early attempts to make sense of the universe, to give it order and relationships, and are foundational to Greek thought.

The Ideas extended beyond the physical universe and into human ethics. Moral behavior requires Ideas of *justice* and *goodness*, timeless constants that are able to withstand politics and other human interventions. Without the Ideas of justice and goodness, we have amoral relativism. Continuing with this line of thought, Plato extended knowledge of reality into the aesthetic. Collectively, the intellectual (True), the moral (Good), and the aesthetic (Beautiful) represent aspects of the ultimate reality. (Plato often linked light, truth, and goodness.) This is yet another perspective on the nature of *wisdom* mentioned above.

The Ideas are foundational to scientific inquiry and moral behavior. They represent the beginnings of Western rational thought and intellectualism. Plato did not trust knowledge acquired through the senses, which he saw as constantly changing and subjective. In the *Republic*, Plato distinguished between authentic knowledge of reality and the impressions or "opinions" about the appearances of things. The structure of the universe is revealed through intellectual thought, which is ascertained by direct experience of the

Ideas. That is to say, the basis of reality lies in the transcendent realm of Ideas, which provide the necessary essence, order, and structure.

There is a certain ambiguity in Plato's thinking, at least when it came to the gods. Plato's teachings were often colored with mythological construction. The artistic and religious imagination is as important in the pursuit of knowledge as is the logical and mathematical. Plato embodied the Greek tension between myth and reason.

Plato, as with Socrates before him, believed in a rational teleology as the basis of existence. The irrational was associated with the senses and the physical world, whereas the rational was associated with the creative mind, the transcendental, and the spiritual. There is the thread of intelligence running through the human mind and the universe, forever linking them.

In terms of a formal (or more formal) definition of knowledge, Plato asserted knowledge is "justified true belief" (in the *Meno, Phaedo, and Theaetetus*). *This definition has been widely accepted by many scholars in KM literature (e.g., Nonaka), but by no means is it considered definitive or complete.* Also, Plato's definition is itself a synthesis, which we discuss in more detail later.

1.11.3.2 Aristotle

Aristotle's contribution may be considered an extension to the idealism of Plato. Aristotle agreed with Plato on the transcendent nature of the human mind, but he refined the positions of Socrates and Plato with precise logic and language. For Aristotle, the essence of an object lies in its structure, but that structure is *intelligible* in its own form, embodied in itself (not as Plato suggests, embodied in corresponding Ideas). That is to say, though immanent, an object's essence is a consequence of its form (itself). Furthermore, form gives an object its developmental dynamic in addition to its distinctive structure.

Biology was Aristotle's chosen science, not mathematics. Aristotle emphasized nature's processes of growth and development where each object followed its own impulse to realize its form. It follows that Aristotle developed a formal teleology where objects moved from a potential to a realized or actual state, from the immature to the mature. The nature of a thing is to realize (or actualize) its inherent form (or its potentiality). Furthermore, no realization is possible without an *already actual being* that has already realized its form. For example, a seed is necessary for a plant to realize itself, but a mature plant must have produced the seed in the first place. We can see here the naturalistic teleology of Aristotle: of things unfolding from an already enfolded form—this is known as a *formative teleology* and will be discussed in Chapter 3. However, to fully account for causality and the complexity of reality, Aristotle proposed a supreme Form, an already existing actuality; it is this supreme Form that is the only Form or Idea entirely separate from matter.

Without entirely discarding their existence, Aristotle diminished the validity of the Ideas, which he replaced with naturalistic and empiricist elaborations.

Reality lies in perception through the senses. For Plato, the material world was a derivative of the universal; for Aristotle, the universal was a derivative of the particulars. (Aside: This is somewhat similar to our conceptualization of the *knowledge worlds* of science and art, respectively, discussed in Chapter 2.)

As with Plato, there is a certain ambiguity to Aristotle's thinking. Apparently, Aristotle believed that the mind's cognition is more than empiricism or rational elaboration of sensory experience; on the contrary, it is eternally active, divine, and immortal. Presumably, the supreme Form played a role in this line of thought.

The central tension between Plato and Aristotle lies at the heart of Greek dualism, which, in turn, underlies Western synthesis and thought. The first set of assumptions arises from a subsynthesis within the Platonic impulse, between a rational analysis of reality on the one hand and the existence of a transcendent intelligence on the other. The material world manifests both the rational and the mythic. Furthermore, a deep understanding of reality is simultaneously intellectual and spiritual in nature.

The second set of assumptions rests on the belief that certain human knowledge arises only from reason and empirical observation, through human experience that is immanent rather than transcendent. Furthermore, causality is entirely impersonal and physical; theory building requires verification in empirical particulars. Human knowledge is never settled; it is relative and must evolve in light of new evidence and analysis.

1.11.3.3 Leibniz and Hume

We introduce both Leibniz and Hume together because they represent opposite poles in the search for certain or objective knowledge.

Leibniz is a leader of the rationalist school where it is asserted that reason is the exclusive basis of certain or true knowledge. Such knowledge is independent of a particular person's point of view, uncontaminated by an individual's beliefs and experiences. The "understanding" of the world (i.e., certain knowledge) contains within it essential principles from which axioms describing the world can be derived. No experience is necessary to uncover the principles—the mind alone is sufficient.

Hume's vision is the opposite of Leibniz: Hume took empiricism to the extreme, with the assertion that all human knowledge is grounded in sense experience. Furthermore, human experience is exclusively phenomenal, of sense impressions—there is no way to look behind the impressions to uncover causation. It is true, the mind alone can speculate and perform logical and mathematical operations; however, without sensory evidence, the mind alone will not lead to true knowledge.

Hume distinguished between sensory impressions and ideas. While the former is the basis of knowledge, the latter are faint copies of impressions. That is to say, every idea has a basis in a corresponding impression. Moreover, causation is not grounded in sensory experience. If each thread of knowledge is grounded in corresponding discrete sensations, then explanations and ideas arise from the mind itself. The mind associates ideas (faint impressions) and brings forth explanations and causal relations—but none of this has a basis in sensory impression.

Originally, the key motivation for Hume was to refute rationalism and deductive logic. To Hume, there are two kinds of propositions: one based on sensation and the other on intellect. The former concerns particulars or concrete facts. The latter concerns relations between concepts. However, the truths of pure reason (e.g., mathematics) exist in a self-contained system with no mandatory reference to the external world. Hence, reason alone cannot reveal the underlying nature of things. That is, empiricism and its logical foundation, induction, now appear unjustified. Ironically, Hume ended with doubts of empiricism: if human knowledge is based on empiricism (sensory impressions), yet induction is not logically justified, then man cannot obtain true knowledge. For Hume, in contrast to Plato, all human knowledge is opinion, and forecasts based on induction are not warranted.

1.11.3.4 Kant

In 18th century Europe, the intellectual community was faced with a growing dilemma of contradictory beliefs. On the one hand, we have the claim of science to certain knowledge; on the other hand, we have the claim of philosophy that certain knowledge can never arise exclusively through experience. Similarly, on the one hand, we have the claim of science that nature is guided, even governed, by unchanging laws; on the other hand, the claim of religion that man is morally free. Immanuel Kant, arguably the most influential philosopher of the modern era, provided a complex and brilliant solution, so much so that Kant was forced to create a new vocabulary to explain his ideas.

Kant attempted to resolve the conflict between Leibniz and Hume. Kant's core conception of knowledge is that of a synthesis between *reason* (which gives knowledge its *form*) and *experience* (which give knowledge its *content*). Either without the other is not sufficient; knowledge bears the mark of both thought and practice. Is such knowledge genuine? It is objective in that it may describe reality, but it is animated through our experience. That is to say, we can never fully know a "thing in itself" that is independent of all human features. Hence, we may develop concepts independent of our point of view, but what we conceive arises from that point of view. These ideas are elaborated in Kant's classics, *Critique of Pure Reason* and *Critique of Practical Reason*.

Knowledge is drawn from two sources: *sensibility* and *understanding*. Sensibility arises from the faculty of intuitions, including our senses; understanding from our faculty of concepts, which must be applied in judgments. Kant argued against the two perspectives as separate entities, commenting that one side "intellectualized appearances," while the other "sensualized the concepts of the understanding." Both the faculties of intuitions and of concepts are necessary *together* to produce knowledge.

1.12 A Brief Comparative Perspective and the Knowledge Triangle

In 1933, Dr. Hu Shih (1934) of the National Peking University presented a series of lectures at the University of Chicago, entitled *The Chinese Renaissance,* published as a book by the same name in the following year. Among the topics was intellectual life in the three great cultures of the world. He observed that the intellectual traditions of India, China, and Greece have concerned themselves with similar endeavors; however, the differences we see today (to which we make reference earlier) follow from an early emphasis magnified over the centuries: India developed great religious traditions, the Chinese moral and political philosophies, and Greece the exploration of nature through mathematics, geometry, and mechanics. The three intellectual lines of thought form our *intellectual* or *knowledge triangle,* the beginning of our exploration into knowledge. It is figuratively the workings and the exchange within the K-triangle that is largely responsible for the rise of knowledge in the new Pangaea.

1.12.1 Knowledge Triangle

The K-triangle serves as a vehicle to organize our thinking and to better understand how knowledge manifests itself in three great traditions. We use this concept as a means to associate these traditions and to create a springboard for further development of our conception of knowledge. Here we discuss several elements of the K-triangle, briefly restating certain ideas discussed earlier, but within the context of the great traditions:

- Whereas a Supreme Being (or the deities) exists in all three traditions, the Supreme Being does not play a central role in their corresponding belief systems.
- None of the traditions advocates a central dogma.
- We believe it valuable to note that the three great traditions, though differing in their respective outlooks, link knowledge with conduct, moral behavior, and ethics, but with a characteristic nuance: Indian thought fixes on a socio-spiritual reformation of society; Chinese on a moral and political transformation; Western on a logical exploration of physical reality.

We put Indian tradition at the top of the triangle because it embraces both the spiritual and empirical; the Chinese emphasizes the practical and experiential; the Western the observational and experimental. This is illustrated in Figure 1.10 where the traditions are linked in the workings of the K-triangle.

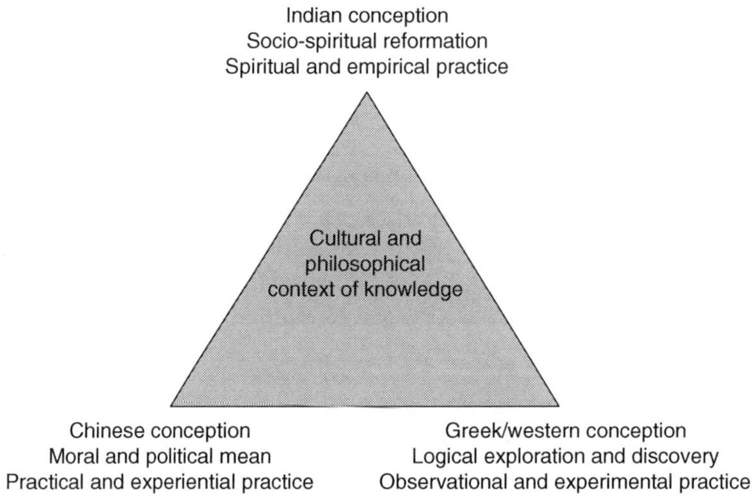

Figure 1.10 The knowledge triangle of the three great traditions.

(Aside: In his famous television series, *Cosmos*, Carl Sagan observes that the earliest and most sophisticated conceptions of the cosmos come from India, including the idea that the universe itself is being born and reborn, thereby recreating itself in cycles of creation and destruction. In this conception of an oscillating universe (making, by the way, the universe a closed system), knowledge in one universe is lost at the transition point to the next. The first observation is a matter of historical fact; the second is supported by recent advances in astronomy.)

In our discussion of the three traditions, our intention has not been merely to provide a brief exposition but to impart an understanding of how knowledge emerged as part of the intellectual tradition of the world. We can readily see that knowledge and culture are intimately connected, even inseparable: Indian, Chinese, and Greek/Western modes of thinking, conceptions of reality, morality and ethics, politics and the social sciences, art and religion, and so on—all are embodied in their respective conceptions of knowledge and vice versa.

1.13 Conceptions of Knowledge in Practice

1.13.1 Justified True Belief (Plato)

Plato is credited with defining knowledge as *Justified True Belief*. This definition is clearly in the nonreligious, philosophical tradition of Greek thought. Figure 1.11 is our illustration of Justified True Belief. Observe that we have

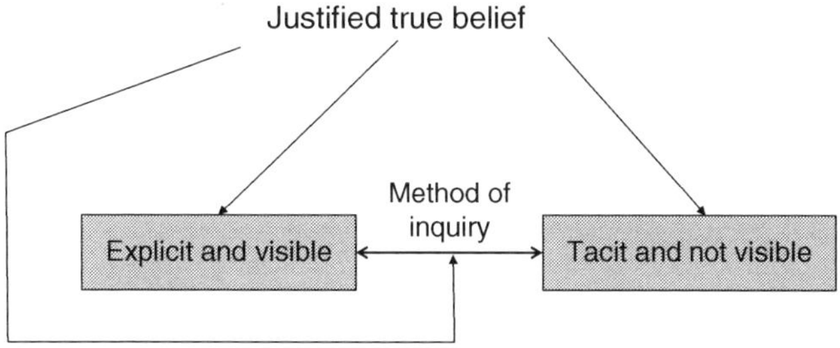

Figure 1.11 Justified true belief.

separated this phrase into three key elements: the visible or what we now call *explicit*, the not visible or *tacit*, and the *justification* that connects the tacit and the explicit. Observe too that the term *true* corresponds to the explicit world, whereas *belief* corresponds to the cognitive apparatus of the mind. Furthermore, what is true interacts with what is in the mind, and knowledge (as opposed to superstition) arises when the interaction is *justified*; hence, the single but composite phrase "justified true belief."

Let us illustrate how these elements work. If I am sitting on a coastline, say 500 hundred years ago, I will notice that a ship sailing away from me gets smaller and eventually disappears. It is reasonable to conclude that the ship fell off the edge, and thus the earth is flat. What has happened? My cognitive system, interacting with the external world (through my perceptions), has come to a rational conclusion that is justified—justified through observation. That is to say, knowledge requires my tacit belief to be justified; otherwise, it is mere superstition or ideological dogma. Justification, in turn, arises from the *method of inquiry*, such as observation and the scientific method, which again takes us full circle into an interaction between the world and my mind.

Continuing with our example, suppose that I now use a telescope to subsequently observe a ship sailing away from me. I will notice that the ship gets smaller but very slowly—and eventually disappears. Now, it is reasonable to conclude that either the world is flat but very large because I reached the limits of observation or the world is round. This new belief is different than the first but equally justified. Additionally, this new belief overturns a previous belief, illustrating the necessity of both courage and desire to pursue the truth on the observer's part.

This sequence of observations is typically the way knowledging works: knowledge constantly revisits itself, as long as the method of inquiry is impartial and the inquirer is honest in his or her pursuit of the truth. (Aside: This example also illustrates the importance of tools and technology to enhance our ability to interact with the world.) Figure 1.12 summarizes our characterization and description of knowledge.

- Constantly changing
- Is a function of time
- Is open to interpretation
- Is synthesized over time
- Is about beliefs and commitment
- Is a function of experience and its assessment
- Is a function of causes and doing—*Action*
- Is always associated with people, their feelings, and sensitivities, and value judgments
- Is about meaning and context (also for info)
- Unlike information, K may be nondeterministic and paradoxical
- Is responsible for changing its own context

Figure 1.12 Observations on knowledge.

1.13.1.1 Gettier Cases

In 1963, philosopher Edmond Gettier challenged the definition of propositional knowledge as justified true belief. (Propositional K is K of a truth or fact and therefore describes some aspect of the physical or conceptual world.) He identified several cases in which a person has a belief that is both true and justified yet that belief is not K. This is because a belief that something is true may be accidental, even if the belief is justified by evidence. For example, if I have reason to believe that it is 12 noon and verify it by looking at the classroom clock, which indicates 12 : 00, my belief is justified that it is 12 noon. However, if it turns out that the clock lost power at 12 midnight, my belief, justified by the clock reading, was accidental, and therefore is not knowledge.

Gettier (1963) cases have been the object of debate among epistemologists for decades. Most such cases describe a situation where there is a degree of luck, fallibility, or accident in the circumstances of the case. The problem created by the Gettier (1963) cases has never been adequately solved. It has been suggested, for example, that there may be a fourth (yet to be determined) element to the definition. Or, perhaps K is not fully definitive.

We would argue that, though not sufficient in the strictest epistemological sense, Plato's justified true belief is both a practical and useful definition of propositional K. Note that in Figure 1.11 we have included *method of inquiry* as a dimension of justification. By structuring justification along an established line of inquiry, we may carry out a finer grain of analysis and expect a reduction in accidents.

As we shall see, not all K is propositional K. K may arise in various ways, such as through senses, faith, intuition, emotion, or imagination. Accordingly, we conceptualize K as elastic and necessarily characterize K with a variety of definitions and models.

1.13.2 Dynamic Cognitive Experience (Polanyi)

Figure 1.13 is adapted from Jerry Gill (2000) and illustrates Michael Polanyi's perspective on human experience and knowledge creation.

We observe that the human mind may be conceived as having two poles: *explicit knowing* and *tacit knowing*. (Note the active verbs—i.e., knowing as opposed to knowledge.) Each pole is connected through two dimensions: *awareness* and *activity*. We may think of awareness as inputs to the human mind and activity as its outputs. Note too that each dimension has a two-way arrow, meaning that each dimension operates through an ongoing interaction between the two poles; that is, there exists a constant give-and-take between explicit knowing and tacit knowing. In addition, observe that each dimension has two subdimensions associated with each pole. Awareness consists of *focal* and *subsidiary* awareness, whereas activity consists of *bodily* and *conceptual* activity. (Aside: Recall from earlier that the Chinese word *t'i* means, "body"; that is, *active personal experience*—note the correspondence with *bodily activity* in the Polanyi model.)

Finally, observe that there is a third dimension, *cognitivity*, which is a resultant dimension of the workings of awareness and activity. It is represented as a one-way arrow from tacit knowing to explicit knowing, hence *the common belief that all knowledge is rooted in individual tacit knowing.*

1.13.2.1 The Knitting Metaphor

Let us illustrate the workings of this model with an example. When we first learn to drive an automobile, we are overwhelmed with the complexity of the

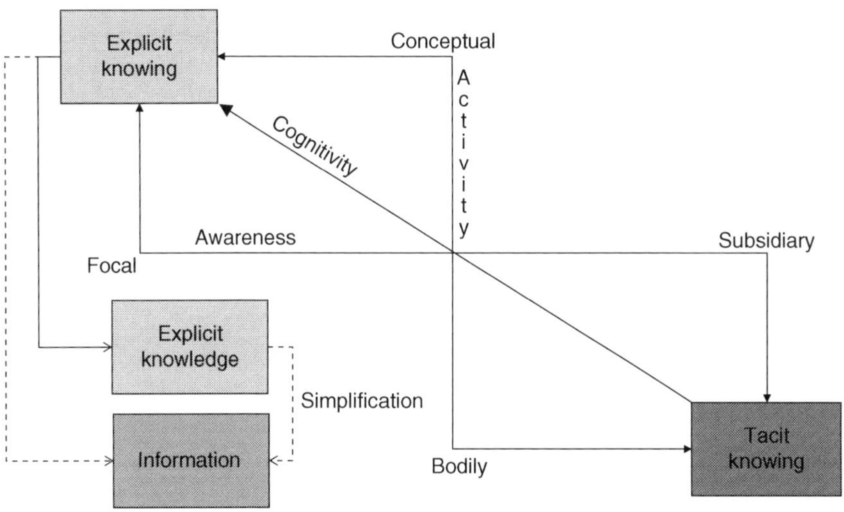

Figure 1.13 Extension of Polanyi's dynamic cognitive experience—Adapted from Gill (2000).

task. We must deal with environmental inputs (e.g., signs and traffic lights), instructions from a driving teacher, and the actual mechanical operations of the car itself. We are inundated; that is, focally aware of dozens of separate input threads. But something happens with practice (i.e., experience): we *knit* (figuratively) the separate threads into a fabric that we may call subsidiary awareness. The separate threads lose their individual identities (in the context of the fabric) and are replaced with a woven fabric that is stronger than the individual threads separately. The fabric may be conceived as a network of threads forming a distinctive pattern. The more we practice, the more we knit, the thicker and broader the fabric becomes. The ability to make sense of an array of complex inputs increases with fabric thickness. The same may be said in reverse: activity starts with a broad bodily activity, which can be taken to mean bodily activity in a physiological sense or as active personal experience. In either case, bodily activity yields separate explicit threads of explicit knowing, or concepts. Furthermore, the knowledge embodied in a single thread still exists, along with the fabric in which it is embedded, somewhat analogous to the *knowledge network* in the Indian cultural thread discussed.

Finally, observe that we have added two additional constructs: *explicit knowledge* and *information*—both of which are absent in the original model. The interaction of tacit and explicit knowing may be externalized and articulated (in a formal sense) into explicit K – a visible expression of explicit knowing. Information, in turn, is only indirectly related to explicit knowing. Both explicit K and information come into existence after explicit knowing is transformed into something concrete, such as a written formula. So, now we ask, what is the difference between explicit knowledge and information? Let us illustrate with another example. We ask our students, is Einstein's formula "$E=mc^2$" information or a thread of explicit knowledge? The answer, from our perspective, is that it depends on whether there exists a tacit fabric or not. If the explicit thread extends into tacit knowing (our fabric), the thread is explicit knowledge. If it does not then it is essentially unconnected to deep understanding—and is information. This is profoundly important: as the original model illustrates, new knowledge comes into existence only through the interaction of explicit knowing and tacit knowing; information, in and of itself, does not yield new knowledge. However, it is possible to recontextualize information; for example, through extensive study (say, following a graduate degree in atomic physics) we knit, and the explicit thread of "$E=mc^2$" shifts from information into explicit knowledge; now, we have the opportunity to create additional threads of explicit knowledge in the field of physics.

Let us summarize Polanyi's perspective: Explicit knowledge represents *conscious* reasoning that can be demonstrated through *explanation* (as in a written exam), usually in a well-structured *language*. Tacit knowledge (or should we say, tacit knowing) represents *unconscious* reasoning or a *skill* that can only be demonstrated by *activity*. It is best characterized by observing the activity in which the *knowledge is embedded*.

Observe too that tacit knowledge has two modes: (1) unconscious reasoning or a mental model, such as an understanding of atomic physics, or (2) a skill, such as riding a bicycle. Nonaka and Takeuchi (1995) have referred to the former as the *cognitive dimension* of tacit knowledge, whereas the latter is its *technical dimension.*

1.13.2.2 Knowledge as Interconnected Patterns

We may represent tacit knowing as *degrees* of fabric thickness. This is illustrated in Figure 1.14, which combines the information continuum with the work of Joseph Novak (1998). In this figure, tacit knowledge is seen as a "pattern of patterns," starting with constructs and working our way up to a worldview, each level representing increasing thickness. In this model, a concept is defined as a perceived (explicit) pattern in data or records of data. A construct, in turn, is its tacit equivalent and is at the bottom of the tacit hierarchy at the right in the figure. Each higher element represents a pattern of increasing complexity and pattern interconnectedness. As an example, a worldview, at the top of the hierarchy exists in the area of investment banking, where we have "bulls" and "bears," the former representing an optimistic worldview whereas the latter represents a pessimistic one. Each worldview, in turn, guides investment advice.

Novak's (1998) work represents a learning perspective, where the lowest level represents simple rote learning and the highest level represents meaningful learning. We move up the hierarchy with a *distinct* knitting pattern. That is, tacitness is an *individual* characteristic, which is why we have different principles, theories, philosophies, and worldviews—and people!

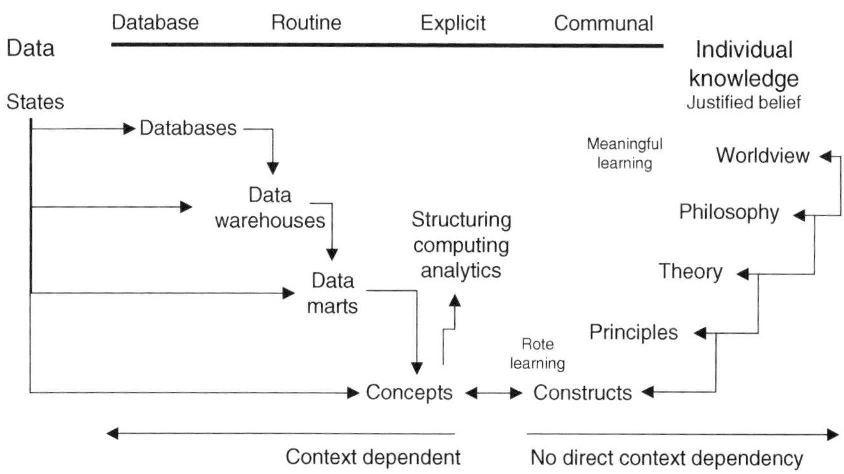

Figure 1.14 Data, information, knowledge interaction—Adapted from Novak (1998).

1.13.3 Spiral of Knowledge

Nonaka and Takeuchi (1995) have made a significant contribution to KM by, in effect, elevating the Polanyi model, which is at the individual level, to the organizational level. This celebrated model is known as the SECI model and describes four modes of knowledge creation: *socialization, externalization, combination,* and *internalization.* Each represents a knowledge conversion and a corresponding tacit-explicit interaction: tacit-to-tacit, tacit-to-explicit, explicit-to-explicit, and explicit-to-tacit, respectively. Knowledge is created as it successively "spirals" through each mode. This is illustrated in Figure 1.15 (illustrated by one of our students). *Organizational design for knowledge creation, then, becomes an exercise in facilitating the four modes of interaction.*

1.13.4 Toward a Working Definition of Knowledge

Finally, let us come to the *definition of knowledge* by Takeuchi and Nonaka (2004):

Definition of Knowledge
Knowledge is a dynamic human process of justifying personal belief toward the truth

Observe that this definition combines Plato's "justified true belief" with Polanyi's "dynamic cognitive experience." We will use this as our working definition of knowledge, with the understanding that its primary utility is technical in nature. The more complex dimensions of knowledge—moral behavior, altruism, social responsibility, justice, and goodness—will not be discarded but will be included in certain aspects of designing K organizations and systems.

1.13.5 Knowledging and Pattern Formation

Knowledging is more than knowing; rather, it is best described as an active process of pattern discovery, construction, and interrelating (Savage 1996).

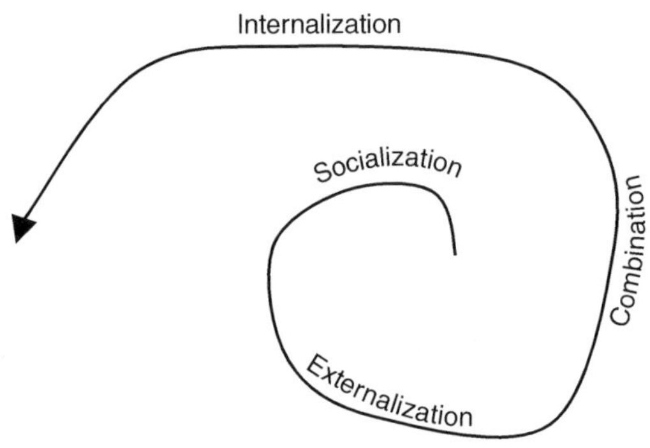

Figure 1.15 The knowledge spiral.

Recall from Figure 1.14 that Novak (1998) views learning or tacit knowledge as a “pattern of patterns,” starting with constructs and working our way up to a worldview. The active process of “knitting” (our term) is essentially interrelating existing patterns with new patterns—this leads us up the hierarchy (learning) in Figure 1.14 or down the hierarchy (unlearning).

Savage (1996) suggests that knowledging looks at both the known (information) and the visionary (what could be). It is the *imagining of possibilities*—connecting the known with an envisioned future—that is the distinguishing attribute of knowledging and what the K organization is all about. The knowledging process will lead us to explore many new concepts, including “intellectual capital,” “organizational intelligence,” and “collaborative intelligence.”

Thus, knowledging is a dynamic and ongoing process that involves our human capabilities to see existing patterns and at the same time to envision new patterns—this is the basis of wealth in the knowledge era. As we will see in Chapter 2, building new ideas is about “leaps” between designs (e.g., patterns) and not the designs themselves.

1.14 The Relationship among Different Perspectives

Figure 1.16 combines the different perspectives of knowledge considered in our discussion. We start with Plato’s philosophical perspective, progress to Polanyi’s individual model, and move to Nonaka’s organizational representation. Note that each perspective builds on its predecessor. Finally, observe that true knowledge is captured, transformed, and routinized into information,

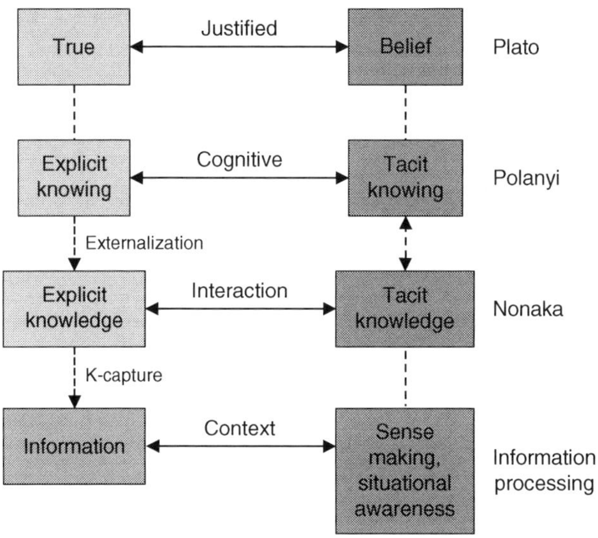

Figure 1.16 Various perspectives on tacit–explicit interaction.

which then becomes part of the information processing perspective. We may add many other knowledge dialectics, such as {reason–experience} or {understanding–sensibility} conceptions of Kant.

These ideas are reinforced by our philosophical review discussed earlier. For example, recall the Chinese terms *t'i-yen* and *t'i-jen*: each may be interpreted with both the Plato and Polanyi perspectives. "Personally testing" refers to the exercise of the tacit; "personally understanding," in turn, may be interpreted as the intellectual, the explicit. Finally, recall too the word *t'i* translates to "body," emphasizing *active personal experience*; again, this has been described by Polanyi as *dynamic cognitive experience*—the correspondence is virtually identical.

1.15 Intangible Assets and Organizational Response

Recall that Figure 1.2 illustrates three architectural viewpoints along the information continuum. We may reinterpret the continuum with, say, a learning imperative: learning has progressed from an agrarian, to an industrial, to a knowledge orientation, where each successive progression requires increasing levels of education and where educational responsibility broadens from the individual to the group and to society.

The learning progression mirrors a corresponding progression in the source of wealth. The agrarian society was distinguished by ownership of land; the industrial society by physical assets or capital (which purchases physical assets); and, finally, the knowledge society is distinguished by knowledge itself.

1.15.1 Ownership and Contribution to the Organization

This shift in assets has raised all sorts of interesting questions—for example, who owns the assets of an organization? In the agrarian society, it was simple; whoever owned the farm owned the land, the chief source of wealth. In the industrial society, ownership of a company is accorded those who provide capital, the stockholders, though they may have no idea as to what the company produces or how it operates. This is our current organizational environment, at least in terms of the legal system and management practice. In the knowledge era, human beings own the source of wealth—knowledge. That means both an organization's current employees and its past employees. How do past employees contribute knowledge? Because they already have, in the form of the systems and practices they developed, and in terms of the culture they helped form. Recall from our discussion of the machine-art metaphor that the machine—the industrial organization—is primarily tangible, meaning its assets get diminished with use. On the other hand, knowledge—the knowledge organization—is imaginative and intangible, meaning its assets actually enlarge with application over time. Hence, former employees may still be making contributions (through the systems they created and culture they formed) to the organization long after their departure. Also, not every employee contributes the same amount or type of knowledge. Furthermore, the workers of an organization now extend to employees, consultants, vendors, and so on. All of this makes the concept of "ownership" slippery in the knowledge era.

How are we responding to this dilemma? Henry Mintzberg's (1983) idea of the adhocracy is one organizational configuration where K workers come together to accomplish a task, such as making a film, and subsequently disband after task completion. Each member of the adhocracy shares in the wealth produced differently, depending on their type of contribution—a famous actor, for example, may receive a percentage of the box office receipts. Peter Drucker (1988) sees the knowledge society moving to a "new society of organizations," implying an adhocracy-like network configuration. However, such organizations may be relatively large and more persistent than a film crew. We see something like adhocracy-like functioning with the growing use of *lateral relations* (e.g., teams, task forces), but the organization's configuration is, at best, some sort of bureaucracy with an "adhocratic addendum" and does not really address the issue of knowledge and ownership.

On the other hand, relatively small, knowledge-oriented organizations are making better strides. A consulting organization is, in effect, a knowledge organization that sells its "know-how" in the marketplace. A common practice among such organizations is to hire their employees right out of school and put them in a training program or on a practice track. Employees are periodically required to meet certain milestones in order to progress; if any employee fails to meet the objective, he or she is asked to leave the firm. If the employee makes it through the last milestone, he or she is made a partner. Law firms operate on a similar "up-or-out" policy, as do universities where junior professors are placed on a tenure track, with the goal of eventually receiving tenure. What these firms have in common is that each employee—a knowledge worker—is an owner (i.e., partner) or on a track to become one. The university is similar: though being a not-for-profit organization, the tenure professors don't "own" the university—they simply run it.

1.16 Valuation, Intangibles, and Intellectual Capital

Daniel Andriessen (2003) has defined the "intangibles," as follows:

- Intangible Asset. A claim to future benefits that does not have a physical or financial embodiment (e.g., patents, copyrights and trademarks, business methodologies, intellectual property). Note that this term was codified by the International Accounting Standards Committee.

- Intangible Economy. A description of the economy that stresses the growing importance of intangible resources in creating added value.
- Intangible Resources. Nonmonetary resources without physical substance that in combination are able to produce future benefits for an organization (e.g., educated workers, brand, location).
- Intellectual Capital. A subset of intangible resources that includes all resources that are based on intellectual capabilities and activities.
- Core Competence. A bundle of intangible resources that enables a company to provide a particular benefit to customers (sourced from Hamel and Prahalad 1994).

Let us elaborate on the last definition. First, a competence must be difficult to emulate; otherwise, it would not be core. Furthermore, a healthy mix of explicit and tacit resources is necessary; a system or organization relying primarily on explicitness will eventually lose its distinctiveness. (We may add, an organization that relies on its explicit embodiments no longer "thinks" for a living.) Continuing, Dorothy Leonard (1995) has stated that a core capability must embody proprietary knowledge that is superior (or should be) to those of competitors. A *core technical capability* includes (1) physical technical systems, (2) employee skills and competence, (3) management systems, and (4) values and norms. Continuing with this line of thinking, Erik Sveiby (1997) characterizes *individual competence* as (1) explicit knowledge, (2) skills, (3) experience, (4) value judgments, and (5) social network.

The work of Leonard (1995) was successively adapted, which resulted in Andriessen (2003) identifying *five categories of intangible resources*, as follows:

- Skills and tacit knowledge, including know-how and competencies
- Technology and explicit knowledge, including patents, manuals, and procedures
- Primary and management processes, including leadership, control, communications, and management information
- Collective values and norms, including client focus, reliability, and quality
- Endowments. This category surrounds the other four and includes customers, brand, suppliers, network of talent, and ownership of standards.

In reviewing Andriessen's 25 models of intangibles, we have selected four illustrative models, each representing a specific emphasis on some form of knowledge capability.

The first is a classic from Erik Sveiby (1997) and is shown in Figure 1.17. Note that the model has two configurations: an intangible model in the form of a hierarchy showing the intangible constituents and a table with variables for each constituent. An intangible model has several organizational uses: the first

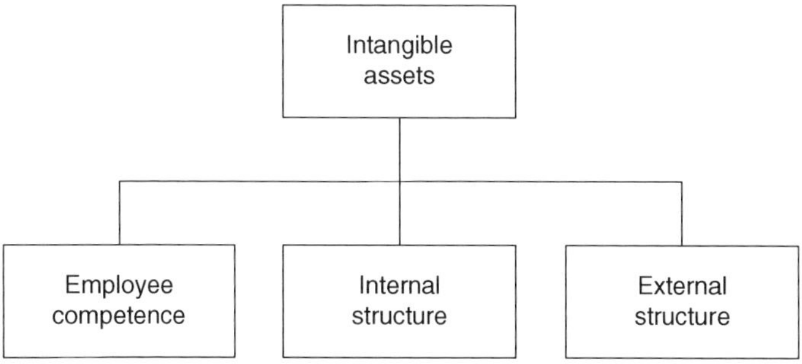

Perspective	Employee competence	Internal structure	External structure
Growth and renewal	Number of years in profession	Investment in information processing systems	Profitability per customer
Efficiency	Proportion of professionals	Sales per support person	Satisfied customer index
Stability	Average age	Rookie ratio	Proportion of large customers

Figure 1.17 Intangible asset model—Sveiby (1997).

is a guide in developing a K strategy, including conducting an inventory and an audit (we will revisit this in Chapter 7), the second is to determine value, and the third is to use the variables to determine an organization's state, performance, and to provide a guide for improvement.

Figure 1.18 is a model of "know-how" provided by the Konrad Group. Examples of know-how companies are consulting and advertising firms. Each sells its expertise or know-how in the marketplace. Also, this model is the first to introduce the concept of structural capital, which includes the attitude of employees toward the company (shaped by hiring and training, among other experiences), the firm's problem-solving ability sold to the client, and customer relationships. Observe the use of the term "structural" comes close to "organization" or "infrastructure."

Figure 1.19 illustrates two models of IC. The first is known as the IC index while the second is the holistic value approach. Observe that in the IC-index model, "relationships" are a subconstituent of structural capital, while in the holistic value approach "relationships" are elevated to a stand-alone constituent, "relational capital."

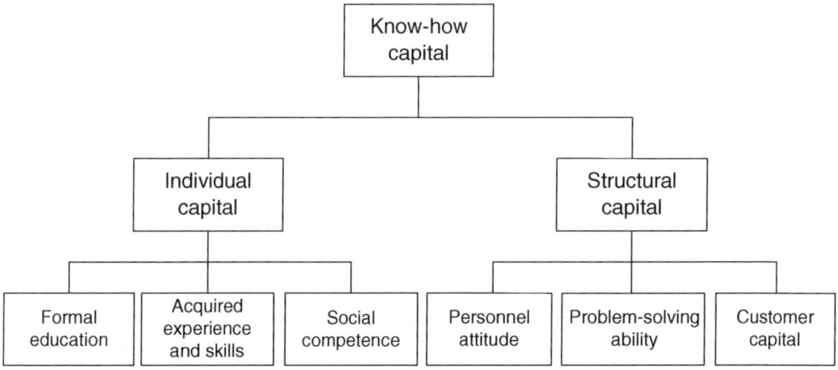

Figure 1.18 Know-how capital model—Konrad Group.

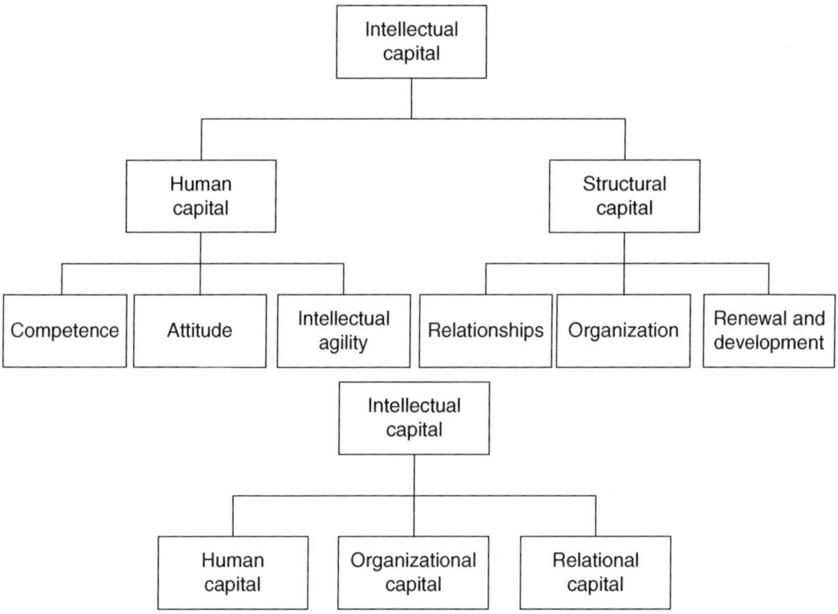

Figure 1.19 Two intellectual capital models—IC index and holistic value.

1.16.1 Student Observations

In discussing these models with our students, several observations come to light. The first is that hierarchical level indicates relative importance; for example, in the holistic value approach relational capital is of greater value than in the IC-index model. The second is that effective models should be of sufficient depth to reveal the underlying workings of the constituents. Also, if possible, constituent variables should be identified for measurement, as in the intangible asset model of Sveiby (1997). Finally, the models we have been discussing represent a particular view and interpretation: the models for actual organizations need to be handcrafted to the organization and industry at hand. This is shown here.

1.16.2 Student Exercises

Figure 1.20 illustrates two "intangible" models of the same large bank prepared by respective student teams at an on-site class (i.e., the students were employees and included managers). Each student team was asked to best represent its bank. We will not discuss every constituent; however, certain interesting observations may be made. First, the teams chose different orientations: the first selected "intangible assets," while the second selected "intellectual capital." Several constituents appear at different levels, suggesting differences in perceived value. Relationships, for example, are at the third level in the first model, while they are at the second level in the second model (this is shaded to highlight the differences). Students consider this difference to be very important: a service or K-oriented organization, such as a bank, may emphasize relational capital more so than an industrial one. Other constituents appear in only one model: "information" and "process" in the first, "product and technology" in the second, and so on. Also, each of the models in Figure 1.20 is a high-level representation: each contains several layers and corresponding variables.

For example, Figure 1.21a represents a full elaboration of the "intellectual capital" leg on the first model, whereas Figure 1.21b is its corresponding (selected) variables.

1.17 Closing Remarks

Two KM Divides

We are leaving the industrial era—the age of the machine—and moving into the knowledge era, one with more in common with a fine work of art than with a machine. However, we are carrying with us the luggage of the industrial era—the performance review, for example. Designing the K organization is more like painting a picture than designing a machine. This is why we need new and creative ideas in our study of knowledge: a suitably rich understanding of its cultural and philosophical development as well as an appreciation of its varied practice and design. The next great innovation may well not be technical, but organizational: the refinement and improvement of K worker productivity including the social and transcultural environments of knowledging.

The first KM divide is characterized by that luggage associated with the industrial era: a collection of management and technical principles and

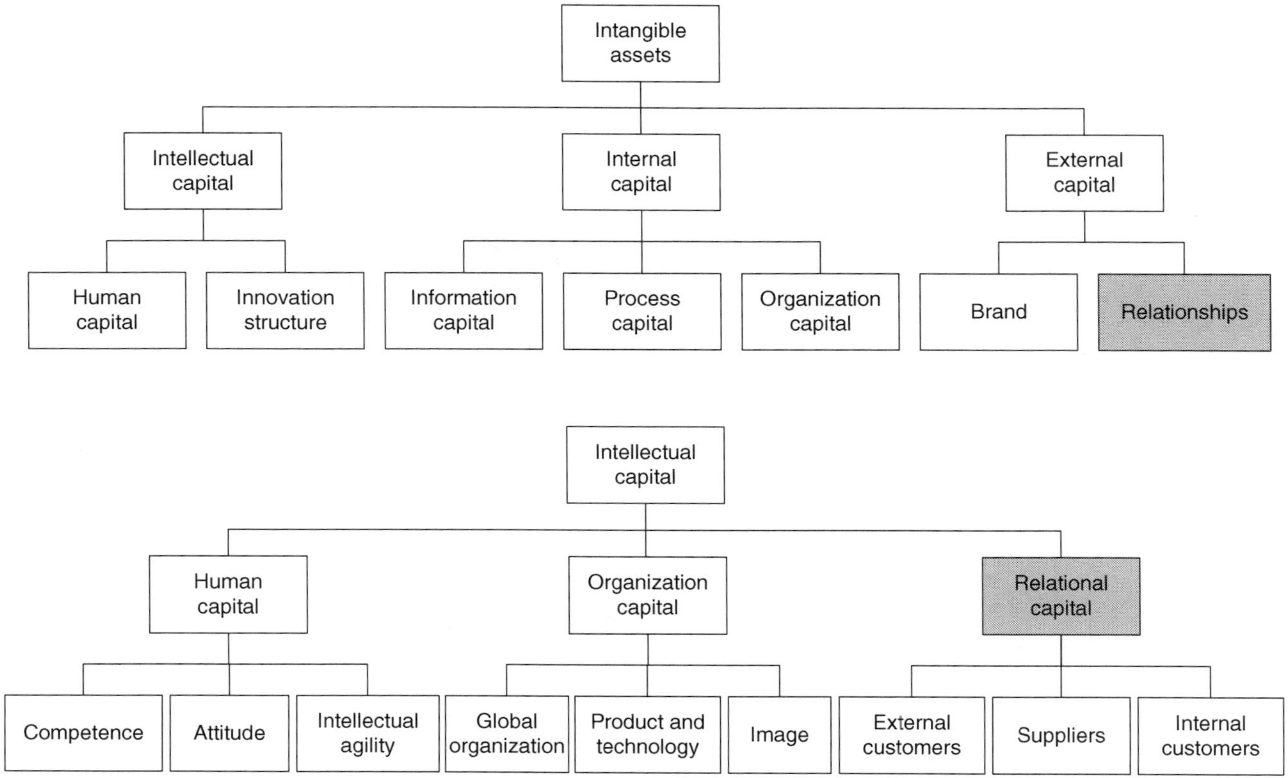

Figure 1.20 Student exercise: two views of the same financial services company.

(a)

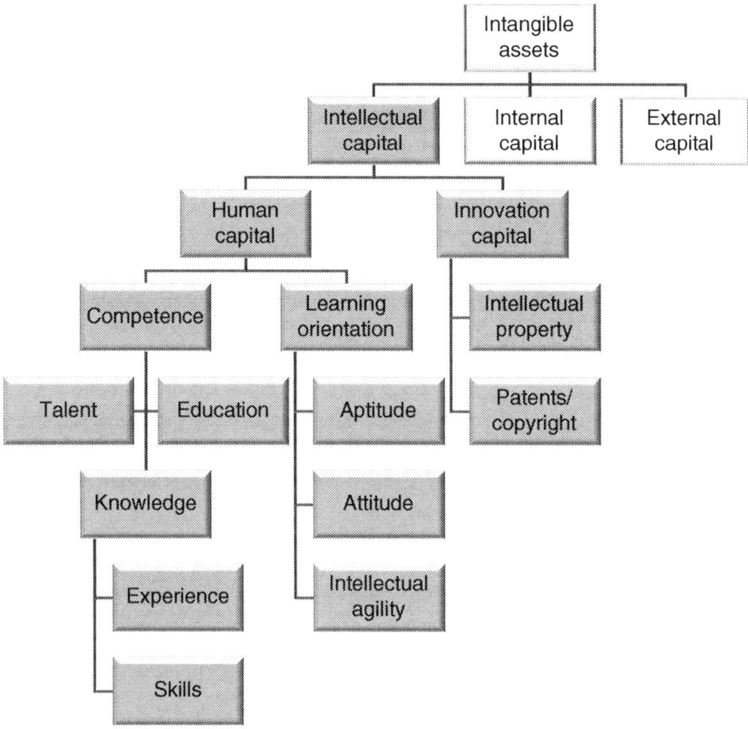

(b)

Perspective	Skills	Experience	Know-how (institutive ability)
Cost benefits	• Cost to train • Time to train (opportunity cost)	• Years in profession • Industry certifications • Industry reputation • Client reputation	• Forecasting and trending skills • "Feels" the market
Value proposition	• Optimal skill selection • Supportability (whether or not needs extensive employee backup) • Ethical decision making	• Experience/value add ratio • Distribution of experience	• Timely decision making • Relationship to external environment
Efficiency index	• Resource usage • Attrition risk	• Self-documenting processes in wiki • Recognition of faulty processes	• Ability to navigate confusing processes • Break up disruptive process loops
Learning and growth	• Limitations of growth • Adaptability to parallel processes	• Ability to teach/pass on others	• Skills adaptability • Teaching others

Figure 1.21 (a) Student elaboration of intellectual capital. (b) Student details of selected variables.

methods. It is the first divide that merely extends the traditional management practices that have developed over the decades. "Modern" or updated approaches create yet another layer in what is essentially an information processing perspective of the organization.

In contrast, the second divide demands a more creative and integrated approach. In this perspective, knowledge manifests itself everywhere, physically, socially, and psychologically; hence, knowledge requires a necessarily concept-rich approach for successful design and implementation. The drivers of this second divide similarly come from a variety of sources, such as knowledge intensity, democratization, networking or linking technology, miniaturization, and the effects of globalization.

Drivers of the K economy
Knowledge intensity Democratization Networking technology Miniaturization Globalization

Historically, industrial revolutions have been driven by both technological and organizational innovations; for example, transportation and electrification, and R&D and Scientific Management, respectively. The same may be said of the current revolution, for instance, miniaturization and networking technology along with democratization and globalization. However, the next organizational innovation will be integrative in nature, a way of harnessing and combining current advancements with knowledge workers and improvement in their productivity.

Cultural Integration

The civilizations of India, China, and Greece associate knowledge with conduct, moral behavior, ethics, and politics. Recall, for example, that the Chinese conception of knowledge includes moral preparation and social action. Further, the four Indian aspects of ethical behavior (Dharma, Artha, Kaama, Moksha), the Plato ideal of justice and goodness, and the Confucian thread of conscientiousness and altruism are additional examples of how the great traditions intertwine knowledge and wisdom. *Every one of these lines of thinking elevates life and leads in practice to knowledge, righteous behavior, and spiritual wisdom.*

We have introduced the K-triangle as a metaphor for understanding the historical development of knowledge and wisdom. Another way of grasping the interplay of the three great traditions is to imagine the points of the

K-triangle to represent the *Self* (Indian), *Social* (Chinese), and *Physical* (Greek) aspects of life and reality. Each is essential to the structure and workings within the triangle. We cannot untangle this essential character of knowledge and wisdom.

We believe the *weaving together of knowledge and wisdom* to be critically important. We de-emphasize the common practice in both KM and information systems literature of representing knowledge and wisdom as the upper two levels of a pyramid with (moving up) data, information, knowledge, and wisdom as being outdated and oversimplified. Moreover, it is not very useful outside of a technical domain. Knowledge does not work that way. Knowledge is an *integral construct* and cannot be separated from the imagination, from the flesh and blood of the human body, from the social structures of our family and society, from our history, from our laws and customs, from our art, and so on—it cannot be reduced to a picture.

Information Continuum

We have created an information continuum to illustrate the interaction and flows of D-I-K. This is an artifact with any number of themes (e.g., architecture), each with several levels of granularity. Each organization will have a unique, instantiated information continuum. Further, we separate the continuum from "Knowledge As Such" located at the knowledge pole with two barriers that isolate knowledge from the rest of the continuum. Here, within *knowledge as such*, there is an explosion of ideas and representations sourced from human history, philosophy, religion, and science and indeed from the human activities and disciplines illustrated in Figure 1.1. Our discussion of the K-triangle, art metaphors, tables, and graphics on knowledge (e.g., Polanyi) are but the beginning as we unravel the idea explosion that is knowledge, illustrated in Figure 1.22.

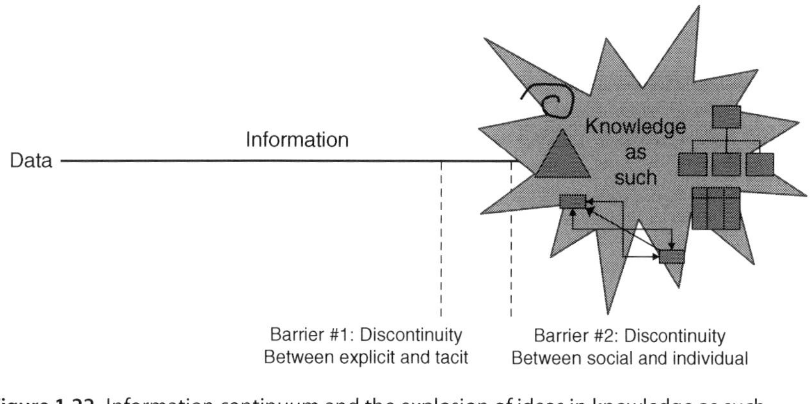

Figure 1.22 Information continuum and the explosion of ideas in knowledge as such.

(Aside: Throughout this book, we will make reference to the elements of the K-triangle and to specific writings from the traditions themselves. For example, in our subsequent discussion of knowledge work, we ask our students to create models of K work using the {sensibility–understanding} conception of knowledge from Kant; we will also discuss the student solutions and provide a commentary.)

Models of Knowledge

In addition to a moral and social perspective of knowledge—what we suggest underpins wisdom—the three traditions emphasize external and internal, visible and not visible, aspects of knowledge. These fundamental ideas have been carried into our *working models of knowledge and KM*: aspects of D-I-K (Figure 1.6), the information continuum (Figure 1.7 and other variations), the K-triangle (Figure 1.10), Plato's justified true belief (Figure 1.11), Polanyi's dynamic cognitive experience (Figure 1.13), Nonaka's spiral model (Figure 1.15), and their interrelations (Figure 1.16). Another important contribution is our *knitting* metaphor, which creates an image of how visible and nonvisible threads of ideas and experience intertwine to create the fabric of knowledge.

At the core of these models are the ideas of tacit and explicit knowledge (or knowing) within the individual and the organization as well as their corresponding externalization into explicit knowledge and information (see Figure 1.13). Also included is the idea of morality and justice (i.e., wisdom), which is part tacit and part explicit. Finally, we make note of Takeuchi and Nonaka's (2004) definition of knowledge, which combines Plato's justified true belief and Polanyi's dynamic cognitive experience: *knowledge is a dynamic human process of justifying personal belief toward the truth.* Each of these ideas and figures will serve as a springboard for further discussion throughout the book.

Intellectual Capital

We have seen the application and practicality of these conceptions to organizations: in our class exercises on intangible assets and IC, students identified both internal and external dimensions of organizations, arranged them hierarchically, and included corresponding measures. The teams had different representations of their respective organizations, even those teams that work for the same organization. Furthermore, each team defined certain intangible aspects, IC, for example, with unique components and measurement variables—a consequence of their collective image of their organization. Lastly, each team assigned different levels of importance (seen in the relative hierarchical position) to the same intangible asset, relationship capital, for instance. All of this suggests that configuring an organization's IC model is a function of perception, context, and strategy.

We are taking a successive or layered approach in our discussion. In this chapter, we have discussed "knowledge as such," the K-triangle, and introduced several models of knowledge. Our students then distilled several K-triangle attributes into real-world IC models of organizations. Subsequently, we will continue our development of models, including a knowledge molecule and an epistemology model, among others—all necessary for successful knowledge design.

The School of Athens

In 1505, Renaissance artist Raphael painted the fresco *The School of Athens*. In this painting, Plato and Aristotle are surrounded by Greek philosophers, scientists, and artists. Each figure represents a thread of knowledge necessary for ascertaining the Truth. Plato is pointing a hand upward to the transcendental, whereas Aristotle is pointing downward to the material. We are reminded of similar paintings and sculptures of Buddha, whose right hand is pointed to the sky and whose left hand touches the earth. Each portrays a balance between the spiritual and the concrete and is essential to discovering certain knowledge. Continuing with our metaphor, the K-triangle is an updated image of this, which seeks to balance the {self–social–physical} aspects of life and the {spiritual–experiential–experimental} behaviors into a unified whole.

1.18 Class Exercises

1 Create a table comparing a machine to a painting. Use as a starting point our assertion that a machine is *of the imagination*, but a work of art is *in the imagination*.

2 Create an intellectual capital model of your organization, university, or school. Include both a hierarchical graphic and a table identifying variables for each component in your model.

3 If every worker in the 21st century is not going to be a K worker, what social and political solution do you recommend? Consider the role that education and its availability play in your answer.

4 Using the material in this chapter, create your own definition of knowledge.

5 Select a tradition (e.g., a specific religion or culture not discussed earlier) and discuss its conception of knowledge.

6 Distinguish between knowledge and wisdom, noting the different perspectives of the great traditions.

7 Using the K-triangle, create a table characterizing your company or school in terms of the three great traditions.

8 Defend or challenge the assertion that robots will replace most human work, even K work. What is your prescription for the future of work?

References

Andriessen, D. (2003), *Making Sense of Intellectual Capital*. Routledge.

Chan, W.T. (1967a), "Chinese Theory and Practice." In *The Chinese Mind*. Ed. Hughes, C.A., The University Press of Hawaii.

Chan, W.T. (1967b), "The Story of Chinese Philosophy." In *The Chinese Mind*. Ed. Hughes, C.A., The University Press of Hawaii.

Chan, W.T. (1967c), "Synthesis in Chinese Metaphysics." In *The Chinese Mind*. Ed. Hughes, C.A., The University Press of Hawaii.

Chen, K.K.S. (1964), *Buddhism in China: A Historical Survey*. Princeton University Press.

Davenport, T.H. (2005), *Thinking for a Living*. Harvard Business School Press.

Davenport, T.H., and Prusak, L. (1998), *Working Knowledge: How Organizations Manage What They Know*. Harvard Business School Press.

Despres, C., and Chauvel, D. (2000), "A Thematic Analysis of the Thinking in Knowledge Management." In *Knowledge Horizons: The Present and the Promise of Knowledge Management*. Eds. Despres, C., and Chauvel, D., Butterworth Heinemann.

Drucker, P.F. (1988), "The Coming of the New Organization." *Harvard Business Review*. January–February 1988.

Drucker, P.F. (1999), *Management Challenges for the 21st Century*. HarperBusiness.

Earl, M., and Scott, I. (Eds.). (1998), *What on Earth Is a CKO?* IBM Survey. London Business School.

Friedman, T. (2005), *The World Is Flat: A Brief History of the Twenty-First Century*. Farrar, Straus and Giroux.

Gettier, E.L. (1963), "Is Justified True Belief Knowledge." *Analysis*, 23: 121–123.

Gill, J.H. (2000), *The Tacit Mode: Michael Polanyi's Postmodern Philosophy*. State University of New York Press.

Gray, E. (2013), "The Original Genius Bar." Institute for Advanced Study. *Time*. July 22, 2013.

Hamel G., and Prahalad, C.K. (1994), *Competing for the Future*. Harvard Business School Press.

Hughes, E.R. (1967), "Epistemological Methods in Chinese Philosophy." In *The Chinese Mind*. Eds. Moore, C.A., The University Press of Hawaii.

Leonard, D. (1995), *Wellsprings of Knowledge: Building and Sustaining the Sources of Innovation.* Harvard Business School Press.

March, J.G. (Interview by Coutu, D.). (2006), "Ideas as Art." *Harvard Business Review.*

Mei, Y.P. (1967), "The Basis of Social, Ethical, and Spiritual Values in Chinese Philosophy." In *The Chinese Mind.* Ed. Hughes, C.A., The University Press of Hawaii.

Mintzberg, H. (1983), *Structure in Fives: Designing Effective Organizations.* Prentice Hall.

Nonaka, I., and Takeuchi, H. (1995), *The Knowledge-Creating Company.* Oxford University Press.

Novak, J.D. (1998), *Learning, Creating, and Using Knowledge.* Lawrence Erlbaum Associates, Publishers.

Radhakrishnan, S. (1929), *Indian Philosophy, Volumes I and II.* The Macmillan Company.

Savage, C.M. (1996), *Fifth Generation Management.* Second Edition. Butterworth-Heine.

Scruton, R. (1982), *Kant: A Very Short Introduction.* Oxford University Press.

Shih, H. (1934), *The Chinese Renaissance: The Haskell Lectures, 1933.* The University of Chicago Press.

Sveiby, K.E. (1997), *The New Organizational Wealth.* Berret-Koehler Publishers, Inc.

Takeuchi, H., and Nonaka, I. (Eds.). (2004), *Hitotsubashi on Knowledge Management.* John Wiley & Sons (Asia) Pte Ltd.

Tarnas, R. (1991), *The Passion of the Western Mind.* Ballantine Books.

Thurow, L. (2009), *Building Wealth.* HarperCollins.

Von Drehle, D. (2013), "The Robot Economy." *Time.* September 9, 2013.

2

Designing Knowledge Systems

Chapter Preview

Introduction

In this chapter, we discuss the transition to the modern worldview. Designing K organizations is mitigated by our concept of knowledge creation. Nonaka's work is pioneering in the field of knowledge management (KM). We explore his ideas, which we consider to be the basics of organizational K creation. Central to the Nonaka and Takeuchi (1995) approach is the K spiral and the SECI (socialization–externalization–combination–internalization) model of organization K creation. In the second half of this chapter, we introduce new complementary concepts to support the next phase of KM and the organization. For example, the concepts of *thin* and *thick*, borrowed and elaborated from the field of ethnography, are central to our design approach. Each set of notions—Nonaka's and our additions—are conceptual in nature but underlie the techniques and systems of KM. On the other hand, we include several practical examples throughout the chapter.

Food for Thought

1) How do you know what you know? How do you feel about what you know?
2) What is the difference between knowledge of facts and knowledge of one's self?
3) Is a hunch or intuition a type of knowledge?
4) What kind of knowledge is best learned on the job as opposed to knowledge that is learned in a traditional class?
5) In your opinion is knowledge a relative or absolute concept?
6) How do you know when you "know enough" to take on a job with confidence?

Designing Knowledge Organizations: A Pathway to Innovation Leadership, First Edition.
Joseph Morabito, Ira Sack, and Anilkumar Bhate.

Topic Layout

The figure below illustrates the topic layout of Chapter 2. The left section briefly describes the transition to the modern worldview that is essential to KM. We inform the organization on the idea of a K world that may be used to distinguish science and art and the nature of creativity. This leads to inquiry systems and the search for knowledge. The center of the illustration shows the central concepts in design for the organization based on Nonaka and Konno (1998). The right side shows our extensions to design in the K era.

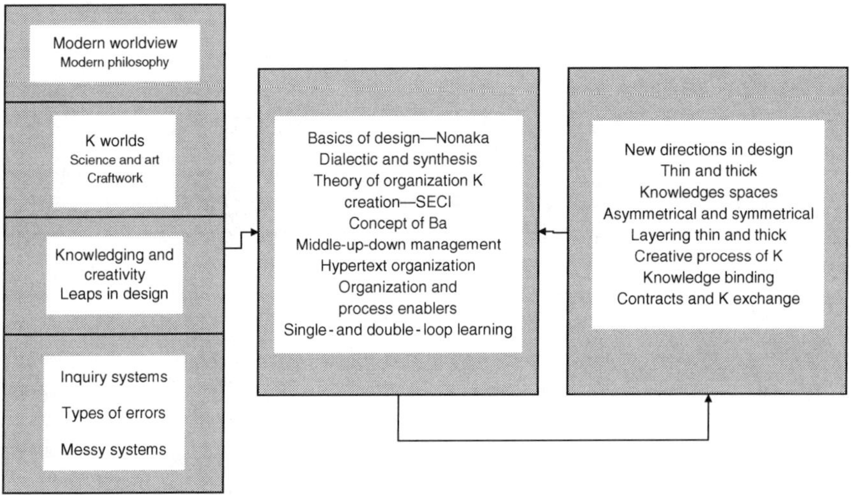

Concept roadmap to Chapter 2.

2.1 Perspectives of Knowledge

2.1.1 Transition to the Modern Worldview

The progression from the ancients to the moderns is characterized by a series of syntheses. In both India and China, we see an accommodation of Hinduism and Neo-Confucianism, respectively, with modern Western thought. Let us recall Professor Chan from Chapter 1: "There can be no doubt that Chinese philosophy will be baptized by Western science, logic, and epistemology." The same may be said of Indian thought and, in fact, world thought generally.

Modern Western belief systems arose in a long series of accommodations between conflicting lines of thought, as new paradigms challenged old-line thinking. We will not list every intellectual movement, but Tarnas (1991) has identified several major currents in Western thought, including Classical

Greek, Christianity, the Scholastic movement, and the development of the university system (e.g., the Papal charter given to the University of Paris in 1215, which in turn triggered university-centered, independent inquiry), the medieval and high middle ages, the quick succession of the Renaissance, the Reformation, and the Scientific Revolution, and, finally, Romanticism and the Enlightenment. While this long evolution (about two and one-half millennia) is responsible for the essence of Western thought, it was the sequence of events beginning with the Renaissance that largely shaped the modern world.

We observe that the Scientific Revolution has had two major phases. Initially it concentrated on the physical world: the science of Copernicus, Kepler, Galileo, and Newton. In turn, the 19th century saw a different type of science arise, with a shift from the physical to the life sciences. Here, Wallace and Darwin created no less a stir—one still in play today in the development of the new science of evolution and the descent of man in particular. Religion has had several centuries to accommodate itself to a heliocentric world, where (as Tarnas (1991) comments) the Earth is seen as only another planet. Today, we wrestle with the idea that man is only another animal. There can be no doubt that science has won the day!

2.1.2 Modern Philosophy and Knowledge

Today, the philosophy of knowledge is called epistemology. However, not very long ago (several centuries ago) it used to be called natural philosophy and was considered separate and distinguishable from moral and metaphysical philosophy. By natural philosophy was meant the enterprise of the human mind as it attempts to study and understand nature in all of its aspects. However, it was limited to answering the question "what?" instead of answering the questions "why?" and "how?" that was under the purview of the earlier-mentioned moral and metaphysical philosophies, respectively.

After the Renaissance in Europe, much of natural philosophy had to be revised simply because the advent of science completely modified the older conception of nature. The new perspective, then, was very similar to that of the revolutionary change that was brought about by Immanuel Kant sometime around 1760 when he read a paper by Euler that argued that space was indeed absolute, as Newton had suggested, instead of being relative, as Leibniz had suggested. In between came the philosophy of Hegel as depicted in his Science of Nature and Science of God or Mind. Hegel developed the overall philosophical "system" that later became the forerunner of much of modern Western philosophy.

Subsequently, Einstein's theory of relativity and the development of quantum mechanics brought yet another sea change in our understanding of nature as such, and that became the foundation stone for knowledge regarding nature. However, during the postmodern era, the earlier-mentioned mechanistic view of knowledge once again got revised when the biology of the modes of perception were taken into account. Furthermore, science has revealed the connectivity of things; that is to say, there are no two things or events in the universe that are totally independent—anything is related or tied to any other thing in some way. In the world of physics this may be called "simultaneity" or "entanglement." But our point here is that the same fact is also true regarding organizations as well. As gravitation, or the concepts of simultaneity or entanglement play the "role of connector," we claim that in the case of an organization, the same role is played by knowledge.

2.1.3 Hegel

Hegel advanced a conception of reality that later would become foundational to modern philosophy and science. Hegel's central idea is that every aspect of reality can be conceived as an unfolding dialectic in which every state of existence brings forth and interacts with its opposite. The interaction of each—*thesis* and *antithesis*—produces a new and higher level of existence through a *synthesis*; in turn, this brings forth yet another interaction and corresponding synthesis and an even higher order of understanding.

From an organizational perspective, leaders are no longer required to choose "one or the other" possibility but to embrace "both-and." Contradictions are not dilemmas but rather opportunities for new knowledge. This is foundational to the creation of new knowledge, as illustrated through the interaction of tacit and explicit knowing discussed in Chapter 1. The same basic dialectic idea extends into every area of the organization—short-term versus long-term strategy, low-cost versus high-quality production, individual versus organization, and so on. Additionally, the K-spiral mentioned in Chapter 1 (see Figure 1.15) may be described as a sequence of successive syntheses, as knowledge is created and refined through consecutively higher levels of abstraction.

We make another observation: in practice, a given synthesis is not necessarily a "little of each" but is asymmetrical where one side predominates. For example, we see this unfolding as religion gives way to science and is compelled to conceive a "two theory" worldview to maintain its standing. As another example, the accommodation, say, between the organization and the individual has shifted in emphasis; during the industrial age the organization predominated, but today, in the knowledge era, the individual drives the synthesis.

2.1.4 Popper and Kuhn

The modern worldview saw an important shift in the assumptions underlying science. Whereas Newtonian physics essentially resolved physics into a set of absolutes, the rise of relative theory and quantum mechanics introduced uncertainty into the natural world. Thus, there arose the idea that scientific knowledge is qualified by human perception and interpretative structures, and the truths of science are neither absolute nor objective.

With the ideas of Kant as a foundation, Karl Popper reinforced this perspective with his assertion that science can never produce certain knowledge. Recall from Chapter 1, Kant originated the idea that we can never understand a *thing-in-itself*. Without a capacity to understand the essence of things, the best strategy is to continually refine, to learn from our mistakes, and to fearlessly search for the truth. Science is ultimately a rational endeavor of systematic criticism through the falsification of existing theories.

Thomas Kuhn further qualified scientific inquiry. Kuhn agreed that science is subject to human perception and interpretation but questioned the actual practice of science. Rather than falsification, Kuhn argues that scientists confirm prevailing theories, reinterpret conflicting data as necessary, and in general conduct confirmatory research that actually strengthens current thought. It is as if existing paradigms are governed by their underlying tacit mental models rather than their corresponding explicit threads, and, as such, observations and actions are filtered through those mental models that, in effect, are self-validating. Only when conflicting data reaches a threshold of sorts is there a sudden conceptual synthesis (i.e., a change in mental models). Furthermore, the synthesis is not rational but a function of the cultural, psychological, and social factors of the scientific and wider communities. Science thus advances as a series of radical, nonrational shifts in vision. Let us be clear: the issue is not science as such but scientists who bring with them the ambitions, biases, and fears of all human beings and societies.

2.2 Knowledge Worlds

"'How then,' comes the question, 'can we ever establish anything finally and completely and for sure, even the most obvious truism and the most cherished credos?' And the answer is, of course, that we can't ... neither by logic nor any other means can we prove something from nothing. We have to start with some premises and principles, and there are no absolute and incontrovertible certainties available. But that does not mean that we must start from careless guesses. We follow our confidences and convictions, which are subject to strengthening or weakening or even reversal as we strive to build right versions or worlds on the basis of these. No starting points or ending points or points along the way are either absolute or arbitrary" (Goodman 1984). Thus, each of us—as scientist or artist or casual human being—creates our own *world version* or *worldview*. And each version is as right as the next—intent and truthfulness determine rightness. Furthermore, rightness cannot be verified by reference to a version-independent world; or as Goodman (1984) puts it, there are no world features independent from all versions. Hence, we make versions; "worldmaking" is a matter of mind.

The knowledge world (K-world) is our construct that operationalizes the thinking of Goodman (1984) by combining several ideas of modern philosophy

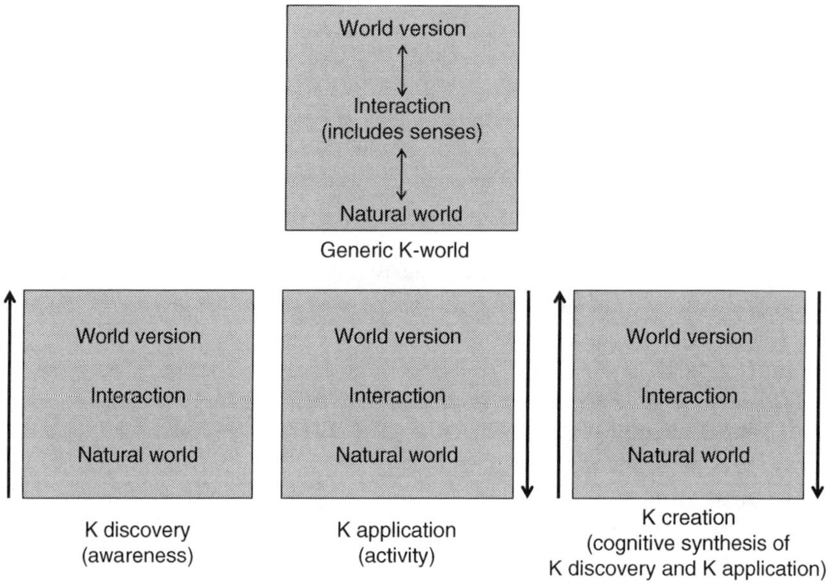

Figure 2.1 Knowledge worlds and modes of K interaction.

with those of KM. Specifically, we include the observer in our representation, with the Mind of an individual as the primary structuring agent in discerning reality. This is illustrated in Figure 2.1.

A K-world is a type of hierarchy, where the human entity's Mind is at the very top, the particulars of reality are at the bottom, and between them are intermediate levels of abstraction. Each person's Mind—individually—interacts with the natural world of particulars along several *modes of interaction,* or knowledging. Looking up the hierarchy, we discover natural particulars through the lens of our individual Mind; this is somewhat analogous to Polanyi's conception of awareness. In contrast, looking down the hierarchy, we operate on nature and apply the Mind's knowledge to the natural world; this is analogous to the activity dimension in the Polanyi model. Finally, the two-way give-and-take is a synthesis of discovery and application, of awareness and activity that brings forth new knowledge or cognitivity. Note too that the human mind—our personal K-world—must be receptive to interaction and new knowledge. *Knowledging is an affirmative activity.*

2.2.1 The K-Worlds of Art and Science (and Management)

Plato was right; the world *is* constituted of archetypes of patterns and laws. However, unlike Plato, science tells us that these archetypes are not transcendent, but mechanical and immanent in nature. Modern thought colors the mechanical view by introducing the observer as an active participant—this is represented by the K-world construct. The challenge of modern society is to uncover and represent the archetypes—we go about this through science and art. (In turn, religion is increasingly less concerned with the natural world and more internally focused—on moral and social behavior, transcendent mind, and depending on the religion, the afterlife.)

The process of revealing nature differs between scientists and artists. Both science and art deal with reality—the same generic K-world—but the navigation paths yield profoundly different goals, results, and other characteristics. That is to say, science and art each have a characteristic K-world, the workings of which bring forth *unique conceptualizations of reality*; as Dewey (1934) states, "science states meanings; art expresses them." The generic K-world and its science and art variations are illustrated in Figure 2.2.

Building on the work of Bayles and Orland (1991), we can generalize the K-worlds of art and science as follows:

Science. (1) Always starts at a high level, and (2) seeks explanations at lower levels, and (3) does not in principle involve the observer (i.e., mechanical analysis), and, therefore, (4) must be repeatable (or it is not science), and hence, (5) the high-level hypothesis is validated or undermined with every iteration.

Art. (1) Always starts at a low level, and (2) seeks generalizations at higher levels, and (3) involves the observer (i.e., thoughtful analysis) who is thus changed by the process, and, therefore, (4) can never be repeated (or it is not art), in which case, (5) there is an internal transformational change within the artist (between each iteration) and his or her art pieces.

Let us introduce a third hierarchical perspective described by Bayles and Orland (1991)—that of "craftwork" (i.e., handicraft).

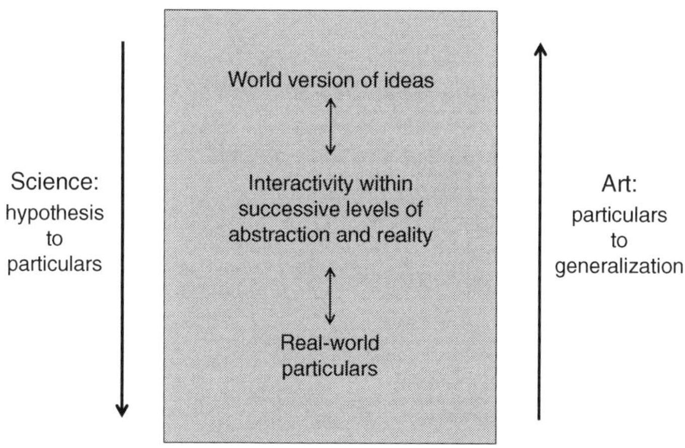

Figure 2.2 The K-worlds of science and art.

Craftwork. (1) Always starts with a high-level template, which is (2) refined at lower levels, and (3) involves the observer in a superficial way, who is thus changed very little by the process and therefore, (4) can and is repeated until (5) a "perfect" work is realized.

Observe that perfection is possible (theoretically) with craftwork and science, whereas it is never possible with art. Scientific inquiries into the same problem should yield the same result; differences between craft pieces are small; differences between art pieces are large, even transformational. Today, we see that even science is relative and conditional on factors of context and perception of the observer. Art is more about process than structure: "art lies embedded in the conceptual leap between pieces, not in the pieces themselves."

Further, as Bayles and Orland (1991) observe, science and craftwork share the characteristics of being "elegant" and "polished"; in contrast, fine art is much less polished, even rough, but always inventive (i.e., the transformational process between pieces and designs).

2.2.1.1 Knowledging

Finally, let us briefly revisit knowledging and pattern formation. Knowledging is about interrelating and building complex patterns. Savage (1996) reminds us that "idea transformation" processes are as critical as "material transformation" or more so. Knowledging, as with art, is about transformational change—the leaps between material designs—and is where new patterns or ideas arise. This is what is meant by the knitting metaphor.

Essence of Knowledging
Knowledging and creativity lie in the leaps between designs, not in the designs themselves

2.2.1.2 Implications for Organizational Design

As a result of our experience and research, we believe that organizations largely operate within craftwork frameworks that we call *templates*. This is particularly the case in operations. Organizations use the notion of templates and craftwork processes to routinize work. (Routinization of work is discussed in Chapter 3.) When knowledge gets organized, it "grinds out" an algorithm, which subsequently leads to a metaphor that produces a collection or variety of templates. For example, architectures, methodologies, best practices, policies, and so on are effectively templates in which organizational elements (including people) are placed, instantiated, and made to interact. The materials of the organization are templates of a sort, for example, procedures and processes, best practices, benchmarks, scorecards, critical success factors; included in our conception of a template are the supporting technologies, particularly operational and transactional systems. The operational workings of the organization share the characteristics of craftwork discussed earlier; it is only a question of selecting the appropriate template for the situation at hand and refining it as such. Organizations borrow, build, and refine their templates to suit their own needs. Please note that we are not reducing management to a series of prescribed routines; we are only commenting that organizations operate within templates and craftwork processes—all necessary to routinize, structure, and improve their work. Think of it this way: if organizations did not operate with templates and craftwork processes, how would they structure and improve themselves?

2.2.1.3 Implications for Organizational Architecture

The *structural architecture* shown in Figure 1.2 is best represented with the organization's set of templates and structures, whereas the *process architecture* typically operates with craftwork processes (as described here). Similarly, the *essential-cognitive architecture* in Figure 1.2 is best represented with the K-worlds of art and science. Further, the structural and process architectures of the organization should be polished and elegant, made possible with templates; on the other hand, the essential-cognitive architecture of the organization is dynamic and rough, and should be inventive. This manifests itself in successful organizational design: structure and process are engineering-like and precise; the cognitive is art-like and ethnographic.

The workings of an artistic K-world exist locally within individuals and specially designated departments, such as R&D. Even here, however, organizations with no understanding of cognitive architecture sometimes deconstruct prior custom into process-like tasks that strip off the *practice* of underlying scientific or artistic thought. *We use the term practice to characterize the workings of (1) art and science and (2) the knowledging organization and knowledge work.* (K work is not confined to an organization, for it may exist, say, in a professional practice.) We will more formally define practice in a later chapter. Organizational innovation requires both science and art K-worlds that are synthesized into organizational templates and routines, behavior patterns, and ultimately organizational culture. From such syntheses arise the organization's new products and services as well as its K structures, K workings, and K systems. See Figure 2.3.

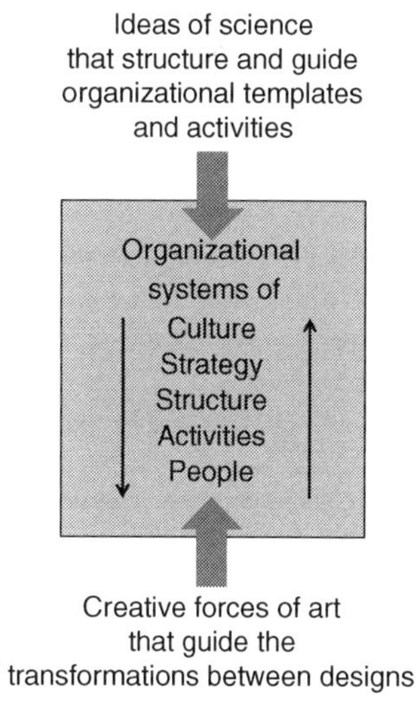

Figure 2.3 The K-worlds of the knowledging organization.

2.3 Inquiry Systems and the Search for True Knowledge

Recall from Chapter 1 (see Figure 1.11) that Justified True Belief implies that the interaction between what is "true" and what is "believed" must be "justified." Justification, in turn, is based on the method of inquiry that constitutes the interaction of tacit beliefs and explicit truths. This is shown in Figure 1.11 where the connection between what is {explicit & visible} and what is {tacit & not visible} is labeled "method of inquiry." Hence, we may extend Plato's definition of {justified true belief} to include an underlying and correspondingly appropriate inquiry system.

An inquiry system is a way of obtaining knowledge about the world. In addition, a given inquiry system corresponds to a particular type of knowledge; that is, the initial conception of what is true in the first place guides the system of inquiry we choose. Here we briefly discuss several well-established approaches described by Mitroff and Silvers (2009) as well as several additional methods necessary for 21st century inquiry. Note that the first two approaches (i.e., expert consensus and one true formula) have been foundational in the scientific and university communities for centuries, whereas the others are either specialized or relatively new methods.

2.3.1 Types of Inquiry Systems

Expert Consensus. In expert consensus, truth is not absolute but a function of consensus among experts. The greater the level of agreement among experts, the stronger the truth. Among the issues are the definition of an expert and the choice of experts. Also, in expert consensus there are typically a limited number of alternatives. For example, admission into a Ph.D. program is typically based on a consensus of selected professors who separately review the submittals of each applicant.

One True Formula. The one true formula approach comes closest to what most people consider science. There is an assumption that truth is absolute and can be ascertained if based on solid principles and laws—that is, rational in a scientific and mathematical sense. This approach is particularly useful in the hard sciences, where a single abstract expression of the truth is presumed to exist. (There is qualified application in the social sciences, where there may be more than one truth.) The key issue is the selection of the realm; for example, are we investigating the very large (e.g., relativity theory) or the very small (e.g., quantum theory)? But this too is subject to the absolute truth idea, as in the search for a "unified theory of everything." The great advantage of this approach, in contrast to expert consensus, is that it allows for the exploration of completely unchartered areas where even the possibilities are not known. We must also keep in mind that, as Kant has stated, we can never know a thing-in-itself. Hence, truth, even in the one true formula approach, is contingent on the human interpretative apparatus; the truths of science are neither absolute nor objective, but relative to the observer, the community, and other factors of space and time.

Multiple Formulas. Multiple formulas attempts to combine the expert consensus and the one true formula approaches. In this method, possibilities are collected into groups, each, presumably, with its own version of the truth. Experts evaluate the possibilities within each group and then reach an overall consensus. The multiple formulas method is premised on the assumption that a complex problem needs to be addressed from several perspectives. This approach helps to facilitate critical thinking, ensure that we are focusing on the right problem or question, and thus avoid "solving the wrong problem precisely." For example, a Ph.D. candidate defending his or her dissertation must do so publicly in front of a committee whose members come from different schools, thus creating a multidisciplinary consensus.

Expert Disagreement. This is the opposite of expert consensus, where a debate or argument is created between two opposing viewpoints and a winner selected by an adjudicator. For example, a criminal trial has two sides of experts who argue a case, and a jury selects the winner.

Systems Theory. The underlying assumption in systems theory is that we can achieve a better Truth if we widen the problem and bring in other forms of expertise in the form of experts, K-threads, or data. This is common in double-loop learning, where we factor in what appear to be external considerations into the problem area. A characteristic of systems theory is that we consider areas often overlooked in the other systems of inquiry, for example, ethical, moral, and aesthetic factors. For example, businesses sometimes replace *ethical* considerations with legal ones. As another example, Apple products are known for their *aesthetic* style in addition to their functionality. Systems theory promotes interdisciplinary inquiry where the participating disciplines *are* affected by the method of inquiry, and thus may create new knowledge. This is in contrast to multidisciplinary inquiry (e.g., multiple formulas), where the participating disciplines are unchanged.

Business and Data Analytics. We have added data analytics as a new and powerful inquiry method. Though analytics is constrained by the availability of data, in the world of big data the issue is not insufficiency but a never-ending *steam* of data. The churning of data and the discovery of patterns characterize data analytics. Within the general area of business intelligence, we may additionally use online analytical processing (OLAP) to discover patterns in a broad sense. The OLAP philosophy views data as structured in the sense that we typically explore a multidimensional data structure where the dimensions are known and designed for in advance, such as time, product, or customer. Thus, whereas the patterns discovered in data by an OLAP analyst may be useful, they will be somewhat constrained by the overall multidimensional data structure. In contrast, data mining and analytics discover the dimensions themselves—dimensions within the data. This may lead to a deeper understanding of, say, products that tend to be purchased together. This is sometimes known as "knowledge discovery in databases" (KDD), though in the era of big data it may be more appropriate to consider streaming data as opposed to stored data. Data mining tools are best used by data mining specialists in conjunction with business experts (see Morabito and Stohr 2010).

Student Case. Several of our students have evaluated a cross-industry data mining process—that is, CRISP-DM lifecycle—and compared it to both the K-spiral and the SECI (socialization–externalization–combination–internalization) model (discussed later). They have found loose correspondence with the socialization mode of SECI and stronger correspondence between internalization in SECI and evaluation in CRISP-DM. This should not be surprising since the data mining process is designed specifically for K discovery in data whereas SECI is designed for K creation; correspondence between the two approaches is more incidental than deliberate.

Knowledge Capture. By knowledge capture, we mean the extraction of knowledge from any number of sources and its abstraction into easily understood facts and rules. In contrast to data analytics, the sources for K capture are not confined to data. For example, we have "tacit K capture" made popular in the development of expert systems, where experts (who possess tacit K) are interviewed, their knowledge extracted, analyzed, and codified, rules developed, and systems built. (This is also known as K acquisition.) Today, we better understand that essential tacit K cannot be so easily captured as it is learned or absorbed, but we still use the term "tacit K capture" when interviewing experts. Similarly, "explicit K capture" refers to the analysis of large volumes of data, in the form, for instance, of a new law or a large document. Accordingly, an organization must abstract and summarize the contents of a new law into rules and guidelines that can be deployed in the organization. Finally, we may have "organizational K capture" (in the whole) when one organization obtains (e.g., buys) the knowledge of another through a merger or acquisition. Note that *K sharing* is closely associated with K capture and will be discussed later in the book.

2.3.2 Types of Problems and Errors

Mitroff and Silvers (2009) have described four types of error that may occur during analysis. The first two are technical, whereas the others are social, psychological, or political. For example, if we are comparing a "new" with an "old" drug, a Type 1 error is one where we conclude that the new drug is better than the old when the old is equal to or better than the new; in a Type 2 error we conclude the old is better than the new when the new is better.

A Type 3 error occurs when "we solve the wrong problem precisely," which typically results from an unintentional failure to expand our thinking.

For example, it may be advantageous to ask whether any drug (new or old) is needed in the first place when a change in diet or lifestyle may be sufficient. Moreover, as described by Mitroff and Silvers (2009), this type of error perpetuates itself in a sequence of self-justifications: having developed the right solution to the wrong problem, we convince ourselves into believing we solved the right problem, and then we stubbornly maintain our belief and perpetuate the problem.

In a Type 4 problem, we *intentionally* solve the wrong problem precisely. At first, this may seem an unlikely situation, but students soon realize that they are surrounded with deliberate misrepresentations and deceits. As Kuhn suggests, people (and let's add organizations) have their own biases and conflicts of interest—perhaps the drug company *wants* the new drug to be better than the old because the latter is going off patent. And let's not forget politicians!

2.3.3 Messy Systems

In today's complex world of business and knowledge, everything seems to be a mess—we even have a description: a *messy system* is one that is composed of constantly changing interconnected problems, none of which can be solved independently (originally described by Ackoff, see Mitroff *et al.* 2012). It is unwise to deal with messes from the standpoint of any single profession or discipline—"a partial solution to a whole system of problems is better than whole solutions to each of its parts taken separately" (Alpasian and Mitroff 2011). The five different kinds of inquiry systems considered by Mitroff *et al.* (2012) together with our own kinds of inquiry may be used as a basis for a coordinated approach to "messy" inquiry. This highlights the importance of systems thinking, but in today's world of big data and social media, systems thinking must embrace and blend all of the inquiry systems mentioned earlier (as well as others to be determined) to deal with messy problems.

We additionally assert that a combination of inquiry systems may need to be *dynamically assembled* in a unique combination to resolve a messy problem—this is characteristic of expert approaches to problem solving (see later). Messy problems can never be completely solved, but the knowledge of the organization must be leveraged to *manage* them. In the context of messes, routine thinking is the enemy of imagination and problem resolution. Furthermore, every problem has a tacit component; hence, soft dimensions, including aesthetics, feelings, values, and ethics, are part of every problem, even if these dimensions are not fully and consciously known or even vehemently denied.

2.3.3.1 Implementation: Education

Most problems in the world today are messy problems, such as climate change, education, healthcare, drug wars, terrorism, and so on. What they all have in common (their messiness) is the following: (1) their complexity, (2) they are ill-structured, (3) no one discipline owns the problem or solution, (4) there is no "legitimate" framework to approach the messiness, and (5) the resolution requires a cognitive and integral architecture.

As an example, consider education. What are the issues? How do we define, for example, a "good or higher education"? How should it be financed, what skills and knowledge should be taught, what type of degrees should be offered, how do we balance the technical, social, and intellectual aspects of education, how (or should) we assess educators, administrators, and students? And many other such questions. Education is so messy that the resolution requires an understanding and deep knowledge of some aspects of virtually everything taught in a school as well as their interrelationships (an interdisciplinary systems approach). The resolution most likely requires satisficing (i.e., "good enough") rather than a clear and complete solution.

2.4 The Basics of Design in the Knowledge Era

This section reviews the basic principles underlying organizational knowledge creation. Given the complexity of KM, we can take this discussion in any number of directions. However, it is difficult to overstate the importance of Nonaka and Konno (1998) in the development of KM, and we therefore describe his (and associated colleagues') key concepts. When we think of the "basics," we think of Nonaka and Toyama (2004). Throughout the book, we will discuss these ideas in more detail as well as provide student solutions to related problems. To round out our discussion, we include additional KM ideas and our own *organizational design implications*, as appropriate.

2.4.1 Dialectic and Synthesis

As mentioned earlier, Hegel introduced the idea of a dialectic and a corresponding synthesis into new levels of understanding. Each higher level is a thread of new knowledge (tacit or explicit) that comes from a synthesis of contradictions. Furthermore, as we discover from an examination of history as well as our experience, a given synthesis is typically not symmetrical, but biased in a particular direction (as suggested by Kuhn). Furthermore, each synthesis is part of a K-spiral (see Figure 1.15) and is an aspect of the SECI model discussed here. This is illustrated in Figure 2.4.

For example, Toyota is well known for synthesizing the apparently contradictory goals of manufacturing low-cost and high-quality automobiles. This is a classic illustration of the workings of this principle: rather than choosing one or the other, Toyota synthesized a new level—a new form of knowledge—that provides a continuing source of advantage. Of course, Toyota had to actually implement the synthesis: we will discuss this here and in subsequent chapters.

Figure 2.4 Asymmetrical K-spiral synthesis.

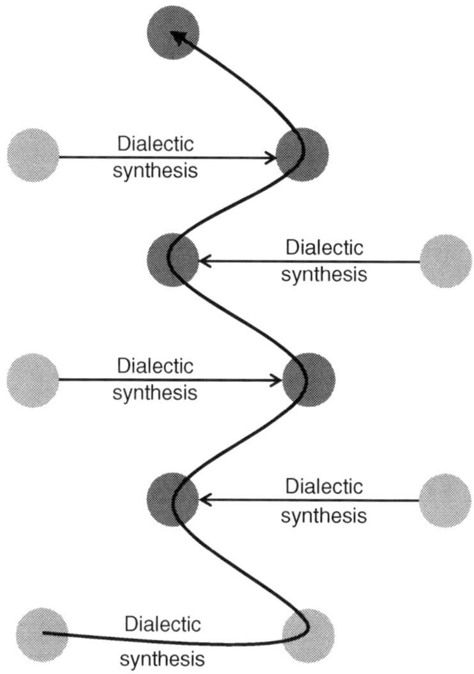

There are several dialectic syntheses that are useful in designing knowledge organizations. Here we briefly describe those made popular by Nonaka and Takeuchi (1995) as well as other scholars:

- Tacit–explicit: This is the foundational dialectic synthesis, initially described by Plato and recently by Polanyi (see Figures 1.11 and 1.13, respectively). The key ideas of Nonaka and Toyama (2004), and those of KM in general, may be traced to the tacit–explicit synthesis. We consider {tacit–explicit} to be a *technical KM perspective*, a viewpoint made popular by Nonaka and favored by organizations. The strengths of a technical perspective include ease of understanding and application in design; on the other hand, a technical perspective overly simplifies knowledge by stripping off the nuanced dimensions of knowledge we believe necessary in the modern world, such as its moral and social applications.
- East–West: A synthesis that is invaluable in the global economy, we consider this dialectic an offshoot of {tacit–explicit} with the East corresponding to tacit K and the West corresponding to explicit K. This is a *cultural perspective*, which we will consider in detail in our discussion of culture in Chapters 6 and 7.
- Body–mind: This is a basic idea in both Indian and Chinese philosophy. We have also seen this in Polanyi's description of knowledge (see Figure 1.13 and the accompanying discussion in Chapter 1). Its primary application in

design is to highlight the importance of experience—even physical experience—in building knowledge. The {body–mind} synthesis represents both a *philosophical and a cultural perspective* and has application in a global economy.

- Both-and–either-or: Related to the {East–West} and {body–mind} syntheses, the {both-and–either-or} synthesis corresponds to an Eastern and Western perspective, respectively. As with the {body–mind} dialectic, this type of synthesis represents both a *philosophical and a cultural perspective* with a global application. For example, recall in Chapter 1 that we discussed the Confucian *Doctrine of the Mean*, which embodies the Chinese idea of both-and, as opposed to Western either-or thinking.
- Understanding—experience; or reason—sensibility: These dialectic syntheses are Kant's description of knowledge and its sources, respectively. This is a *practical and philosophical perspective* that we will use in our analysis of K work in future chapters.
- Hierarchy–task force: This dialectic synthesis attempts to leverage the benefits of the classical bureaucracy and the task force. The bureaucracy is ideal in a stable environment that is characterized by high levels of centralized control and formalization, functional specialization, and work that is primarily routine (or, at least, predictable). In contrast, a task force has the opposite attributes: it is a temporary structure that is dynamic and adaptable, is decentralized with low levels of formality, has high levels of participation, and is organized around accomplishing a task. Nonaka and Takeuchi (1995) call this synthesis a *hypertext organization*, which emphasizes the use of project teams whose members come from functional business units. This is an *organizational perspective* and is discussed in section 2.4.5.

  A related synthesis is that of the {centralized–decentralized} dialectic that leverages the benefits of both centralized and decentralized structures. This is an offshoot of the {hierarchy–task force} synthesis and is sometimes known as a *federated structure*, where responsibilities and roles are distributed in centralized and decentralized subunits. It is important to note that a mere separation of responsibilities (a structural idea) does not necessarily produce a synthesis (a K idea); the organization's culture must promote such a synthesis. Though typically not discussed in KM literature, the federated structure has been successfully implemented in IT departments in large organizations. We discuss the federated structure in Chapter 3.
- Top-down–bottom-up: This synthesis, known as *middle–up–down*, is intended to leverage the benefits of both top-down and bottom-up management practices. While top management creates strategic intent ("what ought to be"), the front-line managers understand reality ("what is"). Therefore, it is the middle line that must rationalize and synthesize both perspectives—it is this synthesis that gives rise to the notion of middle–up–down management, discussed here.

- Individual–organization: This synthesis recognizes the individual, specifically individual tacit knowing, as the source of all new knowledge. The knowledging organization must elicit, convert, and amplify this new knowledge into group and organizational knowledge, which in turn (ideally) becomes embedded in the organization's products, structures, and processes. Knowledge conversion is represented in the SECI model, as is knowledge amplification (K-spiral). This is discussed here.

In the same way that Toyota synthesized low-cost and high-quality manufacturing, organizations must continually search out contradictions as opportunities to create new knowledge. We may extend this idea to virtually every dimension in an organization: {global–local}, {short-term–long-term}, {efficiency–effectiveness}, {continuous improvement–disruptive innovation}, and so on.

2.4.2 Theory of Organizational Knowledge Creation: SECI

The theory of organizational knowledge creation represented by SECI {socialization–externalization–combination–internalization} is the best-known model among both academics and practitioners. SECI embodies several important principles, such as tacit and explicit knowledge, the Polanyi assumption that all knowledge originates as individual tacit knowing, the amplification of knowledge from the individual to group to organizational levels (known as the K-spiral), among others. In fact, SECI is an example of raising the Polanyi model to the organization level. Also, the SECI model captures the notion of dialectic synthesis along several dimensions simultaneously, for example, {tacit–explicit} and {individual–organizational}.

There are innumerable representations of the SECI model. In Figure 2.5 we present a representation used by several of our colleagues in their description of SECI to their students. A fundamental concept underlying the SECI model is that knowledge is created when it moves and is transformed. This is known as *knowledge conversion.* There are four generic K-conversion *modes.*

For example, in Figure 2.5 *socialization* is represented as an interactive person-to-person contact where an understanding is shared; during *externalization* a person or group conceptualizes and creates an explicit conceptual artifact; *combination* brings together the new explicit conceptual artifact with other such artifacts to produce a new explicit system; finally, during *internalization* a person learns from the new system. Note too that in Figure 2.5, one colleague used artificial intelligence concepts in his representation (metaphorically) of people: each person's mind is composed of a neural net and a database, representing tacit and explicit knowing, respectively, or alternatively the processing (neural net) of information (database), respectively. The details of each generic knowledge conversion mode are reviewed here:

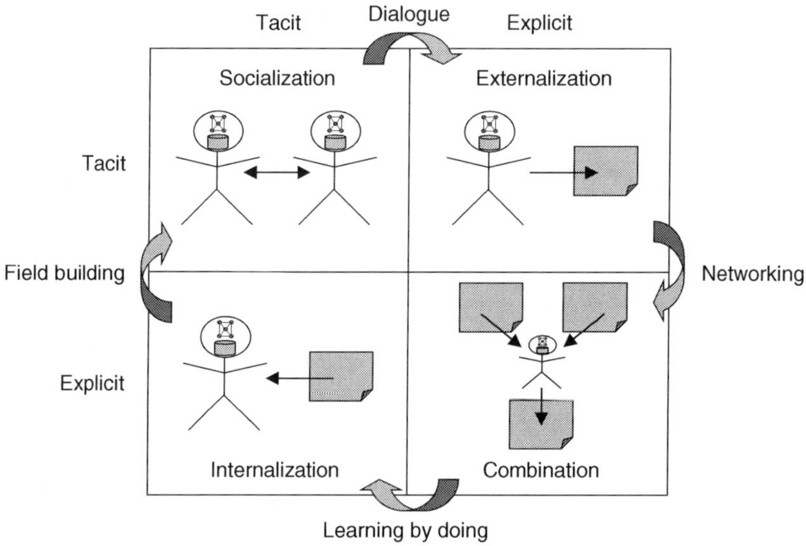

Figure 2.5 SECI knowledge creation model—Adapted from Nonaka and Konno (1998).

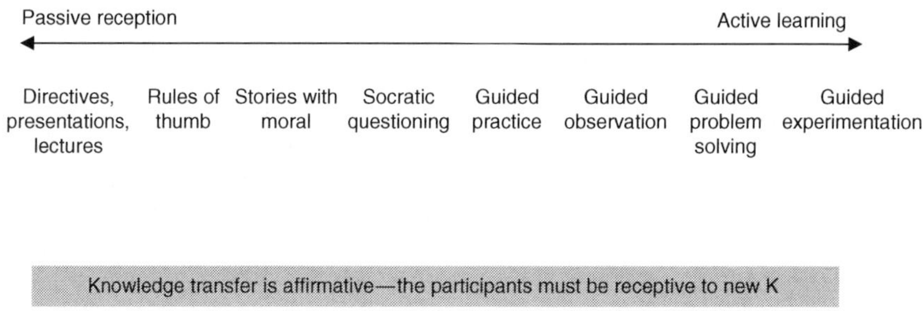

Figure 2.6 Transferring deep knowledge—Adapted from Leonard and Swap (2004).

Socialization. Socialization is the conversion of tacit-to-tacit K. This refers to the transfer of tacit knowledge from person to person, from an originator to another person. Knowledge exists at the individual level. The trigger for socialization is *field building,* or a place where person-to-person communications can take place. This is sometimes known as *learning by sharing.*

Related idea: transferring tacit K. Von Krogh *et al.* (2000) have suggested the following modes for transferring tacit K: direct observation, narration, imitation, experimentation, and joint execution. Additionally, Leonard and Swap (2004) have suggested a continuum of expertise transfer from passive reception to active learning (Figure 2.6). For instance, a team exercise in a classroom is an example of guided problem solving. A single occurrence will not create much of an effect, but repeated practice will more likely create an integrated pattern. As shown in Figure 2.6, the best way to build integrated patterns is through mistakes—that is, guided experimentation. Note too that the transfer of "deep smarts" involves both tacit and explicit K; a look at the continuum in Figure 2.6 suggests the core importance of tacit K transfer.

Related idea: coalescent knowledge. The research of one of our Ph.D. graduates (Morgan 2009) suggests that tacit mental models as such are not shared, but only their underlying assumptions and understandings are shared; he calls this type of shared context *coalescent knowledge.* The patterns that embody a tacit mental model arise from an individual's experience and hence are *unique* to that individual. Coalescent K is the interweaving of underlying assumptions and understandings between people to form new coalescent knowledge that is shared between them—as with deep smarts, this shared or coalescent K may be both tacit and explicit.

Externalization. Externalization is the conversion of tacit-to-explicit K. Once the new tacit K thread is stabilized, it is converted to a corresponding thread of new explicit K (i.e., an explicit concept) through dialogue and collective discussion and reflection. Knowledge has now been elevated to a group level, such as a team or department. The trigger for externalization is *dialogue.* This is known as *learning through conceptualization.*

Combination. Combination is the conversion of explicit-to-explicit K. In this mode, the new explicit K is linked to existing explicit K threads to create a new knowledge system. The trigger for combination is *networking.* Note that a new knowledge system can be anything, from a new product to a simple standard operating procedure. This is known as *learning through networking.*

Internalization. Internalization is the conversion of explicit-to-tacit K. We often describe this mode as the *absorption* of explicit K *back to* tacit K, which is now different from the original tacit K due to its refinement and amplification. The trigger for internalization is *learning by doing.*

Nonaka and Konno (1998) observe a cultural bias within SECI: tacit-oriented organizations (e.g., Japanese) emphasize socialization and internalization, whereas explicit-oriented organizations (e.g., the United States) emphasize externalization and combination. Of course, the lesson of SECI is that both perspectives are necessary for knowledge creation. The cultural dialectic and synthesis, superimposed over SECI, is illustrated in Figure 2.7.

2.4.2.1 Implications for Organizational Design

The K-conversion modes may be applied singularly or sequentially. By singularly, we mean, for example, the socialization that takes place when two people are having a casual conversation without subsequent K-conversion modes. This type of behavior takes place in organizations all the time. In Chapter 5, we will do an analysis of a nonprofit organization working in an underdeveloped country and assess the workings of the K-conversion modes singularly.

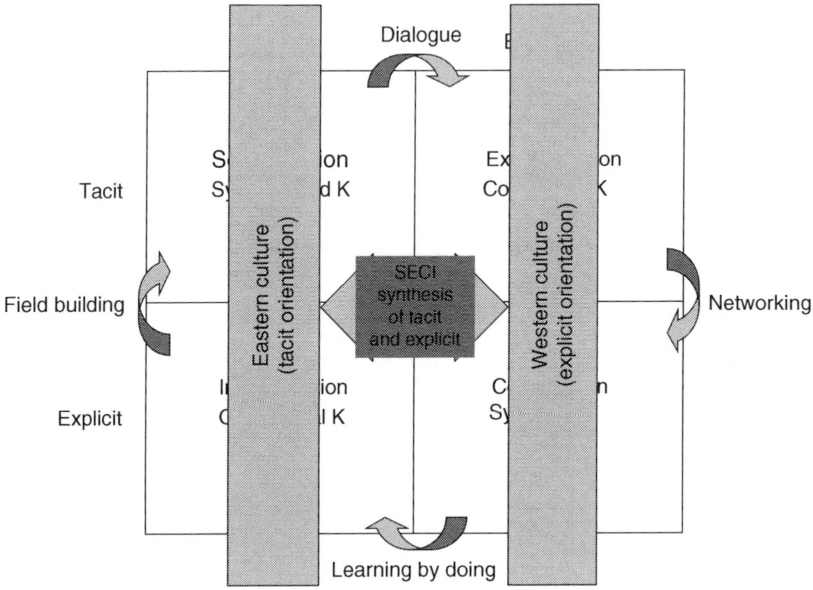

Figure 2.7 SECI and the cultural dialectic.

Presently, we want to illustrate a SECI-inspired new product development workflow proposed by Nonaka and Toyama (2004). As shown in Figure 2.8, the process workflow has a direct mapping to SECI (this workflow is known as the five-phase model of the K creation process). Included in the workflow are administrative tasks, such as a "go no-go" decision gate common in new product development. When we redesign a business process we use a variety of techniques, such as automating routine tasks or replacing sequential tasks with CASE work. If the business process is a K-creating one, then we may use Figure 2.8 as an additional framework targeted to enhancing knowledging. We do this by mapping existing process tasks to each aggregate K-conversion mode in the generic workflow. We would then assess the process as to its K-creating capacity; that is, we may find that our existing process has both strengths and weaknesses with respect to SECI. The final step is to strengthen the comparatively weak and reinforce the strong SECI-related tasks.

For example, as shown in Figure 2.7, a Japanese organization executing, say, a software development process, may be particularly strong during socialization and internalization, but weak during externalization and combination, whereas a US company may be the reverse. We have seen this in class exercises, where we ask our students to name one of their first activities during software development (assuming a US company). Students typically reply "requirements gathering" and then comment that business users are sometimes disappointed

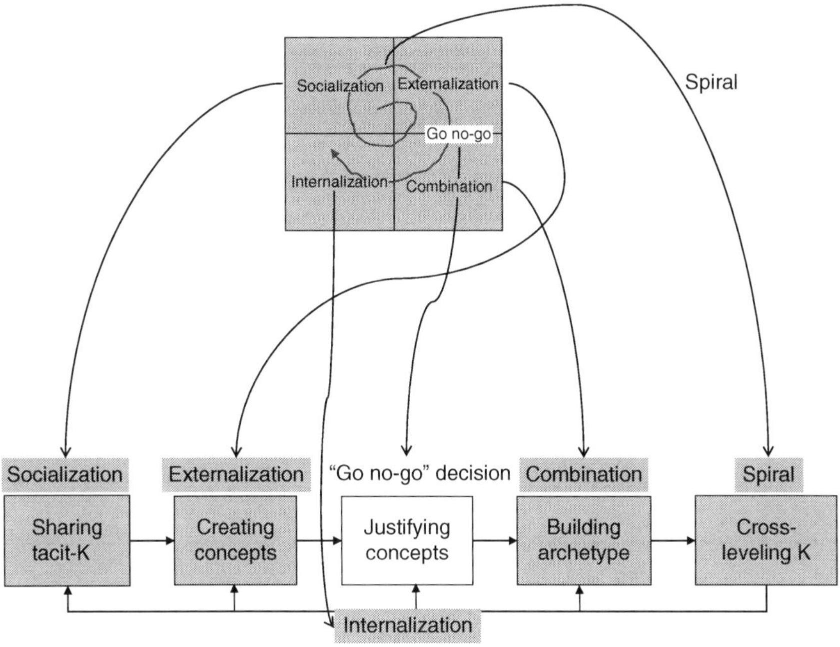

Figure 2.8 Generic process map for SECI (5 phase)—Adapted from Nonaka and Toyama (2004).

with a first version of the new system. Other techniques are employed to compensate for this circumstance, such as using prototyping or agile development. The answer, of course, is illustrated in Figure 2.8 where "requirements gathering" may be mapped to *creating concepts* (i.e., *externalization*), and thus the entire software development process skips *sharing tacit K* (i.e., *socialization*). One solution is to employ suitable socializing activities during the front end of software development, such as strategic offsites. Thus, the reengineered process workflow should strengthen the comparatively weak SECI-related tasks, while reinforcing the strong SECI-related tasks.

2.4.3 The Concept of Ba

Knowledge needs a context in which to be created. Nonaka and Toyama (2004) have called this concept "ba," which refers to the "place" where knowledge is created. There are several dimensions to ba. First, ba is a *context* in *time* and *space* where a "stream of meaning" emerges through interactions and sharing among participants. Second, ba supplies the *energy* necessary for "self-organization" and K-creating. It is the second dimension that elevates ba beyond a mere place to an aspect of culture. Furthermore, self-organization requires "necessary conditions," which in turn, come in many forms, including the organizational and process "enablers" discussed later. Finally, there is a third dimension to ba: the interconnectedness of local ba to produce an integrated whole. That is, a final K product or service requires many streams of knowledge, each embedded in its local ba, to be integrated. Ba integration is typically the function of middle managers (see Nonaka and Konno 1998).

(Note that ba is similar to, but different than, a community of practice (CoP). Ba is an enabling context where new knowledge is created; in contrast, a CoP exists primarily to share embedded knowledge, though new knowledge may also be created. A CoP is tied to the culture and history of a specific group or organization and is relatively stable; in contrast, the context of ba is tied to the participating individuals and their contexts (which may span organizations) and is dynamic.)

Example—The Workings of Ba. As an example of the workings of ba, consider the production of this book. If the authors were to write this book next year, it would look different, perhaps significantly so. The stream of knowledge that is manifest in this book arises from its specific ba and constituents—the simultaneous interaction of people (the coauthors), place (Stevens), time (2016), intention (a KM graduate-level book), culture of academia, and so on. Furthermore, the self-organization apparent in writing this book comes from years, even decades, of collaboration among the authors. Finally, for a successful product to be realized, the K associated with the ba of the individual authors (their respective local ba) must be interconnected, in this case, by the first author. As is apparent, it is sometimes difficult to distinguish local ba from the greater ba of the whole. We think it best to adhere to the basic idea of ba—a stream of meaning in space and time that produces new K. We will subsequently introduce the notion of *K-spaces* for more detailed design.

Example. As an example of local and greater ba, consider the workings of our classes. We encourage team building and problem solving among our students through team exercises (a local, classroom ba). At a higher level of abstraction, Stevens' culture and strategy promotes collaboration and innovation throughout the university (a greater ba). The K and energy of Stevens—its greater ba—arises from the interconnectedness of the local ba in each class.

2.4.3.1 Implications for Organizational Design

If we are to implement SECI in an organization, we must design corresponding ba for each K-conversion mode. We may look at ba from several perspectives or levels of abstraction. The first is generic for each K-conversion mode; the second is context-specific; and the third is connecting ba for knowledge integration and organizational-level K. Another aspect of ba is cultural, such as the promotion of dialogue and conversations throughout an organization. Let us briefly review the key design elements associated with ba.

Originating ba. This is the ba associated with socialization and is best operationalized through a "field" that promotes "face-to-face" interaction (shown in Figure 2.5 as "field building"). This includes, for example, communities, teams, or open office design. Several students tell us that their organizations have experimented with open office design but sometimes find it "too open." We have found that a good solution is a compromise that combines open and private office design, thus promoting both dialogue and private contemplation. In this way, we are facilitating both socialization and the other K-conversion modes. (We will discuss office design in Chapter 6.) In general, originating ba is facilitated through a broad, ecological approach to human interaction.

Interacting ba. This is the ba associated with externalization. This ba is "consciously constructed" and should promote dialogue, conversation, and reflection in a peer-to-peer setting. For example, metaphors and analogies are especially useful in managing conversations and eliciting explicit concepts.

Cyber ba. This is the ba associated with combination. This ba is "virtual" in character and promotes analysis and K reconfiguration through networking. It includes, for example, networking facilities, databases, groupware, and documentation. We thus see a large enabling role for IT in this regard.

Exercising ba. This is the ba associated with internalization. This ba promotes the practice of the new K, or learning by doing. Examples of exercising ba include the use of simulation tools (as in a flight simulator) as well as observational experience and learning (as in being a copilot for a period of time).

Related idea: stickiness. The ease of transferring K is sometimes attributed to the characteristic of K known as *stickiness.* It has been observed that K flourishes best where it originated (e.g., a specific department or group). This observation speaks to the tacit aspect of K, thus making it difficult to transfer. The sequence of SECI-related ba has a varying effect on the stickiness of knowledge. For example, socialization and its originating ba promote close human contact and the sharing of tacit K, externalization less so, and combination even less. This is shown in Figure 2.9 where we observe that stickiness is the most fully developed when person-to-person contact is closest. This has an implication for organization design: socialization is optimal within a department or small group where face-to-face contact is easiest. The efficacy of organizational knowledge arises from the ability of managers to overcome the stickiness of, and connect, the K of all local ba.

Interorganizational ba—an industry example. Almost any context that produces a stream of meaning or the energy necessary to produce knowledge may be considered ba. Our description of creating this book presented earlier is one such example. As another example, consider the nature of knowledge creation across the boundaries of organizations. Ahmadjian (2004) has observed that interorganizational K creation requires context-specific ba, which depends on specifics of the participating organizations, such as their

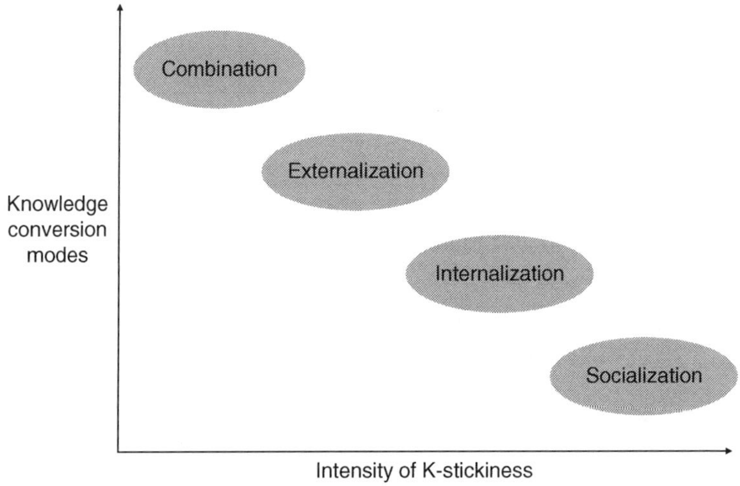

Figure 2.9 Knowledge stickiness versus people distance.

culture and the type of knowledge being created and transferred. Ahmadjian (2004) has proposed two models: a Toyota and a Silicon Valley model. The Toyota ba is centered on Toyota and contains close, long-lasting ties among relatively few organizations. The knowledge that is created is primarily tacit and process-oriented. In contrast, the ba in the Silicon Valley model is not based on an organization, but a place, proximate university and research institutes, and the shared language and experiences of participants. Here the ties are many, fluid, and short term. The knowledge that is created is primarily explicit, is recombined, and product-oriented. In Chapter 6, we will elaborate on the ba associated with interorganizational K creation in terms of culture and social networking theory.

Design in a K organization may be seen as identifying, formulating, and configuring ba. In this regard, recall from Figure 1.2 that traditional design flows from the structure to process to cognitive architectures, but in the K organization it is the reverse. Figure 2.10 illustrates one perspective in terms of ba. In traditional design, process follows structure; in more modern design structure follows process; but in knowledge design, both process and structure (separately or together) follow from the identification, design, and configuration of ba. We will discuss this aspect of organization design in more detail in Chapter 5.

2.4.4 Middle–Up–Down Management

In the K-creating organization, we may visualize the organization as consisting of three components: the top level or the *K officers*, the bottom level or *K practitioners*, and the middle level or *K engineers*. The K officers are composed of the senior managers of the organization and are responsible for strategies and

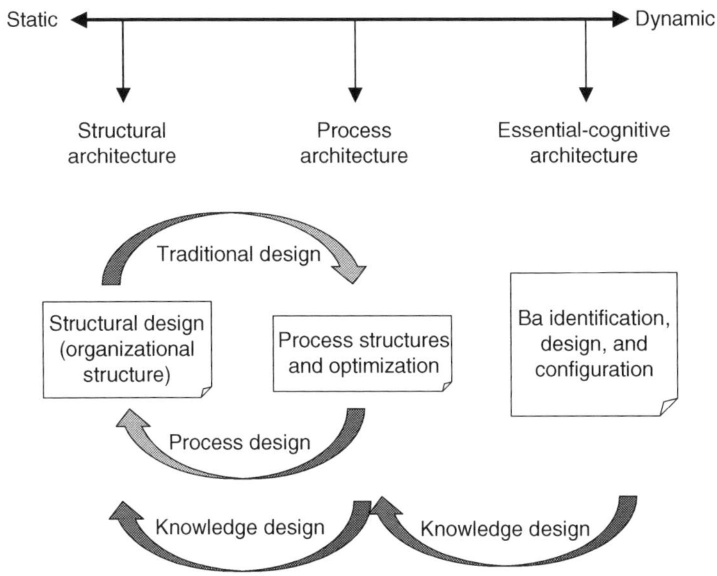

Figure 2.10 Ba and organizational design.

vision. The K practitioners are composed of the front-line employees and managers who are responsible for creating, accumulating, and applying both tacit and explicit knowledge. Finally, the middle line is composed of the middle managers; because of their centralized location between the top and bottom, it is their responsibility to synthesize and rationalize the flows of knowledge between the top and bottom. As K engineers, the middle line typically includes the lead project managers in new product development efforts and is thus responsible for the workings of the K-spiral and the SECI model. This is shown in Figure 2.11.

Note too that in a traditional organization the middle line, again because of their central location, functions as an "information relay." The three components of the K organization and the traditional organization are the same for each (top–middle–bottom). What differs is the philosophy of the organization as to its purpose: K creation in the former and information processing in the latter. It is this difference—that is, their position on the information continuum—that is responsible for the middle line as either K engineer or information relay (see Nonaka and Toyama 2004).

2.4.4.1 Implications for Organizational Design

Let us take this opportunity to elaborate on the structure of an organization. In Chapter 3, we will discuss the five generic components of structure suggested by Mintzberg *et al.* (1998): the strategic apex (corresponding to Nonaka's top level or the K officers), the technostructure (responsible for the

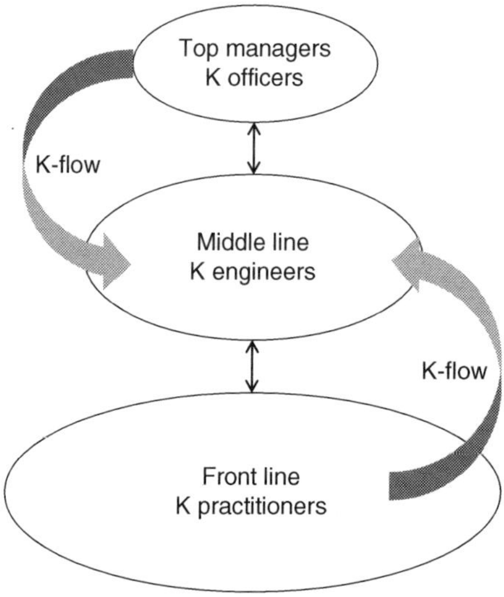

Figure 2.11 Middle–up–down management.

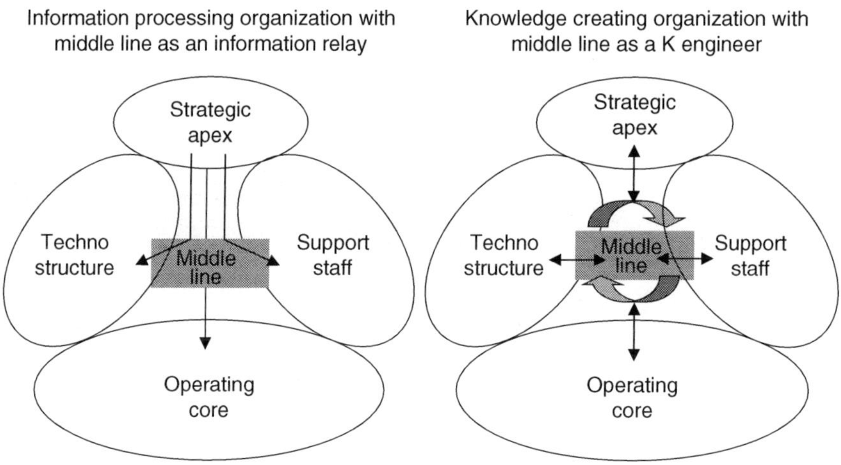

Figure 2.12 Information processing versus K-creating organizations: combining Mintzberg (1983) and Nonaka and Takeuchi (1995).

standardization of work and the development of technical systems), the support staff (responsible for collaborative and support functions), the operating core (corresponding to the bottom level or the K practitioners), and the middle line (corresponding to the middle level or the K engineers). Figure 2.12 illustrates the different workings of the middle line as either an information relay (essentially a flow-through pipe) or a K engineer (essentially fitting the pieces of a puzzle together). In this representation, the K engineer must synthesize K flows from all directions in the organization; that is, from the sides in addition to the top and bottom. Again, the image of the structural elements is the same; only the philosophy and purpose of the organization differs, which in turn affects, more than any other structural part of the organization, the work of the middle line, and with that, the organization itself.

2.4.5 Hypertext Organization

Nonaka illustrates the hypertext organization by comparing the Japanese and US militaries during World War II. Each had an army and navy organized as bureaucracies; however, the United States had the Marines, who developed a new fighting technique—a synthesis that simultaneously integrated land, sea, and air tactics. Nonaka asserts that one important reason underlying US success in the Pacific war was this new type of knowledge-creating structure, in effect, one that produced an "organizational victory" for the United States.

Let us examine the characteristics of a hypertext organization more closely. In the same way that a reader can change contexts in a hypertext document, the hypertext organization is also designed to adapt rapidly to changing contexts. The hypertext organization is composed of three primary structural components: a business system layer, project teams, and a knowledge base. This is illustrated in Figure 2.13.

The business system layer represents the traditional hierarchical organization that is charged with conducting the operations of the organization. The project teams layer is composed of any number of teams engaged in specific knowledge-creating activities, such as new product development. The lower layer is the supporting K-base that contains the organization's full set of knowledge threads, both tacit and explicit. Knowledge threads are typically contained in specifically designed KM systems, such as a K repository (for explicit K) or a

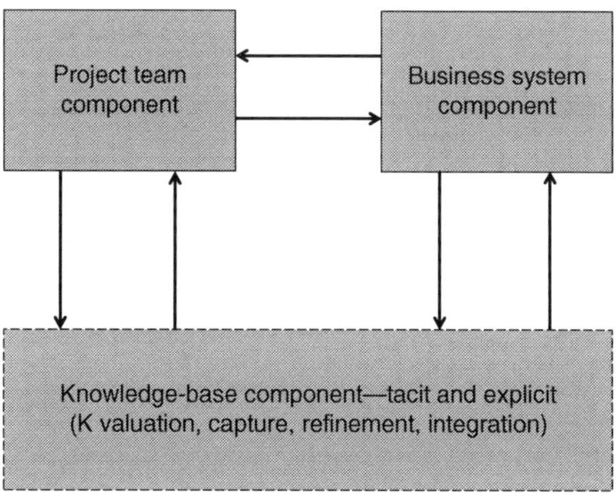

Figure 2.13 Hypertext organizational structure components.

K-directory (for people and links to their tacit K). Examples of tacit threads include the organization's culture and vision.

Observe in Figure 2.13 that the project and business layers are separated. We illustrate the hypertext organization differently than Nonaka in order to highlight the separation in governance of the project and business systems; hence, in our representation we drop the term "layer" and substitute "component." This component separation is in contrast to a matrix organization where the project structure is superimposed over the business system hierarchy. In the hypertext organization, the project team members come from the business hierarchy but report exclusively to a project manager. This too is in contrast to a matrix organization, where a given team member often has two managers: a project and departmental manager. In practice, the hypertext organization compresses communications, speeds decision making, and thus promotes middle–up–down management. Finally, the hypertext represents a synthesis between the task force, which creates and structures knowledge, and the business hierarchy, which accumulates and exploits the new knowledge. This is shown in Figure 2.14. Observe that socialization and externalization structure new K, whereas combination and internalization acquire, accumulate, and exploit new K.

Returning to Figure 2.13, each system component creates a distinct set of knowledge: project- and business-specific. The role of the knowledge base is to capture, refine, and make available both project and business knowledge threads and thus align the project team and business system components. The K-base is not an organizational structure (hence, we draw the K-base with

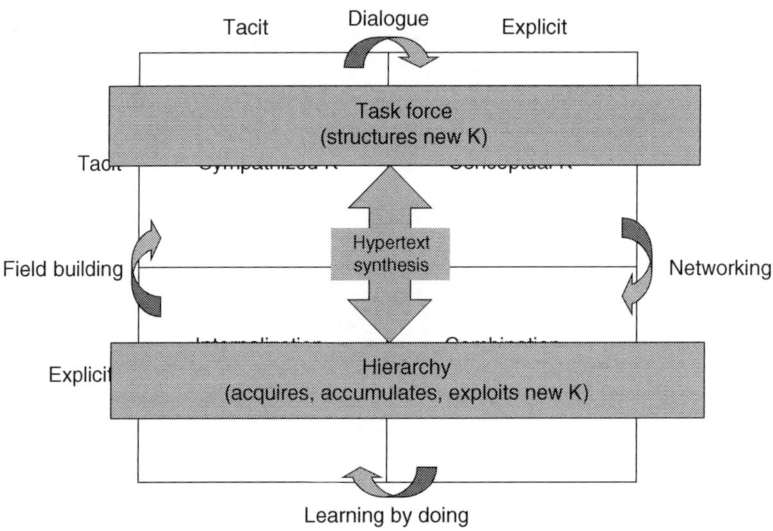

Figure 2.14 SECI and the hypertext synthesis—Adapted from Nonaka and Toyama (2004).

a dotted box), but a set of K-based systems and policies. Included, for instance, are the organization's vision, strategies, and culture as well as specific technology-based systems (e.g., data warehouse) and activities (i.e., data analytics). It is this virtual entity that holds the hypertext organization together.

2.4.5.1 Implications for Organizational Design

We have extended the design of the hypertext organization to include the internal design of the K-base constituent. For the K-base to fulfill its mission and actually work, we believe it necessary for the job function of *K integrator*, or the *K librarian*, to manage the knowledge in the K-base across all constituent systems. The job role of K librarian includes the requirement to assess the value of knowledge threads (in consultation with project and business team leaders), capture and extract the K from its project and business sources, and refine, aggregate, store, and distribute the knowledge. Also, within the K-base component itself, the K librarian integrates separate knowledge threads into composite knowledge structures. It is through such integration that an organizational K-strategy takes shape and that, for example, product design, customers, and marketing and organizational strategies are aligned. Since the K-base is not a distinct structure, there is no single manager; but in the authors' opinion, the K-base should be under the guidance of a *knowledge management officer*. Note that we are presently using emerging categories of jobs, such as the K-integrator or K-management officer. We will discuss these job roles in our discussion of K work later in the book.

(Note: though not directly addressed by Nonaka, certain K roles, responsibilities, and activities should remain embedded in the traditional hierarchy, such as those in marketing and R&D. This should have no impact on the hypertext organization concept.)

Finally, our students ask, "Isn't a project-based organization sufficiently close to the hypertext organization?" Well, it is certainly better than a traditional hierarchy, but falls short of the knowledge facilitating capabilities of the hypertext organization (and of middle–up–down management, for that matter). If we want to migrate from a project to a hypertext organization, two key elements need to be addressed. The first is shifting from a double to a single manager. This represents a not-so-subtle shift in organizational governance that may meet with resistance. As we will see later, more autonomy and less control facilitates knowledging. This goes hand in hand with a shift in responsibility of the middle line from information relay to K engineer. Second, the K-base needs to be organized—staffed and managed—as a single (though virtual) entity from both a technical (e.g., K librarian) and a management perspective (e.g., knowledge management officer). The K-base is the one entity that holds the hypertext concept and, in fact, all knowledge organizations, together.

2.4.6 Organizational and Process Enablers

The organization may pull any number of levers to enable K creation. The enablers come in different flavors, but are primarily organizational or process, and thus constitute the preconditions necessary for ba. Here we list and briefly describe the more popular enablers described by Nonaka, Toyama, and Ichijo. Von Krogh *et al.* (2000) have a particularly good discussion of K enablers:

Autonomy. Virtually every writer on KM states autonomy is a primary enabler. To quote Davenport, "thinking for a living engenders thinking for oneself."

Intention. K creation is driven by organizational intention, or simply strategic intent. This is a core component in middle–up–down management and organizational K creation.

Creative chaos. This refers to the positive tension that arises from a variety of pressures, such as time constraints, ambiguity of intent or job roles, and so on. It is creative chaos that often underlies K-synthesis.

Redundancy. Access to data, information, or knowledge is facilitated through redundancy. Today, we do not need "redundant" systems so much as ubiquitous access through the Internet (or Intranet). Systems correspond to explicit knowledge redundancy. In contrast, job rotation is a good and common way of creating redundant tacit K and know-how.

Requisite variety. K creation is enhanced through a variety of perspectives that is representative of the organization's internal diversity.

Process enablers. Ichijo has described five-process-level enables that correspond to Figure 2.8. These include the following:

1) *Instilling a knowledge vision.* This refers to the translation of business strategies into a specific knowledge vision. A knowledge vision is constituted by the current "organization" mental map, the desired future mental map, and the knowledge necessary to bridge the gap. Note that a knowledge vision is largely tacit.
2) *Managing conversations.* This is perhaps the most important enabler in that it has, more than any other enabler, a primary effect on all the K creation phases shown in Figure 2.8. In this regard, conversations either confirm the existence and content of existing knowledge or aspire to create new knowledge. Von Krogh *et al.* (2000) add that the key principles in managing conversations are (1) actively encourage participation, (2) establish conversational etiquette, (3) edit conversations appropriately (for publication), and (4) foster innovative language.
3) *Mobilizing knowledge activists.* An activist is an important enabler in the beginning of any K-oriented project, for example, the development of new knowledge (as shown in Figure 2.8) or a new business intelligence project. Knowledge activists act as catalysts, coordinators, and merchants. They do not function as knowledge officers—activists enable and influence; they do not control or manage.

4) *Creating the right context.* This refers to the right combination of strategy and structure that enables K creation. For example, it is generally acknowledged that new structures (e.g., teams, teams of teams, communities of practice, self-organizing groups) are required for K creation. By extension, we include culture, social networks, collaborative structures, and the environment as part of the context. That is to say, creating the right context involves structures (and by extension, activities) that foster relationships and collaboration. The very nature of ba is an example of an enabling context.
5) *Globalizing local knowledge.* An important aspect of K creation is the fifth phase of K creation or the cross-leveling of knowledge. Globalizing local K refers to both the K creation process itself and the dissemination of new knowledge throughout the organization. Without cross-leveling K there is a risk of ignoring local units and their accommodations to local cultures that exist throughout the world in a global company.

Care, bestowing, and indwelling. Von Krogh *et al.* (2000) have extensively analyzed and discussed the importance of *care* in K creation. The notion of care elevates the moral dimension of knowledge and inserts it into ba. Organizations with low levels of care operate as information processing companies. With low care, relationships are relatively formal and transactional with little opportunity for K creation. In contrast, high levels of care imply empathy and intimate relationships. At the individual level, high intimacy and care is known as *bestowing* and involves individual socialization and the sharing of insights. At the group or social level, high care is known as *indwelling* where the members of a group live with a concept together. In general, the enabling dimensions of care include (1) mutual trust, (2) active empathy, (3) access to help, (4) lenience in judgment, and (5) courage between members of an organization.

For example, in our case study of a nonprofit operating in the developed world (to be discussed in Chapter 5), volunteers live in a village community for 2 years, share K, and in general indwell. Note that high care involves large reservoirs of tacit K, which may end in specific, explicit actions. Organizations may design for care (this is really designing for tacit K flows). It includes, for example, an analysis of the knowledge vision, identification of necessary tacit knowledge sources (e.g., people), establishment of a caring relationship with each source, and building a K-base with each source. Of course, we may argue that these activities should be universally applied if we are to build a caring culture for any organization.

2.4.7 Single- and Double-Loop Learning

Single- and double-loop learning represent well-established approaches to learning and, to a lesser degree, KM. In single-loop learning, knowledge gathered from within a system or process is used to optimize that system.

Double-loop learning brings new assumptions and ideas concerning the original system into consideration. For example, an automobile manufacturer may refine its lean process by gathering production data and optimizing that process (single-loop learning). It may also gather external knowledge from the environment and factor that into the production process (double-loop learning). Examples of external knowledge include customer satisfaction surveys, auto dealer recommendations, government safety ratings, and so on. Finally, the organization should synthesize both forms of learning into a composite whole (deutero-learning).

2.5 New Directions in Knowledge Design

There are several ways of looking at knowledge. In terms of K creation, the dialectic synthesis embodied in the SECI model is perhaps the best known. Other models of knowledge are descriptive, such as Wiig's (1993) model, which identifies, for example, factual and conceptual knowledge. Other K models have an organizational bent, such as Sveiby's (1997) popular characterization of knowledge as an object versus knowledge as a process or, alternatively, stocks and flows, respectively. The former has an IT orientation whereas the latter has a human resources (HR) perspective. Still other models are attributive, characterizing the key dimensions of knowledge, such as temporality or validity (see Casselman and Samson 2005). Each of these representations says something about knowledge and its design or accommodations. The concepts that we present include these dimensions or attributes, either directly or indirectly. For example, the very definition of ba discussed earlier implies temporality.

As useful as existing ideas may be, if we are to move to the KM divide, we need to explore new concepts that build on and extend those in common practice. We concentrate on the creative design process: K design is like crafting a fine work of art. In effect, *knowledge design is formulating those elements necessary to summon the imagination for the leaps and transformations necessary for knowledging.*

2.5.1 Integrated Understanding of K Models and Representations

Knowledge as such is beyond our understanding. As the reader of this book will see, we have referred to and developed an extensive number of K models and representations to further our understanding. We have started our journey by exploring the ancients (e.g., developing a corresponding K-triangle) and continued through modern thought and practice. This chapter and the remainder of the book will continue to develop K representations (most of the figures in the book) and accompanying discussions—all necessary to understand K and

KM. Our first new direction, therefore, is not an artifact but a process of understanding—a journey for the leader and strategist to develop an integrated appreciation of K in all its aspects. Only with such a deep understanding can the organization succeed in the new Pangaea.

2.5.2 Thin and Thick

All things human—individuals, families and groups, institutions and organizations, universities, culture—may be described by characteristics that range from an outward, surface manifestation to a deep, essential nature. This range, a type of continuum, is a foundational idea in our approach to design.

We may construct a useful model of this continuum by borrowing from the field of ethnography: culture may be understood in terms of both outward appearance and behavior and internal tacit assumptions. Geertz (1983) has described these forms of understanding as *thin* and *thick* descriptions, respectively. Starting from this simple concept, we develop an epistemology of thin and thick organizational dimensions.

We illustrate the concepts of thin and thick by using a city metaphor first proposed by Wittgenstein (1953) in his description of language and later used by Geertz (1973) in his description of culture. As shown in Figure 2.15, the innermost neighborhood is the "old city." The old city is thickly congested: the streets are narrow, irregular, and meandering, where one can easily get lost. The old city is where the original beliefs and values of the city are rooted—its historical origins. By extension the old city is where we find, say, a society's art or an organization's tacit, cultural assumptions. Things move slowly here. Further out

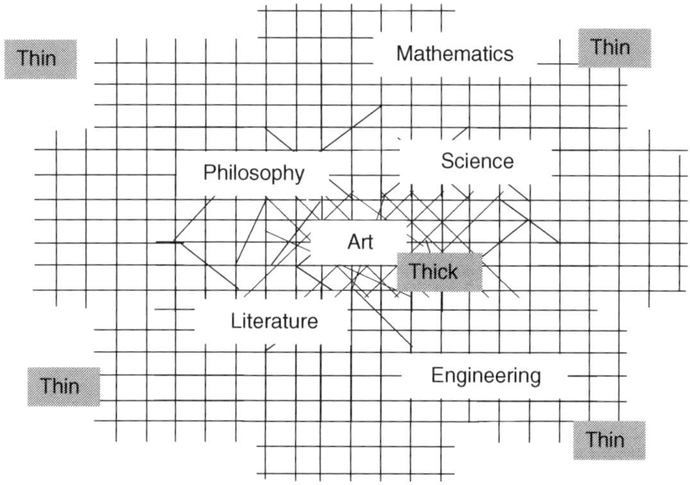

Figure 2.15 Thin and thick city neighborhoods.

from the old city the streets become more regular and clear, a little less thick. Art gives way to science and literature; similarly, an organization's strategy and business processes are part of the middle neighborhoods. Finally, the city's most distant neighborhoods are entirely thin: the streets are wide and perfectly squared. There is no ambiguity and most everything is clear and predictable. The formal discipline of mathematics resides here, as does an organization's policies and procedures, information systems, and routine workflows.

The boundaries between neighborhoods may be thought of as gates that filter what passes between them. Observe that not every street connects directly with every other and that some streets may be one-way and others two-way—that is to say, navigation around the city may be *asymmetrical* where travel is constrained by one-way pathways. In the case of two-way streets, we have more latitude since the pathways are more *symmetrical*, but travel is still somewhat constrained by the presence or absence of adjacent streets. Thus, the gates control specific movement among the neighborhoods and form the particular patterns that are the character or style of a city or organization. All organizational contexts (e.g., strategy, process, structure) have unique identities and specific relationships that shape the identity and style of a given context and the organization as a whole. Furthermore, as a new neighborhood (e.g., a new discipline or technology such as the Internet) is added to the outskirts, there is internal adjustment as new patterns form and the character of the organization emerges yet again as it reshapes itself.

2.5.2.1 Thin and Thick Characteristics

Table 2.1 lists the selected organizational dimensions and their respective thin and thick characteristics. For example, an information processing organization is likely to have thin characteristics, such as top-down knowledge construction (i.e., early K-binding), structure- and process-driven design, where work design is based on engineering analysis. In contrast, the knowledging organization is primarily thick, where knowledge may be created bottom-up (late K-binding) and design is cognitive-driven, where work design follows anthropological analysis. Each of these characteristics may be seen as a design parameter, and they may by applied separately or in combination. Furthermore, thin and thick design characteristics may be applied to every organizational context and design element as well as in every setting. As with explicit and tacit knowing (Figure 1.13), the thin and thick characteristics shown in Table 2.1 are archetypical, interrelated, and may influence each other.

Two Examples

As an example, recall from earlier that deep smarts are hidden—it resides in the head and hands of individuals and thus may be considered thick. It is constituted of both explicit and tacit knowing, includes skilled physical action, and

Table 2.1 Selected thin and thick archetype organizational characteristics.

Organizational characteristic	Thin	Thick
Philosophical orientation	Primarily scientific	Primarily phenomenal
Source of understanding	Primarily analytical	Primarily experiential
Architectural orientation	Structural and modular	Cognitive and integral
Strategy characterization	Deliberate and position-based	Emergent and capability-based
Information continuum	Toward data	Toward knowledge
Structural design	Emphasis on asymmetrical structures and hierarchies	Emphasis on symmetrical structures and communities
Design approach	Engineering-like	Anthropological-like
Knowledge construction	Early K-binding	Late K-binding
Knowledge exchange	Precise and transactional	Relational and psychological
Knowledge dimension	Primarily explicit	Primarily tacit
Operational characteristic	Process	Practice

may be technical or managerial. The importance of thickness in organizations cannot be overstated. Leonard and Swap (2004) make an informing observation on "deep smarts" (i.e., thickness) and its implications in organizational design: "Your best employees' deepest knowledge can't be transferred onto a series of PowerPoint slides or downloaded into a data repository. It has to be passed on in person—slowly, patiently, and systematically." Think about the management implications of this statement with, say, pending retirements. The expert builds an "experience repertoire," achieved through knitting and climbing the tacit hierarchy shown in Figure 1.15. Thick expertise, then, is cultivated—initiated, developed, refined—through extensive experience, often more than 10 years. Whereas thin characteristics change slowly, thick characteristics are dynamic and change constantly—thickness is a function of the accumulated experiences and knowledge of an individual, a practice group, or an organization.

As another example, consider the classroom setting. Upon entering a classroom, we immediately notice thin characteristics, separately in individual students and in the class as whole (i.e., the outward class culture). We see the number of students, their age distribution, where they sit, and so on. More importantly, the professor is faced with a critical educational challenge: to bypass all this thinness and to impart, cultivate, and elicit the necessary thick knowledge from each student, made easier if we additionally create a thick dynamic in the classroom culture.

2.5.2.2 A Painting Metaphor

We further illustrate the notions of thin and thick design with a *painting metaphor* that we have used to distinguish industrial and knowledge work. As children we have seen painting games where a picture is imprinted on a board and each impression is color-coded. The child's task is to accurately fill in a color on a corresponding impression based on the color-coding rules. The vision and knowledge of the task—its imprint and coloring rules—are *engineered by someone other than the child and provided ahead of time*; there is no discretion on the part of the child. Success is determined by how accurately the child follows the rules.

In contrast, a fine artist starts with a blank canvas and only a vision of what is being painted. There are no rules as such. The vision, style, and knowledge of the task's inner workings reside in the *head of the artist*; they are *hidden and revealed only as the picture takes shape.* Knowledge is elicited and created as the task progresses—the in-flight painting and the materials used all have an impact on the future shaping and coloring of the picture. Success is largely determined by the mood, skill, and knowledge of the individual artist and the surrounding environment in which the artwork takes place.

The first way of painting is entirely thin: templates abound, structure is provided by outside experts, knowledge is predetermined, success is easily measured, and so on. The thin design paradigm is that of the industrial organization and is still required for large chunks of any company. In contrast, thick design, which is associated with both the knowledging organization and knowledge work, is characterized by the second painting story, where the necessary knowledge resides in the head of the person performing the task and is only revealed in-flight as the final product takes shape. The structure and content of the final product, though guided by an initial vision, further develops during the process and is not fully known until the end. Nothing is fully predetermined—the psychological disposition of the performer, his or her relationships, and the environment (near and far) all impact the product.

There are several lessons associated with this metaphor—the structure of work, the character and type of people, and so on. Not immediately evident are management issues such as performance reviews. The work in the first example of the child painter is well structured and can be measured; the work in the second example of the fine artist is not very structured and difficult to measure. Yet, performance reviews in organizations are analogous to taking the artifacts used to measure a child painter and applying it to the fine artist. This is one example of the "luggage" of the industrial era imposing itself on the knowledge era—it almost always is counterproductive. We will discuss alternative arrangements in future chapters.

2.5.2.3 Thin and Thick Design

Let us illustrate thin and thick design approaches by looking at work. The design of thin work is characterized by precise specifications where the structure and workings of a process task are elucidated before the task is executed—this is termed *exactification*. Foundational to thin design is the removal of human discretion; hence, design includes detailed control mechanisms and the creation of machine-like environments. Full elucidation is necessary to support, say, process automation. The knowledge required to develop detailed specifications comes from the engineering disciplines and is developed by someone other than the person performing the task, typically an engineer or some other domain expert. This type of process design is sometimes known as *engineering* or *industrial* design and is necessary for the development of large, complex systems.

In contrast, thick work is characterized by ambiguity, where the information requirements and the method of solution are only known generally. In this case, the exact structure and workings of the task cannot be elucidated (and therefore not directly designed)—they reside in the head of the person performing the task (Drucker 1999). Hence, thick tasks can only be *framed* and design is focused on the task environment—one that promotes autonomy, collaboration, and human creativity and innovation. Note too that as work shifts in character from thin to thick, so too must the human specification from one of specific *job roles* to one of *competencies*. The essence of framing or thick design, then, is to design the surrounding context of the task so that the human participant (typically, a "knowledge worker") will be more productive and effective than otherwise.

2.5.2.4 Structuring Thin and Thick Design Spaces

Thin design is characterized by modularity where there is a clear boundary between the design itself (i.e., specification), the person performing the task, and the external environment. This is shown in Figure 2.16 where the thin design space (shaded area) is confined to the workspace. Thick design, in contrast, is integral in character, where the design cannot be separated from the person performing the task, and the environment. This too is illustrated in Figure 2.16 where the thick design space (shaded) extends *inward* into the person and *outward* to the environment. Additionally, the environment has no specific limits and may extend well beyond the boundary of the organization.

Finally, recognize that all work lies on a thin–thick continuum with attributes of each. Let us illustrate with the painting metaphor again: in an organization, almost every task may be visualized as a *partially* filled-in canvas. Those sections of the canvas that are filled in have predefined forms and colors—they represent the organization's embedded knowledge in the form of standards, policies, technical architectures, and sometimes even its culture. Metaphorically, the worker is thrown into the world of the organization. His or her objective is to paint those sections of the canvas that remain blank, while integrating a new portion with the predefined one to produce a unified whole. What varies is the proportional distribution between organization- and worker-provided

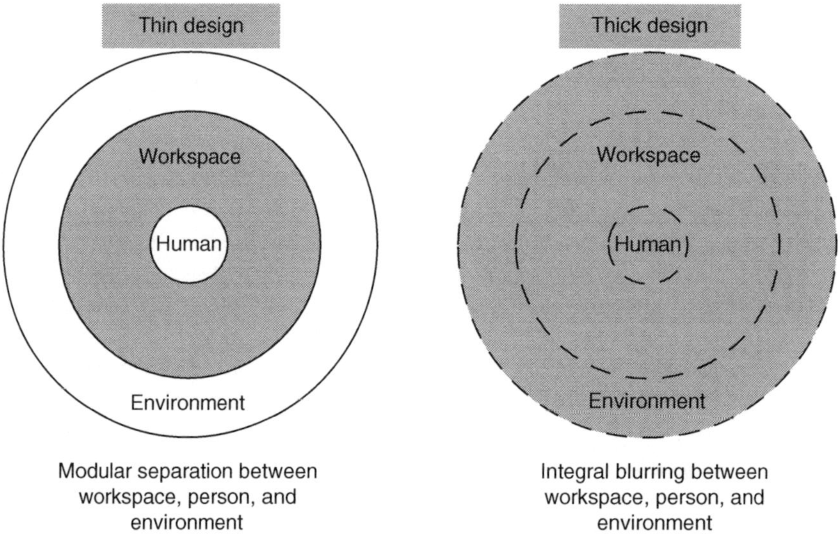

Figure 2.16 Thin and thick design spaces.

content. Organizations, in turn, must design both forms and colors (i.e., structure and content) while providing the right context and enablers necessary for the worker to successfully complete the blank portion of the canvas. *All organization work, then, represents a balance between what is partially thin (someone else's knowledge) and partially thick (the worker's knowledging capability). More broadly, all organizations are constituted with things thin and thick—work, practices, systems, structures, and so on.*

2.5.2.5 The Beginning of Design

Organizations are overwhelmed with the complexity of K design. Do we design, say, for process productivity or innovation? In fact, the organization has a series of design decisions; for example, is the organization primarily an information processing entity or a knowledge-creating one—all organizations have elements of both, it is a question of proportionality. In the spirit of synthesis, we believe it possible to create a thin–thick synthesis, but this will emerge at the organization level. In terms of the nitty-gritty of design, we don't see a synthesis as such, but rather, an intermingling and harmonizing of thin and thick elements with specific modes of interaction. This is where we begin: *imagining, designing, configuring,* and *interacting thin and thick elements*—and then everything unfolds from here.

2.5.2.5.1 Implementation: Music Architecture

During a recent classical guitar workshop given in New York City, Frederick Hand, composer and guitarist for the Metropolitan Opera, revealed to the workshop attendees (among whom was one of the coauthors) the following story. In his later years, the famed concert pianist virtuoso Arthur Rubenstein had stopped concertizing due to declining eyesight. Still he enjoyed listening to other pianists perform the piano literature. During a visit to a well-known music college, he had the opportunity to listen to many young pianists practicing their repertoires. Afterward, he remarked to Hand that he was amazed at the technique and speed of the young players he had just heard. He further declared that at their age he did not have anywhere near their levels of technical skill and speed. Still something vital was missing in their playing, namely, understanding of and expressing the "musical architecture" of the compositions that they just had played.

Music, then, comprises thin and thick interacting elements; however, musical vitality is more than skill and technique, but arises from an understanding of the music architecture itself.

2.5.3 Knowledge Spaces

The idea of a knowledge space is an extension of ba. Whereas ba is a specific context that enables K creation, *a K-space is one of an interconnected series of elements in thick design.* As shown in Figure 2.16, thick design extends both inward and outward from the apparent workspace. Figure 2.17 is an illustration of interacting K-spaces from the very small to the very large; hence, K design requires design-in-the-small and design-in-the large.

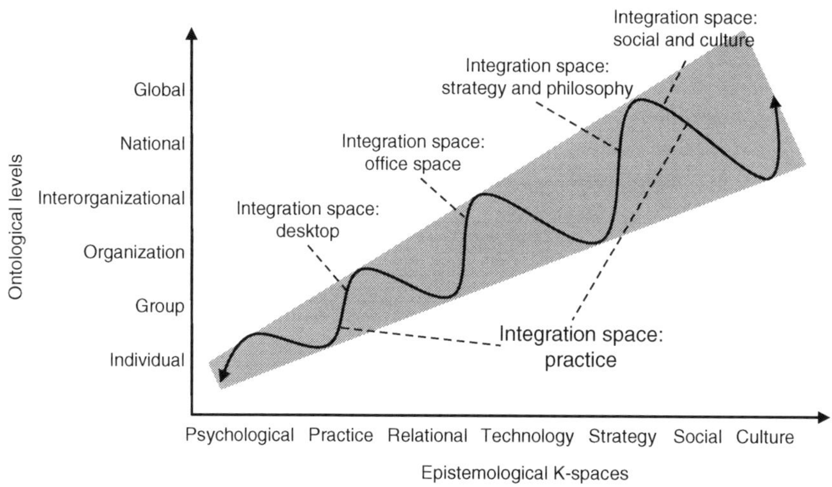

Figure 2.17 Knowledge spaces.

From Figure 2.17 we make the following observations:

- A K-space represents the structure, workings, and motivation (or energy) necessary for knowledging—creation, discovery, or application. (Aside: ba is typically focused on K creation, not knowledging as such.)
- Information flow is embedded between K-spaces. This may be used as a guide in developing supporting knowledge systems.
- We may add any number of *integration spaces* necessary to either highlight or integrate K-spaces for a given design, such as structure, history, organizational philosophy (important, for example, in Toyota), and so on. There are any number of such integration spaces, such as the following:
  - *Desktop workstation* for individuals, work, and technology.
  - *Office space* for individuals, groups and K sharing.
  - *Strategy* and *philosophy* for organizational purpose, planning, and hiring.
  - *Social* and *culture* for social interaction, culture development, and collaboration.
  - *Practice space* for the application of K and K work.
- Each space may "look in" or "look out" to an adjacent space. For example, the technology space may look in and support a given task with an embedded knowledge system, or out to support an organizational system.
- The interaction between spaces is two-way—the small influences the large and vice versa.
- Organizational design follows knowledge requirements. The organization is faced with its need to promote knowledging in all its forms. Guided by Figure 2.10, we can imagine superimposing Figure 2.17; that is, we start with some idea of required knowledge (derived from a K audit, K inventory, a new business or product effort, etc.). We then consider all the necessary K-spaces, select appropriate people, and embed them in existing or new processes and structures. Sounds simple—but that is the essence of design, which we discuss in more detail subsequently.

One of our colleagues has stated, "invention is an individual sport embedded in a social network." We visualize "individual sport" as embodying the expertise and psychological disposition of the individual K worker, whereas the social network extends outward to include his or her immediate relationships, the organization, and, eventually, the entire social space. Also, today the social network extends to the entire global community. The first task of design is to identify and account for those spaces that impact a given design—it is a task of the designer to balance all the influences, some of which may be contradictory.

2.5.4 Asymmetrical and Symmetrical Interactions

In his description of hierarchy theory, Herbert Simon describes the interactions between levels in a given system as belonging to one of two types. The first is *asymmetrical* and characterizes *molecules* as found in biotic systems. The second is *symmetrical* and characterizes *tissues*, such as crystalline or lattice

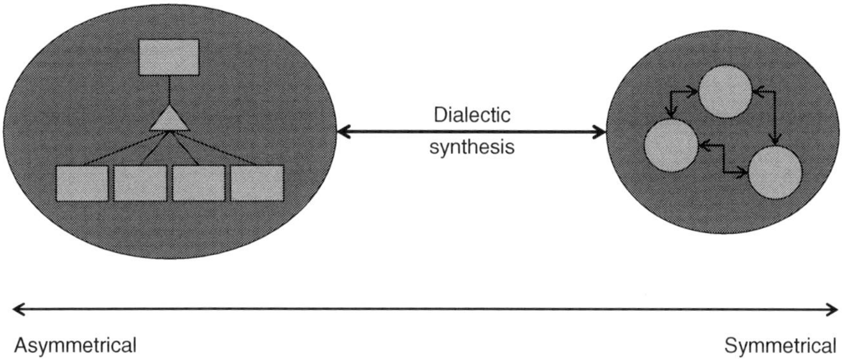

Figure 2.18 Asymmetrical and symmetrical structures.

structures as found, for example, in rocks. By asymmetrical, we mean that the behavior downward is different than the behavior upward between levels; by symmetrical, we mean the behavior is the same in all directions. This is illustrated in Figure 2.18 where asymmetrical structures are multilevel assemblies, whereas symmetrical structures are within the same level and are recursive.

Organizations and their contexts (i.e., subsystems) are primarily asymmetrical, and include, for example, process, structure, strategy, culture, and so on. These asymmetrical assemblies interact to produce what we have termed *organization molecules,* which are representative of a given organizational context, such as a "process molecule" or "information molecule." This has been described in our previous book, *Organization Modeling,* and is reviewed in Chapter 3. However, at a very low level of abstraction, organizational interactions shift in character from asymmetrical to symmetrical where the structure assumes a self-organizing and conversational mode. As importantly, symmetrical structures may be designed, made more formal, and placed throughout all levels in the organization. It is this combination of *interacting asymmetrical and symmetrical organizational elements that we hypothesize is the fundamental (social) architectural structure of K organizations.*

Architectural Structure of the Knowledge Organization
The architecture of the K organization is embodied in the interaction and synthesis of symmetrical lattice-like human interactions and asymmetrical molecular (context) structures

2.5.5 Layers of Thin–Thick Interactions

The essence of organizational knowledge design is the interleaving of thin and thick contexts. As we will presently see, a given context is neither inherently thin nor thick, but it may be designed as one or the other. We architect thin and thick characteristics by creating thin and thick layers of organizational elements, and then establishing interconnections.

Designing the Knowledge Organization
The essence of organizational knowledge design is the deliberate design and interleaving of thin and thick contexts

Figure 2.19 is an illustration of what we mean by the interleaving of thin and thick contexts or layers. In this case, we are focusing on work, but the same idea holds for other contexts such as strategy.

From Figure 2.19 we make the following observations:

- The organization may be visualized as a set of layers or subsystems, each characterized as predominantly thin or thick.
- The lowest level contains individual tasks. These tasks are of two types: knowledge tasks and administrative tasks, the latter being any nonknowledge task (the term "administrative" is used merely for convenience).
- Knowledge tasks are distinguished by a *practice* perspective. The term practice refers to a thick perspective of work, and includes, for example, anthropological-like analysis, high levels of autonomy, an integral architecture where we design the immediate psychological, relational, and practice K-spaces, the standards and professional reference of the practice, and so on.
- Administrative tasks are distinguished by a *process* perspective, where the term process refers to a thin perspective of work. This embodies modular design, which permits, in turn, engineering-like analysis, precise specification using formal terminology such as invariants (business rules that can never be violated), pre- and postconditions (before and after states), and triggers (action initiators). Process typically carries moderate to high levels of formalization.
- The collection of tasks (knowledge and administrative) comprise a thin process perspective (the level immediately above the task level). This includes, for example, a process map or workflow showing task names, their interrelationships and dependencies, business rules, and so on. Every task is well defined with clear boundaries, and, hence, is modular.
- The next level is a thick practice perspective. This level is characterized by an integral architecture where there is a deep, tacit understanding of the entire process. As an example, one of our graduate students (a Ph.D. in biology) who works for a large pharmaceutical company observes a diminishing thick practice level, resulting in more drug safety issues. The thick practice has been replaced with localized thin and thick tasks at the low level; though sophisticated they tend to replace and thus weaken the higher-level practice perspective.

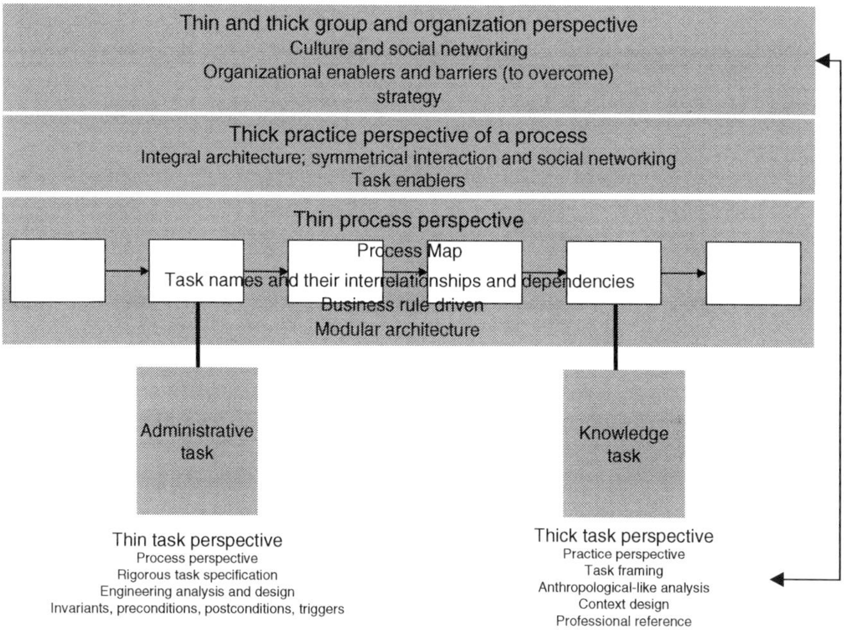

Figure 2.19 Thin and thick layers.

- The top level represents the remaining contexts of the organization, including its strategy, structure, culture, and so on. Each of these, in turn, may be subdivided into their respective thin and thick perspectives.

2.5.5.1 Implementation: An Example from Film and Medical Research

How do we actually implement thin and thick layers? We will use an illustration that we have created, based on the article "The Conspiracy to End Cancer" by Saporito (2013). It has been standard practice that medical research is narrowly focused: a "hero scientist" approach where research is conducted by one researcher, one grant at a time. For a variety of reasons, a group of Hollywood producers thought the research in cancer was moving too slowly and proposed and *implemented* a different model, one based on the film industry. Filmmaking is creative and, perhaps, the process of finding a cure for cancer should also be creative. Figure 2.20 shows a filmmaking organization on the left and a corresponding medical research organization, one derived from filmmaking, on the right. We can make the following observations:

Filmmaking organization. The lowest level comprises individual tasks, some administrative (film crew) and others K-oriented (actors). (These are comparative classifications: the film crew has experts too, but it is the actors who

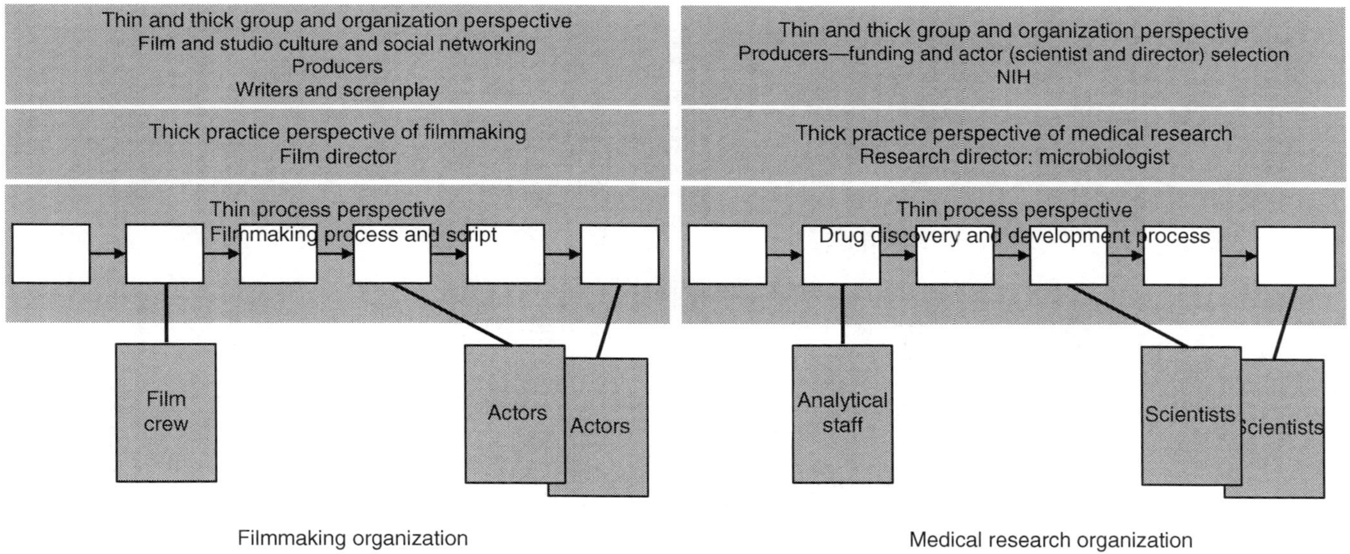

Figure 2.20 Thin and thick layers in film and research organizations.

represent the key thick and creative participants.) The level above the individual task level is the thin process perspective corresponding to the filmmaking process with its task interrelationships and dependencies. The next higher level is the thick process perspective—the film director embodies this level. Finally, the highest level is a mix of thin and thick and corresponds to the work of the producers and the film culture.

Medical research organization. Modeled after the film industry, the lowest level is composed of research scientists, even Nobel Prize winners. The next highest level is the thin research process in which the scientists and analysts participate. The level above is the thick practice level; in this case the director is a microbiologist who provides direction, reviews, and integrates the separate efforts of the individual scientists (the microbiologist is not a manager, but a K director). The highest level is the funding level.

Communications mode. Note too that in both types of organizations the interaction between thin and thick levels is primarily an interactive dialogue between the director and the key participants. That is to say, communications in a knowledging organization is symmetrical, exists at all levels, but often revolves around the thick director.

2.5.5.2 Implementation: An Example from Toyota

In contrast to thin and thick layers, Toyota formulates thin and thick *strands* of "DNA" that are connected by *cross-links* that provide the communications between thin and thick strands. One strand of DNA represents the *Toyota Production System (TPS),* whereas the second strand denotes the *Toyota Way.* The former is the visible production process and may be considered thin, whereas the latter embodies the history and culture of Toyota and may be considered thick. The success of lean and TPS stems from, and is intimately tied to, the Toyota Way; that is, the success of the organization and its thin characteristics are derived from its thickness. The *linkages* between strands include specific practices, systems, policies, and structures, such as quality circles. This is analogous to our design approach—in place of thin and thick layers Toyota uses thin and thick DNA strands. We consider the DNA idea to be an example of what we are discussing and will elaborate in future chapters.

Note too that Toyota is a larger and more complex organization than a filmmaking or research organization. The idea of thin and thick still holds, whether in the form of layers or DNA, but the mode of communications between levels or strands is more varied and formal in Toyota. In general, these modes involve personal engagement, and include, for example, visual management ("go and see for yourself"), quality circles, or policies such as developing and promoting leaders from within the organization (as opposed to hiring outside experts). Coaches, sponsors, process owners, value stream team leaders, and others abound.

2.5.6 The Creative Process of Knowledging: CPK

When a given design (e.g., a piece of art) is complete, it becomes a template to be used throughout the organization—that specific instance of creation comes to an end. However, the creative process continues—we call this the creative process of knowledging (CPK). In particular, the outcomes and learning of the original design should, ideally, produce a transformation within the creative organization (as it does within the artist) and lead to another transformative design that, in time, becomes a future template. When knowledge gets organized, it "grinds out" through the CPK cycle an algorithm, which subsequently leads to a metaphor that produces a collection of templates that can be used to instantiate designs.

The progression of creation follows a spiral-like thick process with a series of thin offshoots that represent given designs (Figure 2.21). The spirals in CPK and SECI are related but nuanced: the CPK represents the hidden pattern of a creative entity, such as an artist, scientist, or organization. The creativity never ends as each innovation builds on previous innovations, sometimes radically so, depending on the transformative leap between designs. This transformation of the imagination lies hidden in the CPK spiral—a kind of incubator.

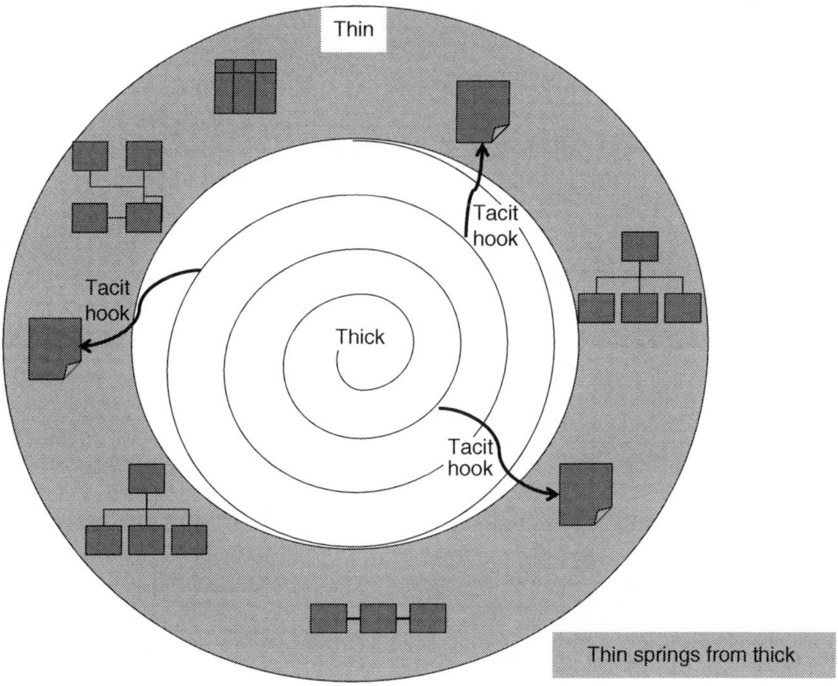

Figure 2.21 The organizational creative process of knowledging.

The CPK cycle represents a historical progression of the imagination—in an individual or an organization. In contrast, the SECI K-spiral represents the cross-leveling of knowledge among increasing communities in a given development effort. SECI represents a given K creation effort, which may be unrelated to other K creations efforts.

The driving force of CPK is the internal thick spiral that represents the organization's leaps and transformations between each new (and past) design. That is to say, *creative designs (or threads of new knowledge) build on history and past K-threads.* This is illustrated in Figure 2.21.

2.5.6.1 Implementation of the Creative Force

How is this accomplished in an organization? There is no simple answer or picture; one purpose of this book is to present ideas that collectively enable the organization to become a creative force. We are presented with the same challenges as in implementing SECI. And many of the solutions are the same, except on a broader scale. Here we highlight several items that stand out as particularly important.

Discourse Structures: Conversations

An essential feature of the creative force in organizations is human interaction. The implementation of symmetrical configurations, drawn from the entire spectrum of K-spaces, will promote conversations. Symmetrical or discourse structures include everything from dialogue and self-organization to collaboration and social networking. Furthermore, we must consciously place these discourse structures throughout the levels of the organization. This is shown in Figure 2.22. Examples include communities of practice, collaborative technologies, open office design, and so on.

Discourse structures take many forms and may be formalized; for example, Toyota promotes thickness through production cells, problem solving (groups and worker), and learning resource centers. Also, discourse structures need not be self-organizing but may be intended and embedded in working arrangements, such as collaboration between research organizations (to be discussed in Chapter 6).

Architecture

The essential architectural feature of the creative force is the configuration of thin and thick layers in the organization (see Figures 2.19 and 2.20). An alternative design is to use thin and thick DNA-like strands in place of layers (to be discussed).

In the same way that K creation requires explicit–tacit interaction, the K organization must promote thin–thick interaction. The "cross-linkages" used by Toyota in its DNA design is representative of how we simultaneously support thin and thick elements (whether in DNA strands or organization layers)—these include specific practices, systems, policies, and structures.

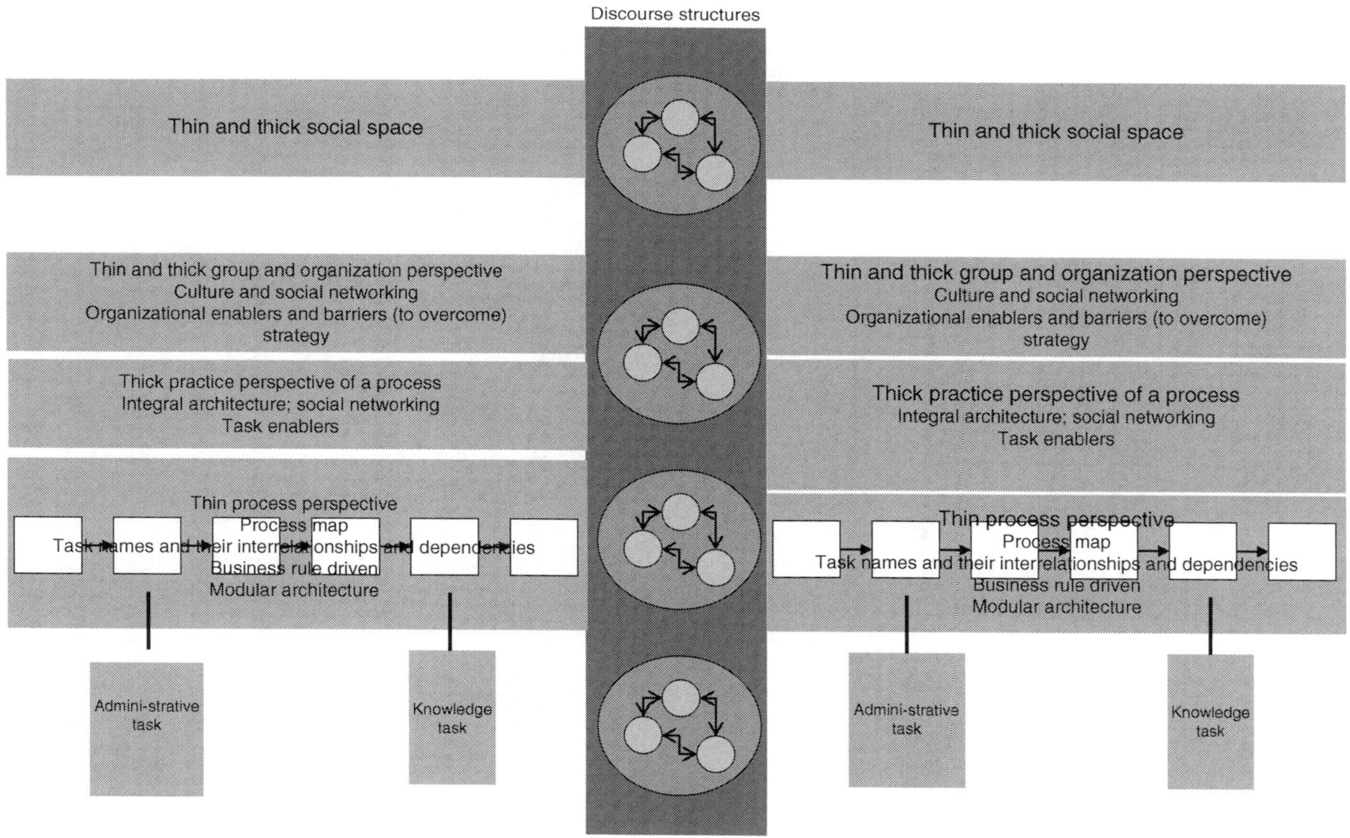

Figure 2.22 Discourse structures and organizational layers.

Finally, integrated design is critical in the K organization. This covers the full spectrum of K-spaces shown in Figure 2.17. We have suggested four integration spaces: desktop, office space, strategy, social and culture, and practice—each will be discussed.

2.5.6.2 Tacit Hook

We sometimes use the idea of a "tacit hook" as a metaphor to illustrate tacit–explicit interaction. For example, it has been observed that practitioner journals often publish a framework or some other artifact that on the surface appears to be useful and has been successfully implemented in a case company. An issue arises when another company tries to implement the same artifact and meets with less than expected results. Without going through something like a CPK, the organization produces meaningless artifacts. All things thin (a framework or some other artifact) should arise from what is thick—a "tacit hook" is necessary. Without the hook, nothing works!

Tacit hook Example. Recall that Toyota uses a DNA metaphor containing a thin and thick strand. The thin strand, known as the Toyota Production System (TPS), contains explicit artifacts (e.g., just-in-time) and is characteristic of lean production. When companies adopt lean techniques, they are emulating those explicit artifacts that lie on the thin TPS strand; however, such organizations often meet with less than Toyota-like results. The reason is explained by Toyota itself: Toyota's competitive advantage does not come from what is visible (i.e., thin), but from what is not visible (i.e., thick). This thick strand is known as the Toyota Way and embodies Toyota's history, principles, and beliefs. Furthermore, the thin and thick elements are *connected* through the cross-linkages that simultaneously connect and reinforce each thin and thick element. (Observe that this representation is not very different than the Polanyi model.) The whole construct—thin (explicit), thick (tacit), and connections—embody the CPK process. Adopting solely thin artifacts produces a temporary impression of success; advantage arises from the entire CPK process that defines and connects thin and thick. To be successful, all thin (visible) artifacts must be supported by thick constructs—this is what is meant by the tacit hook and is one lesson of the Polanyi model. See Liker (2004).

2.5.6.3 Internals of the CPK Spiral

Everything that is thin—all the artifacts of structure, strategy, and process—follows from what is thick. *In turn, thickness embodies the accumulated knowledge and experiences of a person, group, or organization.* What lies in the thick creative spiral? Obviously, the answer is not so easy; otherwise, we would all be geniuses. In particular, we think of the thick creative spiral as a kind of *cultural practice* where the following is a sampling (the list is endless):

- Individual experimentation; that is, everyday activities are seen as an opportunity for an experiment where a hypothesis is advanced and subsequently field-tested through detailed particulars. In addition to the development of a new idea, this is particularly effective at eliciting and cultivating expertise. For example, the internal policies of Toyota encourage individual workers to identify process improvement opportunities that are then tested (see Spear and Bowen 1999). In the same way that Toyota considers it the worker's responsibility to uncover improvement opportunities, we consider the knowledging organization to be one where its workers are expected to experiment, teach, learn, and lead (Drucker 1999). In turn, the K organization should provide the worker with the necessary authority, training, autonomy, care, and trust (Von Krogh *et al.* 2000).

- As implied by the tacit hook, individuals and groups need to actually implement and produce explicit artifacts. This is linked to experimentation and, together, this generates a creative force. We call this energy "late knowledge binding" where knowledge is generated simultaneously with implementation. (See section 2.5.7.)
- Rather than punishment for failure, there are rewards for success.
- Social interaction of all types is encouraged.
- Expert leaders are developed throughout the organization and provide practice guidance.

2.5.7 Knowledge Binding

An organization is a *human construction*. This means that an organization is human-centric and therefore knowledge-centric. An organization's success rests on its knowledge and, as importantly, on the process by which that knowledge is created, gathered, and applied. We refer to this *process of knowledge construction* as *knowledge binding*. Novak (1998) points out that however so constructed, an organization and everything in it is human constructed: people and the organization are knowledge-creating and knowledge-implementing entities.

More formally (i.e., from a modeling perspective), knowledge binding may be seen as the *degree of separation* between specification and implementation. This is shown in Figure 2.23 where we use, for illustrating purposes, an

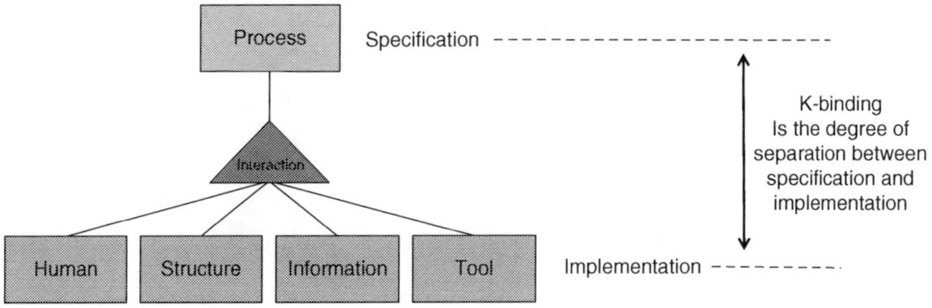

Figure 2.23 Knowledge binding.

asymmetrical process molecule. The philosophy underlying the level of separation between specification and implementation of an organization or system determines its knowledge binding—the process of knowledge construction.

Knowledge may be applied *early*, as is typical in industrial-era organizations with largely structural architectures, or much *later*, which characterizes entrepreneurial or knowledging organizations. In the former case, a specification (i.e., knowledge) is applied well before its implementation, while with the latter there is much less separation. In an extreme case, such as an artist painting a portrait, knowledge binding is *real time*: the final "specification" is realized only when the implementation is fully complete. Table 2.2 summarizes several K-binding characteristics.

The separation or distance between specification and implementation—knowledge binding—is a metaphor for possible mechanisms underlying how knowledge in an organization is actually constructed. For example, when

Table 2.2 Knowledge binding design characteristics.

Organizational dimension	Early K-binding	Late K-binding
Environment	Static	Dynamic
Uncertainty and ambiguity	Low	High
Communications flow	Vertical and asymmetrical	Horizontal and symmetrical
Management processes	Top-down	Middle-up-down
Organizational governance	Hierarchical	Heterarchical
Planning and cycle horizon	Long	Short
Lead time: specification to implement	Long	Short
Epistemological orientation	Analysis	Synthesis
Knowledge type	Explicit	Tacit
Degree of freedom	Constrained and limited	Creative and many
Occurrence of disjoint ideas	Unwelcome and discouraged	Welcomed as possible opportunities
Experimentation and risktaking	Few, predictable with low risk	Many, unpredictable with high risk
Knowledge sources	Top of division of labor	Throughout division of labor
Information requirements	Complete and known	Incomplete and not fully known
Implementation character	Functionally equivalent (testable)	Diverse outcomes (nontestable)
Architectural artifacts	Elaborate and many	Simple and few

strategy is formulated, there is a discernible separation between specification (i.e., strategic plans) and its implementation—this is a classical application of early K-binding. This is characteristic of top-down strategic planning approaches such as the positioning school made popular by Michael Porter (1980). In contrast, when the separation is less clear and formal (i.e., late K-binding), the strategy "forms" and characterizes the learning or emergent schools of strategy (Mintzberg *et al.* 1998). As another example, the constructs associated with the construction of information systems have traditionally contained the artifacts of early K-binding; for example, structured data and process models. Increasingly, however, the need for speed and flexibility has shifted the information planning process to one of late K-binding—from a technical and formal process (early) to a more dynamic and social process (late) (see Earl 1993).

The central theme underlying knowledge binding is straightforward: early knowledge binding is distinguished by the creation of knowledge by relatively few people, typically senior managers or domain experts. That knowledge is subsequently operationalized down the hierarchy through the division of labor. In contrast, late knowledge binding requires the accumulation of knowledge from the division of labor, with rapid, or even unity of implementation.

2.5.7.1 Implementation: An Example from Systems and Software Development

The systems development process has undergone several transformations. In the "early days" of the 1960s and 1970s, software development was informal and craft-like. The systems were comparatively simple and focused on applications that supported the work of a single functional department. As the need for complex and integrated systems arose, there was an attempt to remake systems development into an engineering-like process. This led to the "information engineering" approaches (e.g., business system planning) that were popular in the 1980s. These approaches emphasized architecture and formal integration, but they were extraordinarily time-consuming and expensive and thus suitable to only the most static firm environments. Each of these development approaches are distinguished by the process and artifacts associated with early knowledge binding. In contrast, the "organizational" approach is essentially social in nature and is characterized by learning, experimentation, and other artifacts of late knowledge binding (see Earl 1993).

Today, we recognize that both approaches have their place. The management issue is not necessarily to choose one or the other, but to recognize the benefits and limitations of each, and optimize the selection of either, or some hybrid, depending on the situation. Boehm and Turner (2003) have suggested using risk analysis to define and balance the two opposite approaches of plan-driven (early K-binding) and agile (late K-binding) methods. Table 2.3 summarizes the characteristics of each method (see Boehm and Turner 2003).

Table 2.3 K-Binding example—project characteristics for plan-driven (early K-binding) and agile (late K-binding) software development methods.

Project characteristic	Plan-driven (early K-binding)	Agile (late K-binding)
Primary goals	Predictability, stability, high assurance	Rapid value, responding to change
Environment	Stable, low change, project, and organizational focused	Turbulent, high change, project focused
Planning and control	Documented plans, quantitative control	Internalized plans, qualitative control
Communications	Explicit documented K	Tacit interpersonal K
Requirements	Formalized project, capability, interface, quality, foreseeable evolution requirements	Prioritized informal storied and test cases, undergoing unforeseeable change
Development	Extensive design, longer increments	Simple design, short increments
Culture	Comfort and empowerment via framework of policies and procedures (thriving on order)	Comfort and empowerment via many degrees of freedom (thriving on chaos)

Adapted from Boehm and Turner (2003).

2.5.8 Agreements and the Exchange of Knowledge

We use the concept of a *contract* to describe (or specify) the behavior and exchange of knowledge between people or agents in an organizational context. Contracts are complex. A contract has a recursive character—a given contract contains a mix of tacit and explicit K that, in turn, embodies the knowledge exchange itself. Knowledge exchange takes place between thin and thick contexts and within discourse structures. In order to accommodate the range of thin and thick characteristics of organizations, we have had to draw on a variety of perspectives, including traditional IT techniques, legal standpoints, and psychological perspectives; in turn, this leads to an updated contract continuum to explain the complex behaviors of organizations. *The enablers of K creation and exchange discussed earlier are actually the working of contracts.* As the nature of information or knowledge shifts from explicit to tacit, so too must the assumptions and rules of K transfer change—this is embodied in the various types of contracts.

2.5.8.1 An Organizational Perspective of Contracts

Economic Perspective on Contracts and the Organization

Organizations engage in a variety of relationships that may be described with contracts. Contracts say something about the behavior of the parties participating in the relationship and come in many forms. The appropriateness of a given type is determined by the situation. In classical organization theory the parties to a contract are people and organizations. But any agent that engages in behavior—software, for example—may be described with a contract.

Jensen and Meckling (1976) have proposed a theory of the firm where the organization serves as a *nexus of contracts* among individuals, both within and outside the firm. The contracting relationships serve as a framework in which the conflicting goals of individuals are brought into equilibrium. The firm, in turn, exhibits complex behaviors that are the output of the equilibrium process.

Contracting relationships vary in type and application. Some contracts are formal and explicit, whereas others are informal and implicit; some are long term, whereas others are short term and even immediate (i.e., a spot contract); and so on. Formal contracts may be legally enforceable, whereas informal contracts are often "self-enforcing" through a reference to some standard: personal, professional practice, cultural norms, and so on (Baldwin and Clark 2000). A *contract structure* describes the set of contracts of a firm and includes contracts of all types (Alchian and Demsetz 1972).

The parameters of a contract—its assertions—represent the constraints, obligations, and expectations governing each party to the exchange. It is often useful to use standardized contracts (see Coase 1937)—the assertions are known to the participants, and thus the contract "specification" need not be reinvented. In fact, an entire organization—its contract structure—may be characterized by a standardized form along a continuum of possible forms. Public corporations and partnerships, for instance, represent two standard form contracts, each with unique characteristics that are suitable for the context in which each type of firm operates (Hart 1989).

A Range of Contracts

Industrial-era organizations are characterized by contracts that are, for the most part, mechanical descriptions of work. In fact, such specifications lead to the routinization of work. These contracts are *precise* and have a certain exactness and clarity that leads to engineering-like analysis in the design of a process perspective. In contrast, the knowledge era has brought a new requirement—the anthropological analysis and design of a practice perspective. In this case, mechanical descriptions are not sufficient; rather, *values* and *expectations* are often the primary parameters in the contract. In a modern organization, we must be able to specify both forms—everything from routine workflows to cultural processes—and thus accommodate both precise and rather implicit intentions.

The contract is distinguished by a certain intent, which may be communicated directly with explicit assertions, indirectly through a process of assimilation, or, of course, some combination of them. Thus, the concept of a contract embodies a range of intentions, from explicit assertions such as "business rules" to implicit intentions in the form of strategic intent, cultural assumptions, or personal intentionality.

2.5.8.2 Type 1: Precise Contracts in Fully Explicit Systems

The essential property of a precise contract is that it explicitly defines some behavior or exchange in a system. The semantics of the specified behavior must be precisely and fully defined—there can be no room for ambiguity or interpretation. This requires the use of fully explicit *assertions*—logical statements or functions that define the parameters of the contract. Precise contracts are ideal constructs for specifying the behavior of systems with clearly defined inputs and outputs, such as information systems, most workflows, and corresponding workflow management systems. The formal use of assertions and contracts, in fact, is increasingly common in the development of high quality, complex software (Meyer 1992).

Kilov (2002) describes four categories of assertions: *invariant, precondition, postcondition,* and *trigger.* An invariant is a logical assertion given in the form of rules that the behavior described in the contract must not violate. A precondition lists the inputs and system states that are required to exist before the contract can be satisfied. The third component of a contract is the postcondition that lists the outputs and system states that are expected after the contract is executed. The last component is a trigger, which lists the events and states that initiate execution of a contract. Triggers, pre-, and postconditions are functions of time. In contrast, an invariant is the background of procedures, rules, and relationships that never should change (and hence, an invariant is not a function of time). Collectively, these four constructs constitute a mechanism for the *precise specification of behavior*—and hence are most suitable for fully explicit contracts.

Precise contracts typically exhibit an all-or-nothing pattern of operation: each and every assertion must be logically true if the behavior described by the entire contract is to exist. The all-or-nothing precise contract is typically the type that is used in software. A precise contract may also be used to describe behaviors with a parameterized range of possible pre- and postconditions. Such *relativized* contracts may be used to describe decision-making processes. The business rules of relativized contracts, stated with assertions, are sometimes called *decision rules.* Both all-or-nothing and relativized contracts are explicit in that the assertions may be described with clearly defined statements, algorithms, or functions. Furthermore, relativized contracts may be dynamic in that the assertion parameters may be dependent on a feedback loop, thus promoting a certain amount of learning with each instance or occurrence of the process. Both all-or-nothing and relativized contracts lend themselves to quantitative analysis.

2.5.8.3 Type 2: People and Exchange—Transactional and Relational Contracts

Outside of formal systems analysis, contracts, as are commonly used, involve people and some sort of an exchange or an expectation of an exchange. Such contracts concern human behavior and are simultaneously individual and social in nature. Any discussion of contracts must include the individual, his or her social relationships, and context. Certainly, contracts can be studied in isolation, and for certain analyses this may be useful, but we must always recognize that most such analyses are necessarily incomplete (see MacNeil and Gudel 2001). For example, in our discussion of relativized contracts and decision making earlier, the feedback loop by which most managers operate may be structured in a precise way, but the latitude and interpretation of the manager is largely a function of his or her context, organization, and relationships within that organization. *All contracts involving human beings, even transactional contracts, have a relational and even psychological component.* A key design issue, then, is to know when we can isolate a transactional contract from its relational context in order to solve some restricted analysis, or when we have to recognize that we must accommodate an array of contracts to specify our intention.

Both transactional and relational contracts concern exchange and coordination. In fact, there is a continuum of contracts, where at one end we have the transactional pole and at the other the relational pole. This represents a *continuum from discrete exchanges to ongoing relations.* Discrete exchanges are relatively short in duration and involve limited personal interaction with no entangling responsibilities beyond what is clearly and precisely specified. The parameters of these transactional contracts are fully defined and, in fact, are not very different from that of precise contracts. In contrast, relational contracts tend to be either long duration or ongoing, involve close "whole person" relations, and the objects of exchange typically include both easily measured quantities and not easily measured outcomes. Table 2.4, adapted from MacNeil and Gudel (2001), lists selective differences between transactional and relational contracts. Observe that though derived from the legal profession, transactional and relational contracts represent manifestations of explicit and tacit K exchanges and transfers, respectively.

Example. The transactional–relational poles form a contract continuum that describes human behavior and exchanges. An example of a contract that is toward the transactional pole is a labor contract where, by design, all relational components are removed from the contract. It is essentially an agreement where every anticipated contingency is explicitly stated and accommodated. Hence, most labor contracts are quite lengthy with little or no flexibility. A labor contract is negotiated and covers a finite period time, after which another agreement is required. Its sanctions are primarily legal. In contrast, an employment contract (e.g., an offer letter) is toward the relational pole and

Table 2.4 Contract comparison between transactional and relational contracts.

Concept	Transactional pole	Relational pole
Types of communications	Limited, linguistic, formal	Extensive, deep, not limited to linguistic, informal in addition to or in lieu of formal
Subject matter of satisfaction	Simple, monetizable economic exchange only	In addition to economic, complex personal noneconomic satisfactions, social exchange, nonexchange
Measurability and actual measurement of exchange of other factors	One side of exchange is money; other side is easily monetized, both are measured; no other aspects	Both exchanges and other factors are relatively difficult to monetize or otherwise measure, and parties do not monetize or measure them
Duration	Short agreement process, short time between agreement and performance; short time of performance	Long term; no finite beginning; no end to either relation or performance
Commencement and termination	Sharp in by clear agreement; sharp out by clear performance	Commencement and termination, if any, of relation likely to be gradual
Primary focus of planning	Substance of exchanges	Structures and processes of relation; planning of substance of exchanges primarily for initial period
Planning	Very complete and specific	Limited specific planning of substance possible; extensive specific planning of structures and processes
Tacit assumptions	Inevitably present, but inherently relational and anti-transactional	Recognized aspect of relational planning, without which relations cannot survive
Bindingness of planning	Planning is entirely binding	Planning may be binding, but often some or all of it is characterized by some degree of tentativeness
Incidence of benefits and burdens	Shifting or other specific assignments of each specific benefit and burden to one party or other	Undivided sharing of both benefits and burdens
Transferability	Entirely transferable	Transfer likely to be uneconomic and difficult to achieve
Altruistic behavior	None expected or occurring	Significant expectations of occurrence
Time-sense in view of participants	Presentiation of the future	Futurizing of the present

Adapted from MacNeil and Gudel (2001).

contains both explicit statements and tacit assumptions. Hence, it is fairly small, often no more than one or two pages, and quite flexible. Moreover, it is reinforced or undermined during execution of the contract—the primary means of control are the expectations and interests of the parties to the contract (see Kay 1995). It should not be surprising that the labor contract (transactional) refers to primarily routine work whereas the employment contract (relational) refers to K work.

2.5.8.4 Type 3: Psychological Contracts

While transactional and relational contracts have some level of formalism, psychological contracts describe the relationship between an individual and an organization in entirely unstated expectations. These expectations, in turn, guide and characterize the behavior of both the individual and organization (Handy 1993).

Schein (1980) has described three categories of psychological contracts representing increasing levels of involvement: coercive, calculative, and cooperative. While each exists in an organization, one will predominate and thus characterize the organization's prevalent mindset. This comes closest to our notion of an essential architecture, where the predominant level of involvement in an organization will govern the motivation, cooperation, and inventiveness of its employees, and, ultimately, the organization itself. *The essential architecture of an organization—primarily described by its set of psychological contracts—is simultaneously individual and organizational in nature.*

Coercive, calculative, and cooperative contracts represent a range of attachments and commitments from involuntary compliance to transactional behavior to fully engaged dedication. In practice, calculative contracts are associated with the transactional pole of the contract continuum whereas cooperative contracts are associated with the relational. (The coercive contract does not appear on the legal continuum of Type 2 contracts since there is no "agreement" among participants in a coercive contract, such as a prisoner and a prison.) Whereas the calculative contract appears to predominate in most organizations, cooperative contracts are increasingly necessary if the organization is to flourish in the knowledge society.

2.5.8.5 A New Contract Continuum

Key to understanding the many types of contracts is to recognize that we are describing behavior with *varying levels of K richness,* from information to explicit to tacit knowledge. To accommodate discrete exchange associated with computer processing at one extreme and human inventiveness on the other, it is necessary to extend the legal system's contract continuum (Type 2) beyond the transactional–relational continuum in both directions: we extend to the left to accommodate purely thin behaviors common in computer systems (Type 1), whereas to the right we extend to embrace the psychological

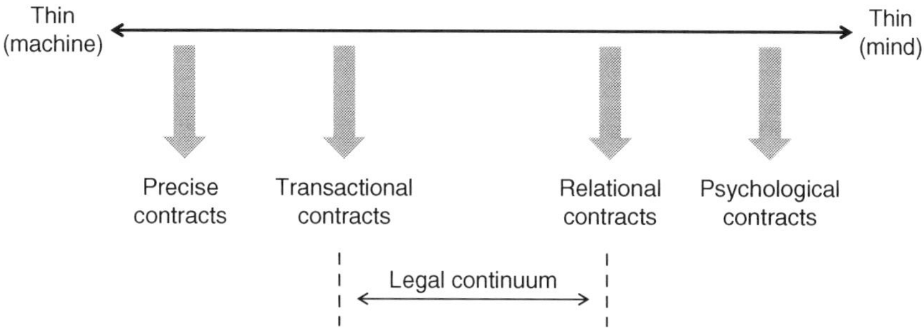

Figure 2.24 A new contract continuum.

orientations of people and their organizations necessary to build a cognitive architecture (Type 3). In Figure 2.24, we construct a continuum based on the concepts of thin and thick, within which the contract types described earlier are placed. We refer to this varying richness framework as a *new contract continuum*.

Observe that the legal continuum between transactional and relational contracts represents the bulk of agreements between people and characterizes the primary mode of knowledge exchange. In turn, psychological contracts represent a necessary precondition for successful K exchange, particularly relational. Relational and psychological contracts are intimately related and contain large reservoirs of tacitness that reinforce each other. On the other hand, precise contracts are necessary for defining explicit exchange and represent the bulk of information systems and operational processing in organizations.

In the contract continuum, contracts shift in character from formal and sometimes legal specifications, to partially formulated (and formed) relationships, to entirely tacit psychological mindsets. As requirements increase in richness, the contract shifts in character from thin to thick: from a "specification" in an engineering sense to a "relationship" in a human context. In practice, this means that the assertions of an explicit contract become less precise and meaningful. The explicit contract is an engineering or legal mechanism for specifying behavior and is not suitable for rich or knowledge contexts. This is where both relational and psychological frameworks come into play. Specifying knowledge work includes developing and reinforcing expectations and the prospect of a long-term relationship. This cannot be done in a formal sense but depends on cultivating trust and social relationships.

The central motivation or requirement of thin design is to satisfy the corresponding (thin) contract specification by reducing organizational uncertainty. Thin contracts completely specify both discrete functions and their interfaces, and easily lend themselves to thin forms of implementation. In contrast, the motivation underlying thick design is to arrive at an acceptable agreement among participants to the corresponding design contract. Thick design is ideal for ambiguous situations where an intention is satisfied through the attitudes and expectations that embody the human constituents of the design—thick contracts focus on relationships, psychological orientations, and the context of the design itself. In turn, context may extend to any number of K-spaces. The people or groups participating in thick contracts determine the design's structure, workings, and output.

2.6 Closing Remarks

Knowledge and imagination are two inseparable aspects of the intellectual experience. The imaginative moment is as creatively central to science as it is to philosophy, poetry, or art. The mind acts upon the sensorial perception of the natural world in the creation of knowledge in the same way as it acts on the elements of human sensibility while composing a poem or painting on canvas or creating a symphony. There exists a constant interplay between the two themes that recur almost like melodies in a sonata and provide intellectual unity to knowledge. They are the role of *conscious human activity* during the creation of knowledge and the *imaginative content* of that knowledge, which we have called knowledging and knowledge, respectively. Lastly, the world of science has fused natural philosophy with moral and metaphysical philosophy—this is the perspective we have taken. We believe this to be the essence of true knowledge: the intellectual, moral, aesthetic (as Plato categorized knowledge) as well as the feelings, emotions, and physiological and sense aspects of flesh and blood human beings.

This chapter presents the key ideas we believe necessary for configuring knowledge and its imaginative content. These ideas include the Nonaka basics and a new set of ideas—*new knowledge principles*—we believe necessary to move the organization from the "routine" of KM to the second divide of imagination. These new ideas are derived from a variety of disciplines from anthropology to the law. This is essential if we are to capture the complexity of K and KM and to integrate the surge in K and system waves that we expect in the coming years.

Central to the new K principles are the concepts of thin and thick. Thin systems are characteristically asymmetrical and may be described with organizational molecules. On the other hand, thick systems are often symmetrical and lattice-like and typically involve relationships and conversations. Design is the purposeful interaction of thin and thick, whether in the form of structure, strategy, work, and so on.

A key design (and implementation) characteristic of a K org is the interweaving of thin and thick elements across K-spaces. From a design perspective, this may be operationalized by interweaving symmetrical and asymmetrical structures, the interleaving of thin and thick organizational contexts, and the superimposition of collaborative and discourse (primarily symmetrical) structures over thin and thick layers, even beyond the boundaries of the organization. It is essential that we create dialectic interactions across the entire sphere of K-spaces.

There is a potentially important, even devastating, aspect to the rise of knowledge in the new Pangaea. As the world generates and exchanges knowledge across cultures, there is a risk of *homogenization*. We witness in cities around the world global fast-food restaurants selling industrialized food replacing local shops selling unique artisan food; similarly, knowledge too may become standardized and common, resisting challenges and new explorations. We too have seen what many have observed—the roots of the global knowledge tree are being refined ever deeper, but there seem to be few new roots taking hold.

In Kant's formulation of K, the *intellectual is twice removed* and easily homogenized, while we wait for yet another Kuhnian paradigm shift to generate a new idea. On the other hand, the *aesthetic is always first hand*, but never homogenized, and will forever take the lead in stimulating the imagination.

(Aside: In the film *Dead Poets Society*, Robin Williams passionately gives the aesthetic the full weight it deserves.

> We don't read and write poetry because it's cute. We read and write poetry because we are members of the human race. And the human race is filled with passion. And medicine, law, business, engineering, these are noble pursuits and necessary to sustain life. But poetry, beauty, romance, love, these are what we stay alive for. To quote from Whitman, "O me! O life! ... of the questions of these recurring; of the endless trains of the faithless ... of cities filled with the foolish; what good amid these, O me, O life?" Answer. That you are here—that life exists, and identity; that the powerful play goes on and you may contribute a verse. That the powerful play goes on and you may contribute a verse. What will your verse be?)

2.7 Class Exercises

1. Identify those organizational activities that are science-like, art-like, and craftwork-like. In what form and where in the organization do they take place?

2. In your opinion, which of the systems of inquiry will tend to yield the least amount of Type 3 or Type 4 errors? Justify your answer.

3. What is the advantage of using more than one system of inquiry to define a problem? Give an example.

4 Describe a particularly messy problem and how your organization or school resolved or dealt with it.

5 Discuss the change in culture and governance that is necessary if we are to shift from a traditional to a hypertext organization. Use your organization as a reference.

6 Select a national leader and discuss his or her thin and thick characteristics and how this manifests in specific policies and actions.

7 Describe the thin and thick characteristics of your school or organization. How has each changed over time? (Hint: an example of a school's thin characteristic is its building layout, whereas an example of a thick characteristic is the type and extent of the school's research.)

8 Describe the thin and thick properties of your office space and what you would change to promote conversations and thickness.

9 Identify and describe the discourse structures in your organization or school and how each contributes to its knowledging. What is the form of the discourse structures and where are they located?

10 Describe CPK in your organization. Identify the tacit hooks that source and make visible new ideas and artifacts. How they are made manifest in the organization?

11 Describe K-binding in your organization with respect to strategy formation. Does the leadership in the organization rely on domain experts, consultants, or are other organizational members included; if the latter, are there certain positions or departments that are preferred? How is the strategy communicated?

12 What are the transactional and relational contract forms in your organization? Discuss their respective context and how they are operationalized.

13 Describe the prevalent psychological contracts in your organization or school and where they are located. (Hint: the type and distribution of psychological contracts typically varies by hierarchical position and departmentalization.)

14 Project. This is an internal organization analysis. Construct and discuss the thin and thick layers in your organization, using Figures 2.19 and 2.20 as a guide. Discuss what policy and structural changes you would make to

(1) enhance the thick presence in your organization and (2) promote K integration using discourse structures. Discuss the content of these discourse structures, how they may vary, and what level you would place each in. Use Figure 2.22 as a guide.

References

Ahmadjian, C.L. (2004), "Inter-Organizational Knowledge Creation: Knowledge and Networks." In *Hitotsubashi on Knowledge Management*. Eds. Takeuchi, H., and Nonaka, I., John Wiley & Sons (Asia) Pte Ltd.

Alchian, A.A., and Demsetz, H. (1972), "Production, Information and Economic Organization." *American Economic Review*, 62: 777–795.

Alpasian, C.M., and Mitroff, I.I. (2011), *Swans, Swine, and Swindlers*. Stanford University Books.

Baldwin, C.Y., and Clark, K.B. (2000), *Design Rules: The Power of Modularity*. The MIT Press.

Bayles, D., and Orland, T. (1991), *Art & Fear: Observations on the Perils (and Rewards) of Artmaking*. The Image Continuum Press.

Boehm, B., and Turner, R. (2003), "Using Risk to Balance Agile and Plan-Driven Approaches." *Computer*, 36(June): 57–66.

Casselman, R.M., and Samson, D. (2005) "Moving Beyond Tacit and Explicit Knowledge: Four Dimensions of Knowledge." *System Sciences. HICSS. Proceedings of the 38th Annual Hawaii International Conference.*

Coase, R. (1937), "The Nature of the Firm." *Economica*, 4: 386–405.

Dewey, J. (1934), *Art as Experience. A Perigee Book*. The Berkley Publishing Group.

Drucker, P.F. (1999), *Management Challenges for the 21st Century*. HarperBusiness.

Earl, M.J. (1993), "Experiences in Strategic Information Systems Planning." *MIS Quarterly*, 12(March): 1–24.

Geertz, C. (1973), *The Interpretation of Cultures*. Basic Books.

Geertz, C. (1983), *Local Knowledge*. Basic Books.

Goodman, N. (1984), *Of Mind and Other Matters*. Harvard University Press.

Handy, C. (1993), *Understanding Organizations*. Oxford University Press.

Hart, O. (1989), "An Economist's Perspective on the Theory of the Firm." *Columbia Law Review*, 89: 1757–1774.

Jensen, M.C., and Meckling, W.H. (1976), "The Theory of the Firm: Managerial Behaviour, Agency Costs and Ownership Structure." *Journal of Financial Economics*, 3: 305–360.

Kay, J. (1995), *Why Firms Succeed*. Oxford University Press.

Kilov, H. (2002), *Business Models: A Guide for Business and IT*. Prentice-Hall.

Leonard, D., and Swap, W. (2004), "Deep Smarts." *Harvard Business Review*, 82(September): 88–97.

Liker, J.K. (2004), *The Toyota Way*. McGraw-Hill.
MacNeil, I.R., and Gudel, P.J. (2001), *Contracts: Exchange Contracts and Relations*. Third Edition. Foundation Press.
Meyer, B. (1992), "Applying 'Design by Contract.'" *Computer (IEEE)*, 25(10): 40–51.
Mintzberg, H. (1983), *Structure in Fives: Designing Effective Organizations*. Prentice Hall.
Mintzberg, H., Ahlstrand, B., and Lampel, J. (1998), *Strategy Safari: A Guided Tour Through the Wilds of Strategic Management*. The Free Press.
Mitroff, I.I., Alpaslan, C.M., and Mason, R.O. (2012), "The Messy Business of Management." *MIT Sloan Management Review*. Fall 2012.
Mitroff, I.I., and Silvers, A. (2009), *Dirty Rotten Strategies: How We Trick Ourselves and Others into Solving the Wrong Problems Precisely*. Stanford Business Books.
Morabito, J., and Stohr, E.A. (2010), "Online Analytical Processing." In *The Handbook of Technology Management, Volume 2*. Ed. Bidgoli, H., John Wiley & Sons, Inc.
Morgan, K.K. (2009), "Does Coalescent Knowledge Exits?" Presented at the *Academy of Management* Annual Conference, Chicago, IL. August 2009.
Nonaka, I., and Konno, N. (1998), "The Concept of 'Ba': Building a Foundation for Knowledge Creation." *California Management Review*, 40(3): 40–54.
Nonaka, I., and Takeuchi, H. (1995), *The Knowledge-Creating Company*. Oxford University Press.
Nonaka, I., and Toyama, R. (2004), "Knowledge Creation as a Synthesizing Process." In *Hitotsubashi on Knowledge Management*. Eds. Takeuchi, H., and Nonaka, I., John Wiley & Sons (Asia) Pte Ltd.
Novak, J.D. (1998), *Learning, Creating, and Using Knowledge*. Lawrence Erlbaum Associates, Publishers.
Porter, M.E. (1980), *Competitive Strategy: Techniques for Analyzing Industries and Competitors*. The Free Press, A Division of Macmillan, Inc.
Saporito, B. (2013), "The Conspiracy to End Cancer." *Time*. April 1, 2013.
Savage, C.M. (1996), *Fifth Generation Management*. Second Edition. Butterworth-Heine.
Schein, E.H. (1980), *Organizational Psychology*. Prentice-Hall.
Spear, S., and Bowen, H.K. (1999), "Decoding the DNA of the Toyota Production System." *Harvard Business Review*. September–October 1999.
Sveiby, K.E. (1997), *The New Organizational Wealth*. Berret-Koehler Publishers, Inc.
Tarnas, R. (1991), *The Passion of the Western Mind*. Ballantine Books.
Von Krogh, G., Ichijo, K., and Nonaka, I. (2000), *Enabling Knowledge Creation*. Oxford University Press.
Wiig, K. (1993), *Knowledge Management Foundations*. Schema Press.
Wittgenstein, L. (1953), *Philosophical Investigation*. Trans. Anscombe, G.E.M. Macmillan.

3

Organizations and Systems

Chapter Preview

Introduction

The overall theme of this book is designing K organizations. Hence, it is expected that the reader has sufficient understanding of organizations to apply knowledging concepts. Accordingly, this chapter reviews the key ideas associated with both organizations and systems.

To organizational science students, we suggest a selective reading rather than a thorough analysis. On the other hand, our review of organizations is not without K in mind. We discuss the importance of *functional design* in the K organization—it is the predominant form in German organizations as well as, say, Google. Additionally, we discuss *symmetrical* structures (i.e., balanced and equal as opposed to hierarchical and unequal) as central to dialogue and the development of K. Closely associated with (symmetrical) dialogue are the local and micro-interactions (e.g., *self-organization*) that characterize a *transformative teleology*. We conclude by comparing key characteristics between industrial, information, and knowledge organizations.

Food for Thought

1) What is the dominant coordinating mechanism in your organization? Is it the organizational chart, culture, teams, information flows, or some other paradigm?
2) How are decisions made in your organization? To what extent are you required to follow the rules?
3) How would you characterize your individual work? Is it highly formalized, repetitive, technical, improvisational, or skill-oriented?
4) How are the departments in your organization configured—around function, customer, product, geography, process, or some combination of them? Can the configuring be improved?
5) How does your organization go about solving complex, difficult problems?

Designing Knowledge Organizations: A Pathway to Innovation Leadership, First Edition.
Joseph Morabito, Ira Sack, and Anilkumar Bhate.

Topic Layout

The figure below illustrates the topic layout of Chapter 3. As shown on the left side of the figure, we discuss classical organization theory (OT), organization and generic structures, and the architecture of organization, particularly one centered on relational contracting. The center of the figure represents the organization in practice: its design considerations and forms. In the same way the left section informs the organization on the issues associated with organizations, the right side informs the organization on system characteristics. This includes our discussion of systems theory and hierarchy theory (HT). We also discuss the evolution of system perspectives through a teleological lens and focus on a transformative teleology as that which comes closest to characterizing a K organization.

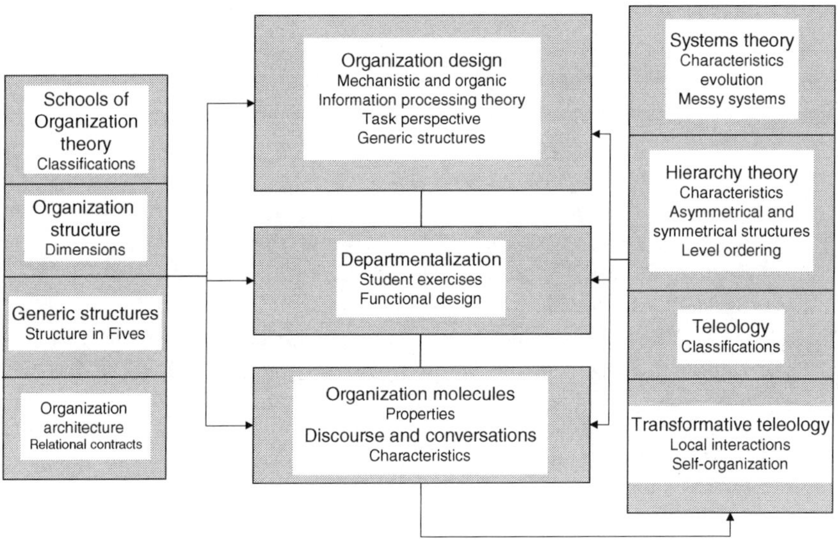

Concept roadmap to Chapter 3.

3.1 Organizations

What is an organization? There are virtually as many definitions as there are books. Robbins (1990) defines an organization as a "consciously coordinated social entity, with a relatively identifiable boundary that functions on a relatively continuous basis to achieve a common goal or a set of goals." Observe that "consciously coordinated" can take many forms: policies and procedures, communications up and down a hierarchy, laterally between peer departments, laterally between subgroups such as task forces or teams, laterally between individuals, and all combinations. Furthermore, the communications may be formal or informal, in any number of modes—written, oral, personal, cultural, or through social media.

The term "relatively identifiable boundary" is increasingly problematic in the 21st century, where communications take place instantly across organizational boundaries and among people, whether or not they are employed by the organization. We may think of a "relatively identifiable boundary" as an emulsion where the elements within it share a kind of *binding force* that serves to connect the elements to the wider suspension (or organization). Furthermore, binding force is a function of distance: the closer (physically) the participants, the stronger the binding force. This is critically important in knowledge design—the binding force directly influences the *stickiness* of knowledge, stickiness indicating the extent and ease with which knowledge is cultivated and shared.

The classical study of organizations has typically been separated by levels of analysis. The macro level is known as *organization theory* (OT), whereas the micro level is known as *organizational behavior* (OB). The former includes broad areas such as strategy and technology (i.e., the work of the organization), whereas the latter concentrates on people, their communication patterns, and their groups (e.g., teams).

3.1.1 Schools of OT

A school of OT represents a particular perspective of organizations. Each school carries assumptions about what an organization is or should be, how it operates, its structure and strategy, its people and their groups, and so on.

As with the definition of an organization, there are innumerable schools of OT. These schools typically follow a historical progression and often reflect advancements in technology, human relations, political trends, and cultural developments. For example, we may view organizations through the following lenses:

1) Classical theory. This theory characterizes the culture of the Industrial Age. We sometimes say that this theory represents the interest of the owner. The underlying design parameter is that of a *machine*. Note that we say "design parameter" and not "design metaphor"—in this approach, organizations largely believe they should be designed and operated as a well-oiled machine.
2) Neo-classical theory. This theory advances the position of the employee (as opposed to that of the owner). Sometimes known as the *humanistic* school, there is an emphasis on motivation and employee involvement.
3) Information theory. Whereas the theories discussed earlier arise from differing perspectives of the division of labor, information theory sees the organization as an *information processing mechanism*.

It follows that several well-known elaborations have emerged, often based on a particular scholar's perspective and research. For example, here we summarize the thinking of Charles Handy (1993). (We will elaborate on several of these schools in this chapter as well as throughout the book.)

1) Scientific management. This school is based on the work of Frederick Winslow Taylor and is often represented as the machine view of organizations. Scientific management includes the scientific design of tasks and the selection and training of people; also included is the separation of planning (management), design (engineers), and execution (labor).
2) Human relations. In response to the mechanistic view of scientific management, Chester Barnard introduced the notion of an organization as a cooperative system, where equal attention is accorded tasks and people. Authority is also accorded (in practice) rather than delegated. Barnard also advanced the notion of the informal organization (group pressure, group acceptance, union pressure).
3) Bureaucracy. Max Weber described the ideal structure of an organization, which came to be known as a bureaucracy. It is characterized by a division of labor, hierarchical decision making, high formalization, and impersonal relationships.
4) Power, conflict, and decisions. This school challenges the notion of rational decision making. Organizations are best understood by looking at power, conflict, and the way decisions are actually made.
5) Technology. This school addresses the organization in terms of its work. There are many contributors to this school, including Burns and Stalker (mechanistic versus organic work), Woodward (unit, mass, and production manufacturing technologies), Perrow (routine, engineering, craft, and nonroutine task technologies), Thompson (long-linked technology or sequential interdependence, mediating or pooled, and intensive or reciprocal). For additional information on Woodward, Perrow, and Thompson, see our previous book, *Organization Modeling*.
6) Systems. This school sees the organization as an open system. This has given rise to new ways of thinking about organizations, including the learning organization and business process reengineering, among others.
7) Institutional. This school advocates the importance of culture, history, and shared assumptions in organizations.

3.1.2 Materials of OT

The "materials" of OT concern the primary structuring elements of the organization: structure, size, environment, technology (i.e., the organization of work), and strategy. More recent additions include the organization's distinctive competencies and core capabilities advocated by Hamel and Prahalad (see Chapter 1). The main goal of OT is to inform organizational design; in classical

OT, this typically means the design of structure. Research by and large addresses the interaction of each context with design; for example, how does strategy affect structure? How does technology affect structure? And so on.

3.1.3 Organizational Structure

What is structure? When we ask this question to our students, most of them answer that it is the organization chart (org chart). This is not wrong, only incomplete. Traditionally, three dimensions of organizational structure have been described as follows.

- Complexity. Complexity deals with the extent of differentiation along a given dimension. For example, complexity includes the extent of differentiation among job tasks, job skills (i.e., job specialization), geographic location, and so on. While these indicate horizontal complexity, vertical complexity concerns the depth of the org chart and the span of control.
- Formalization. Formalization indicates the extent to which the day-to-day operations of the organization can be coordinated and managed on the basis of standardized rules and procedures. Hence, we often say that, for example, a bureaucratic organization has a high level of formalization, whereas a research division has a low level of formalization.
- Centralization. Centralization deals with the way decisions are made in an organization; that is, the locus of decisions. Is it centralized or decentralized? In practice, most organizations have some measure of both centralization and decentralization; differences lie in the extent of each and the distribution of authority in the organization.

3.2 Organizational Design

Collectively, the three dimensions characterize the structure and workings of organizations. For example, if we compare an organic and mechanistic organization, we can make the following observations: an organic organization has low complexity, low formalization, and low centralization; in contrast, a mechanistic organization is the reverse—there are high levels of complexity, formalization, and centralization. Other forms, which we will describe later, have other combinations; for example, the bureaucratic organization typically has high levels of complexity and formalization, but low to moderate levels of centralization. This is shown in Table 3.1, adapted from Hatch (1997).

Organizations often attempt to leverage the benefits of both mechanistic and organic forms. For example, one hybrid design is the *federated* structure, which balances the benefits of both centralization and decentralization. This requires that the organization decide which elements are best suited for centralization, and which are better suited for decentralization—these decisions are a

Table 3.1 Dimensions of structure and three organizational forms.

Item	Complexity	Formalization	Centralization
Organic	Low	Low	Low
Mechanistic	High	High	High
Bureaucratic	High	High	Low

Adapted from Hatch (1997).

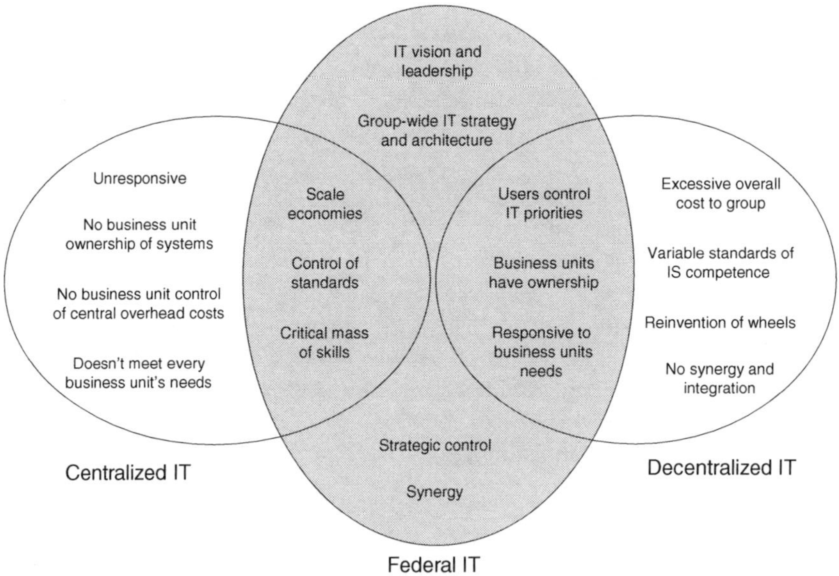

Figure 3.1 Centralized, decentralized, and federated representation—Adapted from Rockart *et al.* (1996).

function of strategy and situational context. For example, John Rockart *et al.* (1996) have described a federated structure for IT departments: infrastructure areas are best centralized to gain economies of scale, establish standards, and for obtaining a critical mass of skills; in contrast, application areas should be decentralized, even embedded within business groups, to facilitate business ownership of IT, responsiveness, and effective understanding of business concerns and requirements. This is illustrated in Figure 3.1, adapted from Rockart *et al.* (1996).

The left circle shows the advantages (right portion of circle) and disadvantages (left portion of circle) of centralized design; the right circle shows the advantages (left portion of circle) and disadvantages (right portion of circle) of decentralized design; the center circle represents the federated structure, which captures within it the advantages of both centralized and decentralized design while avoiding their respective disadvantages. Centralized and decentralized elements are held together through IT vision and leadership, IT strategy and architecture, and strategic control.

3.2.1 Task Perspective on Design

Figure 3.2 illustrates the Perrow task technology model. In this model, knowledge-based tasks are represented along two dimensions. The *X*-axis shows task variability (also known as the number of exceptions) and represents the degree to which one occurrence of a task is different from the next occurrence. The *Y*-axis shows the level of task analyzability, shown as increasing levels of task unanalyzability; this dimension assesses the type of search procedures used to locate successful methods for responding to task exceptions.

The 2 × 2 model is used to classify four types of tasks, as follows: *Routine* tasks have low levels of variability with easy-to-analyze problems; examples include tasks found on an assembly line. *Engineering* tasks have many exceptions, but each exception can be resolved in a systematic, easy-to-analyze manner; examples include design activities. *Craft* tasks have low levels of variability with difficult-to-analyze problems; examples include creating a work of fine art. *Nonroutine* tasks have high levels of variability and difficult-to-analyze problems; examples include research and development. Table 3.2 summarizes several structural and design characteristics associated with each task category.

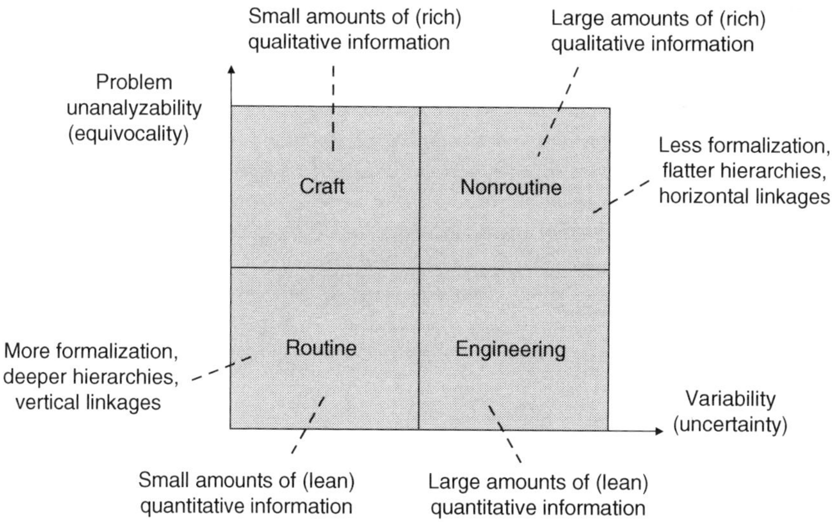

Figure 3.2 Characterization of knowledge tasks—Adapted from Perrow (1967).

Table 3.2 Design characteristics associated with the Perrow tasks.

Task	Formalization	Centralization	Education and training	Worker experience	Information characterization
Routine	High	High	Low	Low	Small quantitative
Engineering	Low	High	High	Moderate	High quantitative
Craft	Moderate	Low	Moderate	High	Low qualitative
Nonroutine	Low	Low	High	High	High qualitative

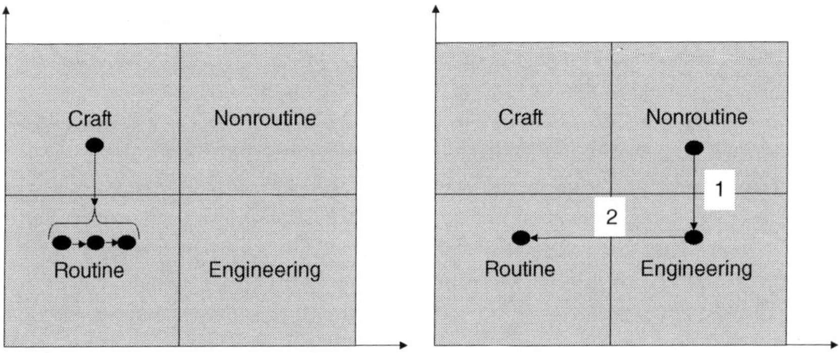

Figure 3.3 Examples of routinization using the Perrow framework.

We can use this model to illustrate the notion of *routinization*, that is, the analysis of a task with a high level of unanalyzability, and then breaking it down to simpler tasks, each with low levels of unanalyzability. For example, in Scientific Management we analyze tasks (typically craft tasks) and discover that they *are* analyzable—they only need to be decomposed, modularized, and designed as such; this is shown on the left side of Figure 3.3. The right side of Figure 3.3 shows routinization as a consequence of learning and experience. For example, a new surgical procedure may initially be nonroutine, but with analysis, publication, and further experience the procedure becomes routinized in stages. In general, organizations prefer routinization for apparent reasons—for example, the people performing the work are less educated, replaceable (i.e., "substitutable"), and therefore, they cost less than would be the case with a more nonroutine task.

Finally, Figure 3.4 was created by one of our doctoral students and illustrates the application of the Perrow framework to the major functions in the pharmaceutical industry. Note the cluster of routinized work with notable exceptions, specifically R&D, marketing, and sales. We will revisit the Perrow model in our analysis of knowledge work later in the book.

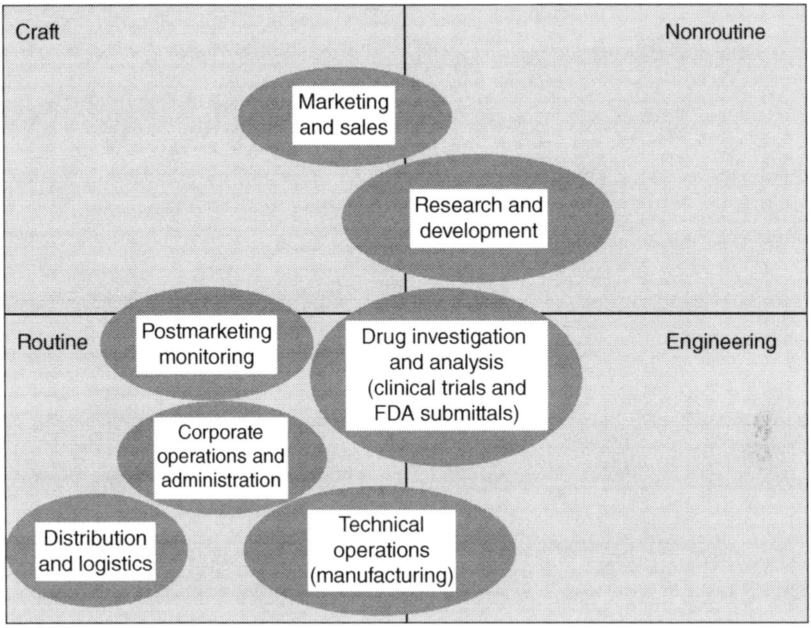

Figure 3.4 Perrow model adapted to the major functions of the pharmaceutical industry.

(Aside: Recall from Chapter 2 that we introduced the term craftwork. The term craftwork is used in a different context and does not refer to craft tasks; rather, it is used in contradistinction with science and art. In fact, craftwork *is* analyzable and is represented by the idea of a *template*. In this sense, both routine and engineering tasks in the Perrow model may be described with the notion of craftwork as we use it. In contrast, the craft and nonroutine tasks in the Perrow model are closer to some combination of craftwork and art or science.)

3.2.2 Information Process Theory and Design

According to Daft and Macintosh (1981), organizations become more effective if they achieve a "good fit" between the *requirements* to process contextual information (e.g., information associated with strategy, technology) and the *capacity* to process information. Continuing with this line of thinking, the capacity to process information is associated with *structural design choices*. Structural design choices typically fall into three categories: vertical information linkages, horizontal information linkages, and departmental grouping. This framework is called the information processing approach to design.

Figure 3.5 is an illustration of both vertical and horizontal linkages, adapted from Daft (1998). Observe that vertical linkages are associated with high

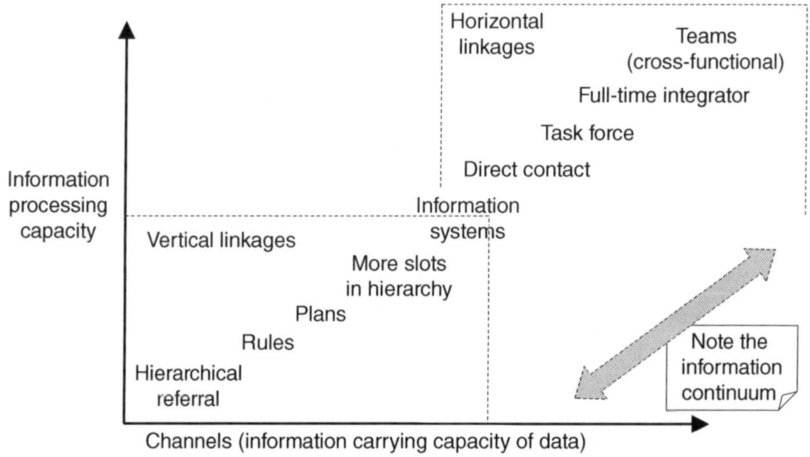

Figure 3.5 Vertical and horizontal information linkages—Adapted from Daft (1998).

levels of formalization and hierarchical communications, whereas horizontal linkages are associated with low levels of formalization and lateral communications in the coordination of work. Note too that the linkages are placed from the least to the richest communications channel, in essence, following the information continuum from the lower left to the upper right of the illustration. Organizational design involves the selection of the proper linkage necessary to reduce the uncertainty associated with a given task or decision. This is the central idea underlying the information processing view of an organization.

As mentioned earlier, a key idea associated with structural design choices is the reduction of uncertainty, where uncertainty is typically defined as the absence of needed information. Goodhue *et al.* (1992) have combined the work of Tushman, Nadler, and Galbraith to produce an updated information processing model, shown in Figure 3.6. Observe the factors that cause an increase in uncertainty: interdependence among subunits (e.g., departments), complex or nonroutine tasks, and an unstable subunit task environment. In response, the organization may employ two design strategies. The first is to reduce the need to process information through slack resources (e.g., reduce the level of performance) or with self-contained tasks (e.g., employ a nonfunctional structure such as one based on product lines). The second strategy is to increase the capacity to process information though investment in computer information systems or with lateral relations. Lateral relations reduce uncertainty and ambiguity by promoting horizontal communications among groups and individuals and are implemented through the horizontal information linkages shown in Figure 3.5. Other examples of horizontal linkages include the process-oriented organization and lateral processes (each discussed later). A more

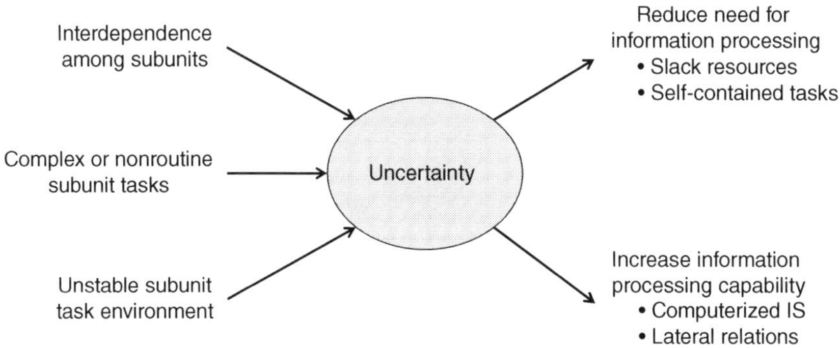

Figure 3.6 Updated information processing model—Adapted from Goodhue *et al.* (1992).

recent addition to lateral relations is social communities, such as communities of practice (discussed later in the book). Observe that the latter examples would, in effect, extend the continuum shown in Figure 3.5 to the upper right.

The third element of design is organizational form or departmentalization. We believe the best way to visualize form is to identify the level immediately below the CEO; if it is functions, then we have a functional structure; if it is product divisions, then we have a product structure, and so on. Galbraith (1995) has identified five generic organizational forms: *function, product, market, geographic,* and *process-oriented.* Each of the five generic structures has strengths and weaknesses. However, organizations have a dilemma: they need to be multidimensional and leverage the benefits of all the structures. Additionally, they must design within the constraints of an organizational hierarchy. Hence, the selection of the first level below the CEO is critically important in organizing the focus of the company; once selected, the other dimensions will be replicated within that first level. For example, if we select a product structure (say, three product categories), then the primary functions (e.g., R&D) will be replicated across each product line. Figure 3.7 shows four functions within the first product line, but this arrangement must be replicated within the other product lines too. If we consider the five generic faces an organization must wear, the complexity becomes overwhelming. The solution is to create a hybrid—more than one form immediately below the CEO, and then employ lateral relations or processes to provide communications and coordination across whatever hybrid structure has been design.

Figure 3.8 is a sample structure defined by our students at a large international bank. Observe that this hybrid structure balances *geographic* (i.e., Japan, Latin America, Europe, Middle East, Africa, North America), *product* (i.e., institutional and global client groups), and *functional* (i.e., operations, technology) design. Lateral processes (not shown) coordinate work across selected areas in the organizational chart.

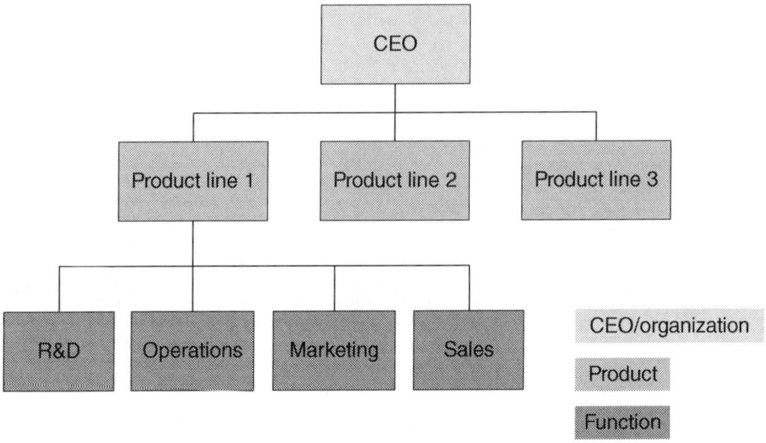

Figure 3.7 Product structure with embedded functions.

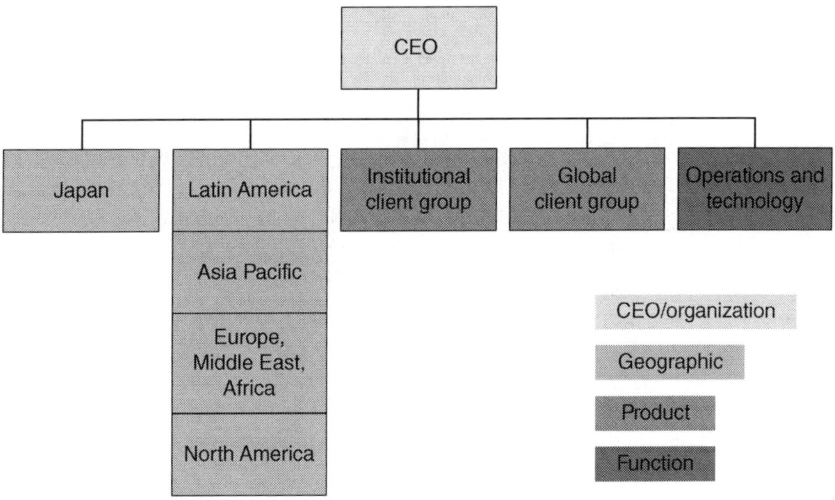

Figure 3.8 Hybrid organization structure at a large international bank.

Lateral processes are permanent structures that coordinate ongoing activities and promote decentralized decision making. Lateral processes may be implemented with a formal group, full-time integrators (a horizontal linkage shown in Figure 3.5), or even within the generic process structure itself. Note that the introduction of lateral processes does not create a matrix structure. A matrix structure is a project-oriented secondary structure where project teams (typically cross-functional teams) are superimposed over a primary structure (one of the five generic or hybrid structures). A distinguishing characteristic of a matrix structure is that project teams are temporary, existing only for the life of a given project. In contrast, lateral processes are fairly permanent structures that provide continuous coordination.

3.2.2.1 Implementation Patterns

Organizational forms are commonly implemented as hybrid structures, as illustrated in Figure 3.8. A hybrid structure is not a haphazard combination of other forms but is considered and designed. Here we highlight several properties of the generic forms, which we may use to construct hybrid structures.

Functional structure. The oldest form, the functional structure is suited to a small organization with a single product line or service. From our perspective, its most important attribute is that it *promotes expertise and scale* within a function (i.e., a knowledge domain).

Product structure. This form is product focused and has the shortest product development cycle of any form; hence, it is one of the most popular forms with manufacturing organizations.

Market structure. The market structure is organized around market segments. This form provides a single point of contact and focuses on customer satisfaction. The market structure is one of the most popular forms with service organizations.

Geographic structure. This form is suitable for large international organizations that want to present a local face. It seeks to exploit local knowledge and relationships.

Process structure. The process structure is the most recent form and promotes flatter hierarchies, processing efficiency, and short cycle times.

Common hybrid forms. Galbraith (1995) describes two common hybrid forms. The first is the product–function structure and the second is the product–market structure (also known as the back–front organization). Each hybrid seeks to combine the advantages of its constituent forms. The first hybrid is essentially a product structure that wants to gain scale in a particular function, such as fabrication or sales. The second is typical in financial services organizations that want to present a single face to customers (front) while speedily providing a suite of products (back).

Note too that, as mentioned earlier, the matrix structure is project-oriented where a secondary structure of projects (temporary entities) is superimposed over a primary structure (one of the generic forms or a hybrid). Also mentioned earlier, in contrast to projects, lateral processes are mostly permanent structures that provide coordination across parts of the organizational structure.

Finally, the actual structure of a given organization, though specific to that organization, reveals the underlying patterns discussed earlier. It is not surprising, therefore, that the large international financial services organization illustrated in Figure 3.8 is a three-way hybrid of {product–function–geography}.

Observations. It is easy to assume that the organizational forms, as presented, represent a progression or evolution as each succeeding form compensates for the weaknesses of previous forms and accommodates itself to the modern world. However, this is not necessarily so. For example, German organizations, though huge with a large number of products, are typically designed as functional structures; perhaps because of the German legacy of science and engineering (see Chapter 1). But this is not necessarily cultural—Google also designs itself along functional structures (Schmidt and Rosenberg 2014). The common denominator is not the industrial attributes of size, efficiency, and so on (though important), but the 21st century attribute of knowledge. *Where knowledge is concerned, think functional design!*

3.2.3 Generic Structures

Henry Mintzberg (1983) has created a framework consisting of five generic components, illustrated in Figure 3.9. For example, a "machine bureaucracy" is one that emphasizes the technostructure and the standardization of work, leading to a deep hierarchy, high levels of formalization, top-down decision making, and so on. Whereas every organization has these generic constituents, each will emphasize one constituent over the others, giving it a particular character, structure, and behavior pattern. Table 3.3 is a summary of the five organizational forms and their selected characteristics, adapted from Mintzberg (1983).

Observe that the simple form is associated with entrepreneurship and has those characteristics typically associated with a young, small organization. In contrast, the machine bureaucracy characterizes an old, large organization; if we expand the machine bureaucracy by further increasing its size, we evolve into the divisionalized bureaucracy, in essence, a collection of machine bureaucracies.

Finally, note the similarity between the professional bureaucracy and the adhocracy. Both require high levels of training and education, low levels of formalization, and insignificant flows of authority. Also, each form—though potentially different in size—is dominated by knowledge workers within the operating core. (The other generic forms have knowledge workers too, but they tend to be located in dedicated departments

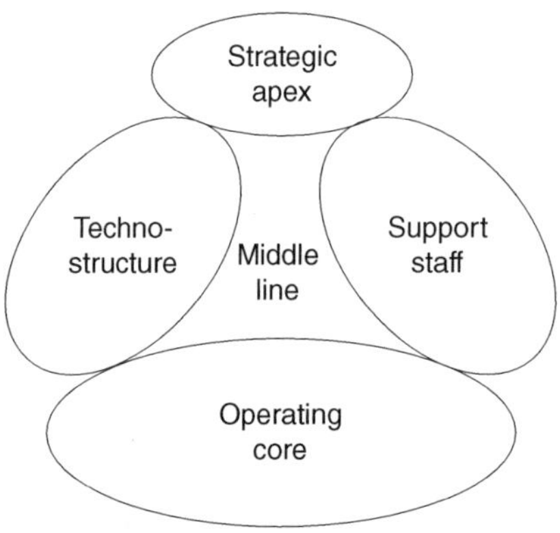

Figure 3.9 Mintzberg's five generic organizational components—Adapted from Mintzberg (1983).

Table 3.3 Generic forms and organizational characteristics.

Item	Simple	Machine	Professional	Divisionalized	Adhocracy
Emphasis	Strategic apex	Technostructure	Operating core	Middle line	Support staff
Coordination	Direct supervision	Standardization of work	Standardization of skills	Standardization of outputs	Mutual adjustment
Job specialization	Little	Much vertical and horizontal	Much horizontal	Some horizontal and vertical between HQ and divisions	Much horizontal
Training	Little	Little	Much	Some	Much
Formalization	Little	Much	Little	Much	Little
Unit size	Large	Large at bottom	Large at bottom	Large at top	Small throughout
Level of decentralization	None, i.e., centralized	Limited horizontal	Horizontal and vertical	Limited vertical	Selective
Flow of authority	Significant from top	Significant throughout	Insignificant	Significant throughout	Insignificant
Flow of informal communications	Significant	Discouraged	Significant in administration	Some between HQ and divisions	Significant throughout
Flow of decisions	Top-down	Top-down	Bottom-up	Differentiated between HQ and divisions	Mixed, all levels
Environment	Simple and dynamic	Simple and stable	Complex and stable	Relatively simple and stable	Complex and dynamic
Age and size	Young and small	Old and large	Varies	Old and very large	Young

Adapted from Mintzberg (1983).

such as R&D.) Furthermore, we see these characterizations in the field: the professional bureaucracy characterizes university and hospital structures and therefore has a large educated workforce. Similarly, the adhocracy is characterized by a workforce with high skill and tacit knowledge, as is typically found in a film company. *Both the professional bureaucracy and the adhocracy come closest to a generic knowledge organization.*

3.2.4 Organizational Architecture

Organizational architecture, as is commonly used, is a particular application of organizational design. It concerns itself with the design of an organization's distinctive capabilities. Here, we briefly describe several approaches to architecture.

First, Nadler and Tushman (1997) view organizational architecture as a way of structuring and coordinating people and processes in order to maximize an organization's capabilities. Their approach is to look at architecture as a sociotechnical system, with an emphasis on information processing theory—matching an organization's information processing requirements with its capacity to process information. In practice, their approach emphasizes the design of structure and process. It includes enterprise-level organizational forms (i.e., a kind of architectural continuum but one that emphasizes organizational structure that encompasses the integrated enterprise, interdependent business groups, and independent business units) in addition to operational designs that correspond with the broader organizational forms. This is essentially a business process design perspective and includes the design of processes, teams, information systems, and people that lead to a "high performance work system."

Second, John Kay (1995) regards architecture as the set of contracts in which an organization participates. By *contract*, we mean agreements, behaviors, and expectations between parties. Contracts need not be formal; in fact, Kay (1995) claims that *relational* contracts (e.g., informal or lightly detailed agreements) are responsible for success. This includes (1) internal contracts between a firm and its employees or among its employees, (2) external contracts between a firm and its suppliers or customers, and (3) network contracts among a group of collaborating firms. Architecture, and thus distinctiveness, rests on contracts as a group: though a single contract may be imitated, the system embodied in the group is too complex to reproduce. The value of architecture derives from the ability of an organization to leverage its contracts to create organizational knowledge that is greater than the sum of its individuals' knowledge, to establish a cooperative ethic, and to implement organizational routines.

There are innumerable IT architectures (e.g., information engineering) and several hybrid approaches that combine IT and business. For example, Ken Kusunoki (2004) envisions architecture as a synthesis between the modular knowledge of IT and the integral knowledge of business. Recall that we have previously used the term *integral* to describe the essential character of knowledge, a characteristic that distinguishes it from virtually every other construct used to describe and design organizations. Note too that the relational contracts mentioned earlier represent one way of operationalizing integral knowledge between people and organizations. We will revisit these ideas throughout the book.

3.3 Systems Theory

The system perspective is a useful framework for understanding an organization—both its design and management. We can visualize an organizational system as a whole consisting of interacting parts or subsystems; hence, design admits of a structural perspective, that is, the individual subsystems, and an internal behavior perspective, that is, the subsystem interactions.

Robbins (1990) defines a system as "a set of interrelated and interdependent parts arranged in a manner that produces a unified whole." Everything from machines such as automobiles to biological systems such as human beings to social systems such as organizations fits this definition. A closed system is one that does not interact with its environment, but this perspective has little practicality. In contrast, an open system is one that dynamically interacts and exchanges energy with the environment in the form of inputs and outputs, while the organization itself acts as a transformation agent as it converts those inputs to outputs. This is shown in Figure 3.10.

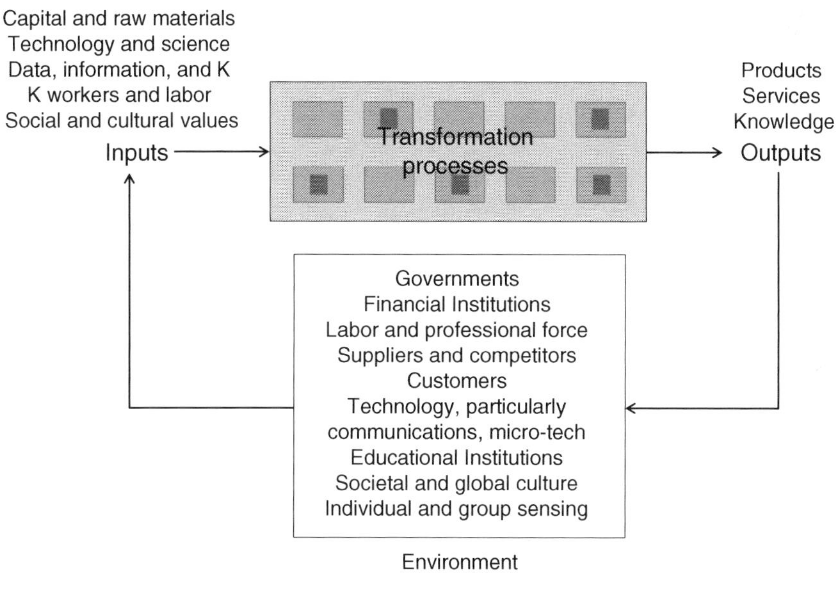

Figure 3.10 An organization as an open system.

Let us look more carefully at Figure 3.10. The figure is an adaption from many such illustrations of industrial organizations. We have generalized the representation to include other factors necessary for both service and knowledge organizations. Principal inputs include both hard and soft assets, such as raw material and technology, cultural and social norms and values, and of course, both labor and K workers. The major outputs include products, services, and knowledge—knowledge is certainly embedded in modern products and services, but it is also an output proper that is available as input to organizations. The environment includes traditional constituents, such as governments (note the plural) and financial institutions, and newer additions such as educational institutions and the values and norms associated with a global culture. Finally, note that the center box represents the transformation processes of the organization: observe the smaller boxes within the larger box; and several of these smaller boxes may contain even smaller boxes, and so on. These boxes represent interacting subsystems within subsystems and exist at different levels within the system as a whole. This is further discussed here.

3.3.1 Characteristics and Evolution of Systems Theory

We will not list all the characteristics of an open system but several stand out. They include *environmental awareness* and *feedback* (Figure 3.10), *homeostasis* or a tendency to move (or adapt) to a steady state, *negative entropy* meaning that the system may repair itself and even grow, *adaptation* as the system adjusts to internal and external demands, and *equifinality* or the likelihood that there exists several pathways to the same goal. The latter is particularly important since it implies that the organization can produce the same output even if the inputs and transformation capabilities of the organization change.

Stacey (2010) describes the evolution of systems science through three primary phases: *general systems theory*, *cybernetics*, and *systems dynamics*. Biologists took the lead in developing general systems, whereas engineers developed cybernetic and systems dynamics. General systems theory introduced the notions of order and stability, whereas cybernetic systems are self-regulating and goal-directed—both types of systems move toward adapted equilibrium. Most organizational subsystems (e.g., planning, budgeting, performance appraisals, incentive systems) are cybernetic in nature: targets are set, actual performance appraised and compared with those targets, the gaps measured and fed back, and necessary adjustments made. Management change initiatives are typically cybernetic too, such as culture change and business process reengineering. There exists an assumption in cybernetic thinking that an organization can be designed as if it were a fine watch and that human activity should be kept to a minimum; from an engineering perspective, humans represent interference with the workings of a fine instrument, such as a watch or an organization. This takes us full circle back to classical theory and the culture of the Industrial Age where design (at least) follows a mechanical-like paradigm. (As we will see in Chapter 6, organizations are typically characterized and designed according to the assumptions underlying the engineering subculture described by Edgar Schein (1996).)

Systems dynamics changes the terms of systems thinking by introducing non-linear equations, interactions, and responses. The implications are significant: (1) the recognition of *feedback*, where feedback is the process of returning part of the output to the input, either to oppose the input (negative feedback) or to reinforce the input (positive feedback), (2) the possibility of *unexpected outcomes* rather than a state of equilibrium; (3) a system no longer self-regulating but *self-influencing*. All three variations of systems thinking follow the notions of efficient and formative cause (see Table 3.4), in which the organization designs its subsystems and takes specific actions toward the achievement of predefined goals (Stacey 2010).

Traditional views of organizational systems (e.g., mechanistic, organic) assume stability and a reduction of uncertainty—a movement toward certainty. One issue with traditional systems thinking is that emergence and novelty are not easily accommodated; in fact, they increase uncertainty. However, organizational systems do have emergent characteristics and they certainly invent—create knowledge—and therefore we need to further our development of systems thinking to account for both evolution and transformative processes. In this line of thinking, systems may evolve or be transformed toward unknowable futures.

The concept of *evolution* introduces a new idea into systems: movement toward uncertainty as opposed to certainty. A given system, such as a biological species, exploits unforeseen incongruities within (e.g., "mutations") and opportunities without (e.g., "environmental niches")—this dynamic interaction between internal and external forces is responsible for the emergence of new and novel species or systems. This follows the adaptionist teleology shown in Table 3.4. On the other hand, evolution is not entirely novel but contains within it a process of progression as species develop from one form to the next and is constrained by previous forms (i.e., history is important). For example, embryonic development reveals its phylogenetic history during the development stages—in the case of birds and mammals, we see visible reptilian structures in the early stages of embryo development.

In contrast to evolution, complexity science suggests *transformative* teleology and the importance of *local interactions* (also known as *self-organization*). Such interactions are amplified into organization-wide emergent patterns of behavior, even as they are formed by those same patterns. Hence, a system is better understood, not as one moving toward equilibrium, but as one characterized by a small number of "archetypal patterns of interaction." Successful managers control the dynamics of such systems by recognizing and leveraging these archetype patterns. (This perspective of systems underlies the notion of

Table 3.4 Systems teleology.

Item	Efficient	Rationalist	Formative	Adaptionist	Transformative
Movement toward a future that is:	A repetition of the past	A goal chosen by reasoning autonomous humans	An unfolding of a mature form enfolded at the beginning	Unknown and dependent on chance	Perpetually constructed by the movement itself as continuity and transformation (known and unknown) at the same time
Movement in order to:	Reveal hidden order; realize optimal state	Realize chosen goals	Realize a mature form already given	Survive as a species	Express continuity and transformation of individual and collective identity
Process of movement or cause is:	Universal, timeless laws, of "if-then" kind	Human reason within ethical universals	Unfolding a whole already enfolded in rules of interaction	Random variation in individual entities; natural selection	Local interaction forming and being formed by population-wide patterns
Nature of variation or change:	Corrective/fitting/aligning	Designed by rational choice to get it right	Shift from one given form to another in stages of development	Small chance variations	Diverse micro-interactions and escalation of small changes
Freedom and constraints:	Freedom through conforming to constraint of natural laws	Through reason constrained by ethical universals	No freedom, only constraint by given forms	Freedom arising by chance, constrained by competition	Both freedom and constraint arise in spontaneity and diversity of micro-interactions

Adapted from Stacey (2010).

organizational learning described by Peter Senge 1994.) We argue that the emergence of local interactions is constrained by the history of an organization (its culture) in the same way that evolution is similarly constrained by the history of a given species. Also, we suspect that transformative organizational change is less random than with evolution since, despite the existence of emergence, managers select (rational or otherwise) desired interactions, whereas in evolution the selection is "natural." We will exploit this idea in our approach to designing K organizations later in the book.

Table 3.4 is a summary of systems teleology, adapted from Stacey (2010). By teleology we mean the processes of stability and change that lead to a specific final cause. The first three forms cause certainty (i.e., reduce uncertainty) and encapsulate the ideas of systems thinking, whereas the last two cause uncertainty and encapsulate the ideas of evolution (adaptionist) and complexity science (transformative). Observe too that we have already seen elements of teleology in our discussion of philosophy. Recall that Aristotle—a proto-biologist—emphasized patterns observed in nature, where each object moves from an immature (potential) to a mature (realized) state. Furthermore, no realization is possible without an *already actual being* that has realized its form. This is referred to as an *enfolded form* in Table 3.4, that is, a form which has no freedom; it is only constrained by internal forces. This corresponds to formative teleology. We will use elements of transformative teleology, HT, and our own research to explain the formation and development of knowledge in organizations and systems of people.

3.3.1.1 Messy System Structures

Recall from Chapter 2 that a messy system is one composed of constantly changing interconnected problems, none of which can be solved independently (Mitroff *et al.* 2012). One conceptualization is to consider a messy system as a high-level superordinate context consisting of any number of interacting constituents. Furthermore, these constituents are largely human constructed and therefore bring all the strengths and weaknesses of people. It is no wonder that the constituents (problem components) contradict and are often in conflict with each other, creating a messy system structure.

Each constituent is a separate knowledge domain with its own experts. The leader—we say leader because messy systems are sufficiently large and complex that a senior executive is often charged with formulating a strategy—is responsible for managing the different, interconnected perspectives. As we will see in the next chapter, individual expertise is characterized by the creation of a unique solution to a novel problem through the dynamic assemblage of separate knowledge threads. However, unlike an individual expert, the dynamic strategy associated with messy problems is *channeled through* a leader, who draws on the expertise of many domain experts to create a unique solution framework. This is similar to systems thinking; however, with messy systems, the leader does not formulate a solution as such, but rather a management strategy or framework that gathers the diverse solutions of experts throughout the organization or systems (i.e., this is an example of late K-binding). Recall from Chapter 2 that healthcare and education are good examples of messy systems with no real solutions.

3.4 HT and Design

HT has its roots in the work of economist Herbert Simon (1973), chemist Ilya Prigogine, and psychologist Jean Piaget. Our description of HT comes largely from the works of Simon (1973), Allen and Starr (1982), Salthe (1985), and Ahl and Allen (1996). Most work on HT has concentrated on biological and ecological systems as well as complex systems theory. Whereas Simon (1996) has focused on systems theory, his writings have never been translated into an explicit design approach for organizations. In fact, the common interpretation of "hierarchy" as a rigid authority structure is a misinterpretation that has caused social and organizational scientists to shy away from the field. Indeed, its adoption encourages those who pursue the biological sciences.

Simon (1973) illustrates a hierarchy by using "Chinese boxes" as an example. Opening a box reveals one smaller box, which in turn may be opened to reveal yet another box, and so on. Such an arrangement is known as a complete ordering. In contrast, a hierarchy is a variant of this theme: opening one Chinese box reveals a *set* of smaller Chinese boxes, opening any one of which reveals yet another set, and so on. Mathematically, this is known as a partial ordering, or simply a *tree*. According to Ahl and Allen (1996), a system is hierarchical if, in addition to being a partial ordering, it is "composed of stable, observable subunits unified by a superordinate relation." Furthermore, hierarchical subunits—the high-level box and any or all of the set of small boxes—need not be tangible.

Simon (1996) also describes the interactions between things in levels as belonging to one of two types. The first is *asymmetrical* and characterizes *molecules* as found in biotic systems. The second is *symmetrical* and characterizes *tissues*, such as crystalline or lattice structures as found, for example, in rocks. By asymmetrical, we mean that the behavior downward is different than the behavior upward; by symmetrical, we mean the behavior is the same in all directions.

Organizations and their subsystems are primarily asymmetrical. These subsystems interact to produce what we have termed *organization molecules*. This has been described in our previous book and is reviewed here. However, at a very low level of self-organization and human conversation, the interactions are primarily symmetrical and lattice-like. It is this combination or *layering of asymmetrical and symmetrical subsystems that we hypothesize is the fundamental architectural structure of K organizations.*

System Characteristics of a Knowledge Organization
A layering of symmetrical lattice-like human interactions alternating with asymmetrical molecular structures

3.4.1 The Role of the Observer

HT concerns levels of organization and the role of the observer in analysis. First, HT is a science of systems that emphasizes the notion of levels and their behaviors and interactions. Second, HT is an *observer-centric* approach to analysis that complements the more traditional mechanical approach that underlies science and engineering. By involving the observer, we change the terms of analysis: from mechanistic techniques to experience and learning, from problem reduction to problem expansion, and from narrowing a question to broadening it. By introducing an observer, we put another wrinkle into systems dynamics as described earlier.

3.4.2 Complex Systems

Ahl and Allen (1996) describe a *complex system* as "one in which fine details are linked to large outcomes." ("Larger outcomes" include emergent properties.) This requires that we address several levels simultaneously and is in contrast to simple systems where low-level details may be ignored. A simple system is typically described using only one level of analysis as the central explanation of the observed behavior of the system. That is to say, a simple system may be understood by looking only at two layers: the phenomenon level displaying the behavior of interest and the explanatory level immediately below it. In contrast, a complex system is one where the effects of fine details amplify at successively higher levels, thus affecting the behavior of the whole system, and therefore requires that we understand several levels at once.

3.4.2.1 An Example of Simple and Complex Systems Perspectives

Ahl and Allen (1996) explain this phenomenon by using a chair. A chair is a simple hierarchical system, where behavior at one level can be explained by understanding the molecules immediately below it. A chair is, for example, solid because behavior of lower-level molecules (e.g., vibration) dissipates and produces an average condition of solidity. Solidity is an emergent property at a higher level that is not a characteristic of individual molecules at the lower level. If we ask at what load will the chair fail, the chair may still be analyzed as a simple system. On the other hand, if we ask, where and how the chair will fail, then we have to analyze the chair as a complex system.

3.4.3 Selected Properties of Hierarchical Systems

Here we discuss several characteristics of hierarchical systems that we will use in our design of K organizations and systems. We start with two fundamental characteristics of hierarchical systems. The first is that a *hierarchy* may be described (metaphorically) as a set of boxes within a box. Second, subsystem (i.e., box) interactions may be either *symmetrical* or *asymmetrical.* Recall too that we asserted that most organizational system interactions are asymmetrical or molecule-like, but *lower-level, local human interactions* shift to symmetrical or lattice-like behaviors.

3.4.3.1 Near-Decomposability

The notion of asymmetry leads to the idea of near-decomposability. A level is nearly decomposable, meaning that it cannot, by virtue of its asymmetrical association, be completely decomposed. Also, a system is nearly decomposable if its subsystems *within the same level* have comparatively weak intersystem forces; hence, "interactions *within* units at any level are rapid and intense in comparison with the interactions *between* units at the same level" (Ahl and Allen 1996).

Near-decomposability implies that a level, or a specific thing in a level (i.e., known as an *entity* or *object*), is not merely a collection of the set of entities at the next lower level. For example, a single molecule of water (at a high level) is more than a collection of one oxygen and two hydrogen atoms at the next lower level; rather, combining entities at the lower level produces certain properties in the corresponding higher-level entity. These properties are *system level*—they are visible in the higher-level entity or level, but depend on the elements at the lower level and therefore are functions of the entire system. The weight of a water molecule, for example, is a system-level property that, in this case, is a summation function corresponding to the weight of individual atoms at a lower level. In contrast, properties specific to an element, such as the specific weight of individual atoms of hydrogen and oxygen, are said to be *deliberate.* Figure 3.11 illustrates a water molecule from a hierarchical systems perspective.

Finally, the interrelated concepts of asymmetry and near-decomposability lead to the notion of *inexactness* as a general property of hierarchies: Simon (1973) observes, "properties of the lower-level, higher-frequency subsystems will 'show through' faintly into the behavior of the higher level, lower-frequency systems." As a consequence, Simon (1973) states that most laws of nature hold only approximately and only if we ignore lower-level details.

3.4.3.2 Modeling Distinctions

At this point, we need to make a fine modeling distinction introduced by Mario Bunge (1977). Both hierarchy and systems theories make reference to

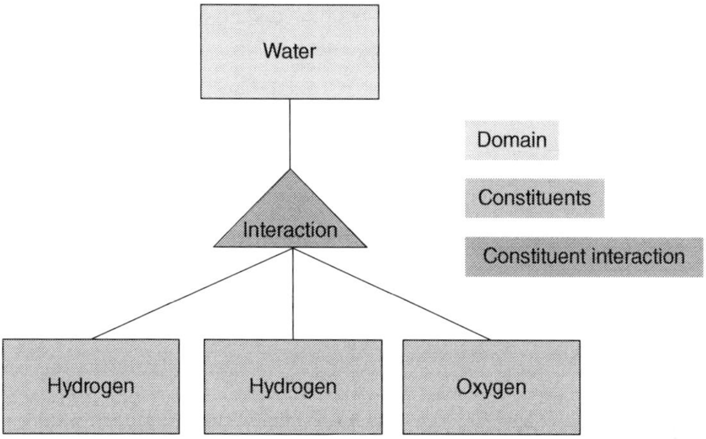

Figure 3.11 A molecule of water from a hierarchical systems perspective.

"emergence," though in the earlier discussion we use the term "system level." System-level properties (also known as *bulk* or *global* properties) are functions of the system itself—that is, its multiple layers—and are of two types. The first are *resultant* properties, sometimes known as hereditary properties, and the second are *emergent* properties. Resultant properties are typically *functions* corresponding to lower-level properties and therefore exist in at least one lower-level subsystem. As an example, the weight of water is a summation function: weight is a property of at least one constituent (in this case, weight is a property of all the constituents), and therefore the weight of water is a *resultant* property corresponding to the weight of its constituents. In contrast, wetness is a property of water that does not exist in any of its constituents but *emerges* through lower-level component interaction.

Let us extend our discussion with another example used by Bunge (1977). *Thinking* is a property of human beings. Is it resultant or emergent? To answer this question, we need to look at our human system as a complex one; that is, we must analyze multiple levels at once. A human is a composite of many organs, one of which is the brain; a human brain, in turn, is a composite of neurons and synapses. Now, where does thinking first arise? Presumably, thinking first arises in the brain and is therefore an emergent property of the brain. However, at the level of the human being, thinking is a resultant property since it already exists (albeit, in a different context) in the lower-level brain. This leads to a general observation: Within a complex system (i.e., a human being or an organization), properties typically first arise as emergent properties but are then carried upward as resultant properties. This corresponds to the idea expressed by Simon (1973) that properties of lower-level systems "show through faintly" in the higher-level system. Finally, it is easy to conclude that properties of complex systems are an entangled composite (i.e., a mess) of innumerable resultant and emergent properties arising from many different levels simultaneously.

3.4.3.3 Loose Horizontal and Vertical Coupling

Generally, a subsystem has most of its strongest interactions with nearby subsystems. Hence, a subsystem is characterized by strong bonding forces with nearby subsystems, but interaction with more distant subsystems is characterized by a comparatively weak bond. Where the subsystems are in different levels their interaction—*loose vertical coupling*—describes the hierarchy itself. When the subsystems are within a level, they may communicate through *loose horizontal coupling.* Loose vertical coupling creates independence between levels and thus provides an opportunity to create a larger and more complex structure (provided subsystem equilibrium properties do not affect the system behavior at higher levels). Loose vertical coupling is largely responsible for the evolution of complex systems, where, despite internal discontinuities and external turmoil, larger assemblages may be incrementally composed over time. Similarly, loose horizontal coupling also permits the creation of a larger structure, but here the key interaction considerations are the inputs and outputs between subsystems at the same level.

3.4.3.4 Frequency and System Behavior

In general, we observe that the vibration (i.e., motion or behaviors) of a constituent high in the hierarchy is lower than the vibration of a constituent in a low level. Therefore, we can say that behaviors of a system that depend on low-frequency levels (high in the hierarchy) are so slow that they will be difficult to observe and may be replaced by *constants.* In contrast, motions determined by high-frequency modes within a subsystem (low in the hierarchy) will be, of course, isolated from that of other high-frequency subsystems. Moreover, they will be so rapid that the corresponding subsystems will appear to be in equilibrium with no internal degrees of freedom. The result is that corresponding subsystems within the same low level will behave (with respect to each other) like *rigid bodies.* After eliminating the very high and low frequencies, then, it is the middle-frequency layers that determine (to a good approximation) the observable dynamics of the major subsystems. This is illustrated in Figure 3.12.

3.4.3.4.1 Example of a Low-Frequency System

Organizational culture sits at the very top of the organization. Hence, its motion is so slow that it may be taken as a constant. This is one reason management is attracted to "culture change" initiatives—changing an organization-wide constant should have a dramatic effect on the organizational system as a whole. Similarly, *organizational learning*—another low-frequency subsystem closely associated with culture—is another initiative that attempts to create the "learning organization." Edgar Schein (1996) argues that such efforts often meet with failure: occupational subcultures (i.e., cultures based on education and work roles)—each with different and contradictory constants known as *tacit assumptions*—work against the communication necessary for learning. Attempts to change very low-frequency subsystems—culture or learning, for example—often meet with difficulty. Low-frequency layers such as culture and learning are largely created through organizational evolution and are resistant to the effects of changes in high-frequency, lower-level subsystems.

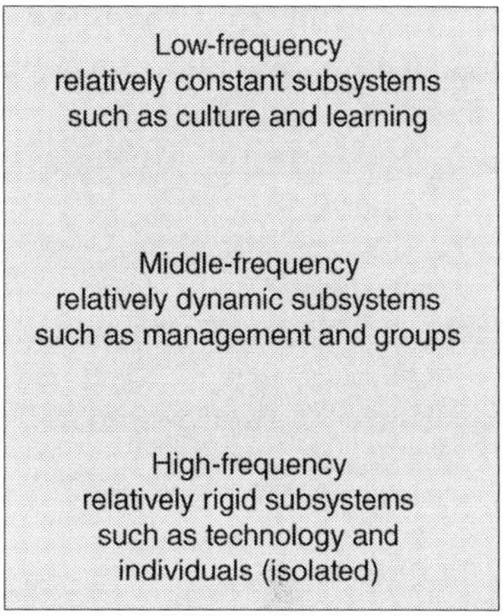

Figure 3.12 Frequency range in subsystem levels.

3.4.3.4.2 *Example of a Middle-Frequency System*

Recall from Section 3.2.3 that Mintzberg (1983) has identified five generic structural components including the *middle line*—the middle-frequency layer in an organization's structure. As shown in Figure 3.9, the middle line is centrally located and manages the flow of information among the other components. In essence, the middle line serves as an *information relay*, but this is true for a firm operating as an information processing mechanism. In contrast, a more modern perspective that reflects the emerging knowledge era is that the firm is a knowledge creating entity (as opposed to an information processing entity). Therefore, the middle line—again, because of its central location—functions as a *knowledge engineer* (Nonaka's term) who rationalizes and integrates knowledge flows from all directions. We will discuss this in more detail later in the book.

3.4.3.4.3 *Example of a High-Frequency System: An Important Departure*

As we show in Figure 3.12, low-level subsystems have comparatively high frequencies and may sometimes be seen as rigid bodies. This observation is characteristic of a traditional or information processing organization and is associated with classical HT. We have identified two low-level, rigid-like subsystems: technology and isolated individuals. However, individuals are rigid bodies only if their interactions remain local and unconnected to other organizational dynamics. In contrast, when individuals start to interact, even if they are located low in the organization chart, there arises the possibility of self-organization, a form of group dynamic that, if leveraged, will have its locally created knowledge amplified into organizational knowledge. This is the underlying premise of the SECI model of Nonaka and Takeuchi (1995) and is foundational to organization-level knowledge creation. *The knowledge organization is distinguished by its local, often low-level conversations, which are amplified into organizational knowledge at a higher level.*

3.4.3.5 Level Ordering

How do we characterize the hierarchy itself? As mentioned earlier, a higher-level entity will have a slower frequency than a lower-level entity. Ahl and Allen (1996) describe a higher-level entity as having a (1) lower frequency and (2) less bond strength and integrity than a lower-level entity. Additionally, a higher-level entity (3) offers context to, (4) constrains, or is a (5) composite of lower-level entities.

3.4.3.6 Complexity Versus Complicatedness

A complex system is characterized by the necessity to understand several levels at the same time. Therefore, if we add levels to our system, we make it more *complex*. But why would we consciously add levels? Recall from Figure 3.5 that one option in promoting the capacity to process information is to add slots; that is, by making the hierarchy deeper we increase its capacity to process information. Additionally, we may say that increasing complexity lowers *complicatedness* and vice versa. For example, organizations look for opportunities to flatten the hierarchy and thus reduce complexity. However, the resulting behavior at any given node in the flattened hierarchy (which now contains many more connections) is more elaborate and complicated—just ask any working manager!

3.5 Organization Molecules

We have synthesized organization principles with those of hierarchy and systems theories to propose building blocks, or *organization molecules*. We deliberately use the term "molecule" to characterize systems between levels in organizations. Recall from Section 3.4.3.1 that a molecule of water may be represented as a set of interacting elements as shown in Figure 3.11. As a *molecule*, this figure is an authentic representation of water in all its forms, as a sold, liquid, or vapor.

Similarly, an organization molecule is an authentic representation and design artifact of organizational elements (i.e., subsystems) in an organization. It is a finite assembly of elements that is constant across organizations and represents an architectural framework for design, specifically specification and implementation. Figure 3.13 is a representation of a generic organization molecule.

Illustrated in Figure 3.14 are generic *process* and *information molecules*. The process molecule, for example, represents a particular assembly describing the organization of work: the {process} at a higher level is a description or specification of work, whereas the lower-level elements of {human, structure, information, tool} define an interaction of implementation constituents

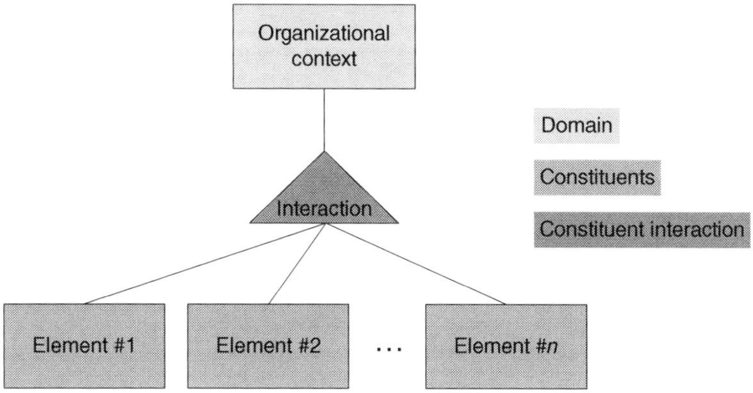

Figure 3.13 Generic organization molecule.

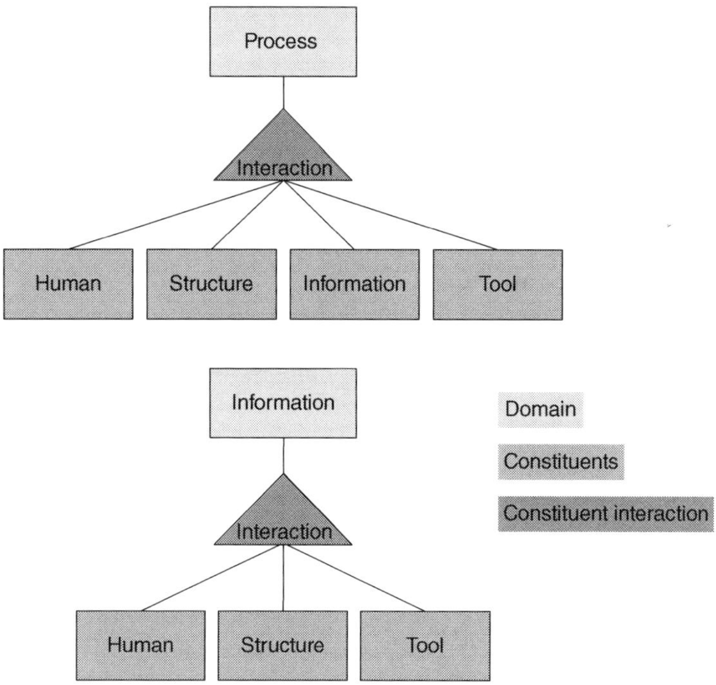

Figure 3.14 Process and information molecules.

immediate below its corresponding work specification. We will revisit this representation later in the book to compare craft, process or routine, and knowledge work. Similarly, the information molecule is characterized as an interaction of {human, structure, tool} elements and underlies our previous discussion of data, information, and knowledge. Note too that the information

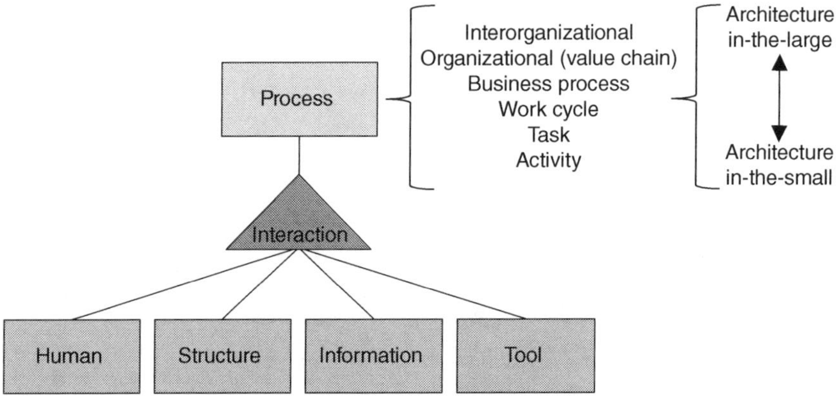

Figure 3.15 Process layering: architecture-in-the-large and architecture-in-the-small.

continuum and its illustrations in the previous chapter represent an application of the information molecule. Finally, any organizational context, not only process and information but also strategy, culture, structure, and so on, may be generalized in the molecule framework.

Abstraction. Finally, each domain and corresponding organization molecule may exist at any level of abstraction. For example, a process molecule (in its entirety) may define (i.e., specify and implement) the notion of work at successive levels, that is, interorganizational, organizational, business process, work cycles, tasks, and activities. This is illustrated in Figure 3.15. Furthermore, we may think of these levels of abstraction in terms of HT, where each level simultaneously defines work at a given level while containing and constraining lower-level forms of work. Thus, (looking up) a given business process is one of many processes in the organization, and (looking down) may contain any number of work cycles, which, in turn, may contain any number of tasks. This gives rise to a continuum of successive levels from architecture-in-the-large to architecture-in-the-small. We may make the same observation for every domain—structure and strategy, for instance.

3.6 Symmetrical Structures, Discourse, and Conversation

In contrast to the asymmetry of most organizational domains such as process, knowledge arises from human interaction that is characteristically symmetrical. The central characteristic of symmetrical interaction is that its mode is lateral rather than hierarchical.

We are using the term discourse in two ways. The first is the formal discourse among peers that includes, for example, the discourse in project teams or task forces. Such interaction may be found at all levels in the organization and typically involves specific outcomes, recommendations, and decision making. The second is informal discourse that is better described as a conversation and that may take place, for example, at the coffee machine. Between the formal and informal, semistructured conversations take place. For example, semistructured interactions may occur in an open office setting (more toward the informal) or part of a collaboration effort (more toward the formal). It is the informal conversation (including the semistructured), often embedded in a community network, which is often identified in KM literature as that where knowledge creation originates. The job of the organization is to encourage informal and semistructured conversation, elicit new knowledge, and amplify it through the more formal symmetrical and asymmetrical structures of the organization.

As an example, the authors have had experience in laboratory environments. The interactions were (typically) initially informal, which, if of perceived value, were elevated and became more formal. If direction came from the top of the organizational hierarchy, the conversations were quickly driven to the lower levels for analysis. These modes of conversation and discourse range from the informal to the formal and are characteristic of a knowledge organization.

Finally, social interaction need not be confined to verbal communications. It may be visual, mere gestures (Stacey 2010), constrained by unconscious assumptions embedded in a culture, or some other psychological or social mode not even known. We use the term *social heat* to describe all modes of social interaction that are foundational to human experience and knowledge creation. Returning to our laboratory experience, it is not that every person in a laboratory setting is constantly producing streams of new knowledge, but, rather, collectively the environment, in whole or in part, provides the social heat necessary for that knowledge which is created. Finally, social heat must intersect with the traditional (asymmetrical) domains of an organization if new ideas are to emerge into new products and services.

3.7 Closing Remarks

The modern organization lives in two realms. The first is that of the traditional organization, one with a hierarchy and the traits of a classical organization. These traits are the subject of organization science and include, for example, the relationship between structure and strategy, the level of formalization and communications patterns in industrial and service organizations, the decision-making process and how organizations reduce uncertainty, and so on. The second realm concerns the global context within which the organization now finds itself. This is a context of high uncertainty and globalization, an educated workforce with its own unique expectations, social and cultural integration, and a rapidly changing technology. However, what distinguishes this current upheaval from others is a revolution focused on knowledge itself. Therefore, it has the power to transform economies, organizations and institutions, cultures, and people in heretofore unknown ways. Furthermore, the second realm is distinguished by knowledge not necessarily confined to the organization but one that is at once individual and social, spanning the cultures of the world.

As the organization wrestles with rationalizing the two realms, it is simultaneously faced with a choice: follow either the first or else the second KM divide (see Chapter 1). The first divide will take the organization down a continuum of management and technology practices, where knowledge is managed not very differently than information. The second divide will lead the organization to seek out knowledge in its many forms and sources and to embrace the new organizational paradigms that will surely follow. Table 3.5 shows a representative progression of ideas as organizations move from the industrial through the information and into the second divide of the knowledge era.

The systems view of organizations derives largely from an engineering perspective of the world. Here, the organization is best seen as a machine that

Table 3.5 A comparison between industrial, information, and knowledge organizations.

Item	Industrial	Information	Knowledge
Social metaphor	Organization	Global	Transcultural
Organization intent and configuration	Automating	Informating	Knowledging
Structure	Formal and hierarchical	Formal and lateral (e.g., teams)	Informal and networked (e.g., communities)
K-binding (K construction)	Early	Early–late	Late–real-time
Organization analysis and design	Mechanical and engineering-like; modular design	Systems and engineering-like; modular design	Anthropological and ethnographic-like; integral design
Key design artifacts	Frameworks and prescriptions	Case studies; empirical roadmaps, architectures, and templates; statistical analysis	Ethnographic studies; discourse and social analyses; organization, social, and integrative modeling
Teleology	Efficient	Rationalist and adaptionist	Adaptionist and transformative

can be designed and managed as such. A complex transformation such as business process reengineering may be mapped to a template consisting of before and after measurements and workflows, gap analyses, improvement recommendations, and all other artifacts associated with a technical view of the world. People tend to be treated as automatons, working in teams, and managed by numbers with a variety of templates. That is to say, the organization develops an essential character as a well-engineered system. Though Peter Senge has exploited systems thinking in his contribution to organizational learning, Nonaka and Takeuchi (1995) make an important criticism of Senge and other approaches to organizational learning: the failure to see knowledge and knowledge creation (KC) as important dimensions of learning, the role of individuals in KC, and the role of tacit knowledge including physical, bodily learning in KC. Examples of the latter include the importance of trial-and-error learning, largely dismissed in organizational learning approaches. Note too that integrated body and mind learning (also dismissed in organizational learning) is one example of a dialectic, a foundational idea in KC.

In contrast to traditional systems perspectives, evolutionary (adaptionist teleology) and transformative systems are unpredictable, at least in the long view. Furthermore, they embody many of the characteristics of a knowledge organization, such as novelty or the creation of new knowledge. However, things are not so easy; recall from Chapter 2 that K organizations draw on many characteristics, often representing an interweaving of all five teleologies and other ideas and artifacts specific to our research such as the thin–thick continuum. See Table 3.4.

An organization molecule is a construct that characterizes an asymmetrical system. It simultaneously configures a subsystem domain to any number of levels while providing its corresponding specification and implementation. The description or formulation of a domain at a given level is a specification for what is below it in the hierarchy, whereas the interaction among corresponding lower-level subsystems represents the domain's implementation. That is to say, looking down, a subsystem specifies what is below; looking up, a subsystem implements (in part) what is above. Additionally, the specification may be informal as in a simple description, more formal as in a contract specification or other system artifacts, or completely formal as with the use of formal methods. Such asymmetrical systems characterize organizational subsystems, such as process.

In contrast, human interactions (formal or informal, at a high or low level) represent tissue- or lattice-like symmetrical structures. It is through such structures that knowledge is created and cultivated. As we will subsequently see, human interaction provides the necessary *social heat* for knowledge to arise. Further, the interaction of asymmetrical and symmetrical subsystems amplifies and promotes knowledge within the larger organization.

3.8 Class Exercises

1 Using Charles Handy (1993) as a reference, characterize your organization in terms of the schools of OT. If necessary, suggest other schools or a hybrid of those discussed by Handy (1993), which may represent a better fit with your organization.

2 Characterize your organization or school in terms of the primary dimensions of structure, that is, complexity, formalization, and centralization.

3 Characterize the major functions of your organization or school using the Perrow framework. Use Figure 3.4 as a guide.

4 Construct the organizational chart of your organization, explaining what it says about your company's organization of work and how it interacts with the environment.

5 Using the five major generic structural forms described by Mintzberg (1983), describe your organization in terms of the characteristics shown in Table 3.3, with the understanding that your organization may be a hybrid of the five forms.

6 Create a series of representative subsystem layers for your organization, identifying the contents of each layer by subsystem name and frequency. Describe or characterize the interactions between subsystems within and between layers. Use Figure 3.12 as a guide.

7 Construct representative organization molecules for your organization, identifying the domain being described (at a comparatively higher level) and its implementation constituents (in correspondingly lower-level subsystems). The process and information molecules are two examples—you may extend these molecules or develop other such molecules, such as a strategy, structure, or culture molecules.

8 Project. This is a comparative analysis. Research two or more organizations from different industries (including your organization or university). Describe and compare each as an open system, using Figure 3.10 as a reference. Include in your analysis the major system constituents, including the TRANSFORMATION processes and the interaction between the INPUTS, ENVIRONMENT, and OUTPUTS. In particular, describe those elements from the ENVIRONMENT that are significant for each industry and how each affects the OUTPUTS and INPUTS. In your analysis, consider how selected environmental elements modulate and harmonize inputs and outputs.

9 Identify where and how discourse or conversations take place in your organization or school.

10 Identify and locate the modes of interaction between the asymmetrical and symmetrical structures in your organization or school. Describe how that leads to knowledge origination and amplification.

References

Ahl, V., and Allen, T.F.H. (1996), *Hierarchy Theory: A Vision, Vocabulary, and Epistemology*. Columbia University Press.

Allen, T.F.H., and Starr, T.B. (1982), *Hierarchy: Perspectives for Ecological Complexity*. The University of Chicago Press.

Bunge, M., (1977) "Emergence and the Mind." In *Scientific Realism*. Ed. Mahner, M., Prometheus Books.

Daft, R.L. (1998), *Organization Theory and Design*. Sixth Edition. South-Western College Publishing.

Daft, R.L., and Macintosh, N.B. (1981), "A Tentative Exploration into Amount and Equivocality of Information Processing in Organizational Work Units." *Administrative Science Quarterly*, 26: 207–224.

Galbraith, J.R. (1995), *Designing Organizations: An Executive Briefing on Strategy, Structure and Process*. Jossey-Bass.

Goodhue, D.L., Wybo, M.D., and Kirsch, L.J. (1992), "The Impact of Data Integration on the Costs and Benefits of Information Systems." *MIS Quarterly*. September 1992.

Handy, C. (1993), *Understanding Organizations*. Oxford University Press.

Hatch, M.J. (1997), *Organization Theory: Modern, Symbolic, and Postmodern Perspectives*. Oxford University Press.

Kay, J. (1995), *Why Firms Succeed*. Oxford University Press.

Kusunoki, K. (2004), "Synthesizing Modular and Integral Knowledge: Business Architecture Innovation in the IT Era." In *Hitotsubashi on Knowledge Management*. Eds. Takeuchi, H., and Nonaka, I., John Wiley & Sons (Asia) Pte Ltd.

Mintzberg, H. (1983), *Structure in Fives: Designing Effective Organizations*. Prentice Hall.

Mitroff, I.I., Alpaslan, C.M., and Mason, R.O. (2012), "The Messy Business of Management." *MIT Sloan Management Review*. Fall 2012.

Nadler, D.A., and Tushman, M.L. (1997), *Competing By Design*. Oxford University Press.

Nonaka, I., and Takeuchi, H. (1995), *The Knowledge-Creating Company*. Oxford University Press.

Perrow, C. (1967), "A Framework for the Comparative Analysis of Organizations." *American Sociological Review*.

Robbins, S.P. (1990), *Organization Theory: Structure, Design, and Applications*. Prentice Hall.

Rockart, J.F., Earl, M.J., and Ross, J.W. (1996), "Eight Imperatives for the New IT Organization." *Sloan Management Review*. Fall 1996.

Salthe, S.N. (1985), *Evolving Hierarchical Systems: Their Structure and Representation*. Columbia University Press.

Schein, E.H. (1996), "Three Cultures of Management: The Key to Organizational Learning." *Sloan Management Review*. Fall 1996.

Schmidt, E., and Rosenberg, J. (2014), *Google: How Google Works*. Grand Central Publishing.

Senge, P.M. (1994), *The Fifth Discipline: The Art & Practice of the Learning Organization*. Currency Doubleday.

Simon, H.A. (1973), "The Organization of Complex Systems." In *Hierarchy Theory: The Challenge of Complex Systems*. Ed. Pattee, H.H., George Braziller.

Simon, H.A. (1996), *The Sciences of the Artificial*. Third Edition. The MIT Press.

Stacey, R.D. (2010), *Complexity and Organizational Reality*. Second Edition. Routledge.

4

Knowledge Work and Technology

Chapter Preview

Introduction

The central theme of this chapter concerns K work and the character of practice. We argue that the notion of a practice, grounded in deep philosophical and sociological roots, is the essence of K work. There are several dimensions to K work, known or unknown, but it is the exercise of practice that defines its character.

We continue our discussion in previous chapters and distinguish between *theoretical* and *practical K* and identify the factors that mediate their *interaction*. That interaction, in turn, forms the basis of practice, and as importantly, leads to the development of *social K*. In a very real sense, then, the exercise of K changes our social settings in large or small ways.

K work in the 21st century cannot be untangled from technology. We briefly review several important K-related technologies and present a student exercise on job analysis and the integrated design of K work and technology. Finally, we discuss the emergence of the robot economy and the factors that influence the automation of work, including K work.

Food for Thought

1) What is the difference between knowledge and experience?
2) What is the primary basis of your expertise? Is it facts, rules, methodologies and procedures, experience, social interaction, or some other factor(s)?
3) What role does social interaction play in the application of your work?
4) What organizational or professional constraints exist in the exercise of your practice?

Designing Knowledge Organizations: A Pathway to Innovation Leadership, First Edition.
Joseph Morabito, Ira Sack, and Anilkumar Bhate.

5) How important is expert knowledge in the kind of work you do or would like to do?
6) What role does technology play in your work? Do you believe that certain elements of your job may be automated through artificial intelligence (AI) and robotics?

Topic Layout

The figure below illustrates the topic layout of Chapter 4. The left side of the figure informs the organization on the nature of K work, K workers, and knowledge management (KM) tools. This includes classifications and perspectives and student exercises. We further discuss expertise with illustrations on piloting aircraft. The right side of the figure discusses the design issues that characterize K work, its architecture, and associated technologies. We also discuss the impact of AI and robotics on automation of work. The center of the figure represents K work in action. We distinguish between process (thin) and practice (thick) work, with an emphasis on the tacit characteristics of K work. We provide an extensive discussion of practice and include implementation guidelines. Also, we discuss design strategies that the advent of big data creates for the organization. Finally, we illustrate integrated job design through a student exercise.

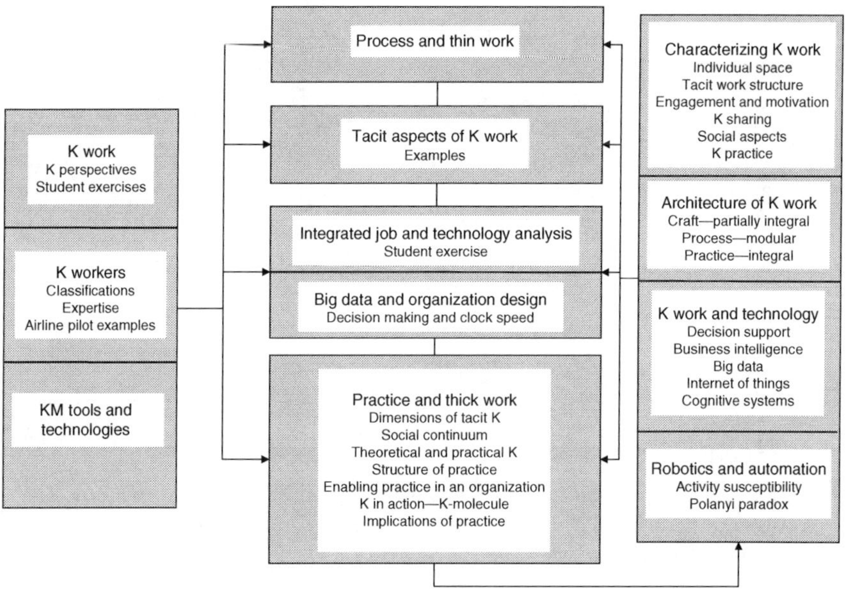

Concept roadmap to Chapter 4.

4.1 What Is Knowledge Work?

When we ask our students what is K work, we receive a wide variety of answers, with a particular emphasis on analysis and decision making. These answers are not wrong, but they are in need of further refinement if we are to understand the essential nature of K work.

Let us look at this in more detail. Is an auto mechanic a K worker? Most of our students respond "probably not," but the answer may depend on a given task: diagnosing the cause of an auto problem seems to involve K and a certain level of understanding, whereas repairing the auto after the diagnosis seems more mechanical in nature. This leads to the necessity of distinguishing work from the worker. Presumably, a K worker engages in a range of tasks, some of which may be more K-oriented than others.

We now ask, "Is clerical work K work?" In fact, is there anything that happens in an office environment that is not K work? Peter Drucker (1999) helps answer these questions. (Aside: Drucker is the first person to coin the terms K work and K worker.) In Drucker's perspective, K work is characterized by the manipulation of information, as opposed to the manipulation of physical things; alternatively, mental work as opposed to manual work, respectively. This means that clerical work is K work, a clerk is a K worker, and, in fact, much of what happens in an office is K work. This explanation makes our students a little uncomfortable and does not correspond to their perceptions of K work and K workers.

Service work often contains the two components mentioned earlier in the auto example: the first is a K-oriented analytical task (e.g., diagnosing declining sales), whereas the second is a little more straightforward (e.g., developing a new sales plan). However, this observation must be qualified; there are different degrees in the range of K tasks. A physician may follow a similar two-part task pattern as the auto mechanic or marketing analyst; however, the second task, say, writing a prescription, *is* a K task (manipulating information), but sometimes less intense than the first initial diagnosis. This leads to the idea of a range of K work, as there is a range of work in general. In addition, Drucker (1999) has classified a new type of K worker (really a subtype), a *technologist*, who simultaneously manipulates both information and physical objects, such as a surgeon performing surgery or a dentist filling a tooth.

As an example, Figure 4.1 illustrates several sample task pairs along a range of thick K-oriented and thin mechanical tasks. Naturally, the details depend on the specific problems identified and the solution chosen (i.e., each identified problem may have several possible plans of action). In general, the pairs follow prescribed templates and corresponding craftwork plans of action. However, the thicker the problem, the more integral is its design and implementation; that is, with thickness, problem analysis and its solution are more

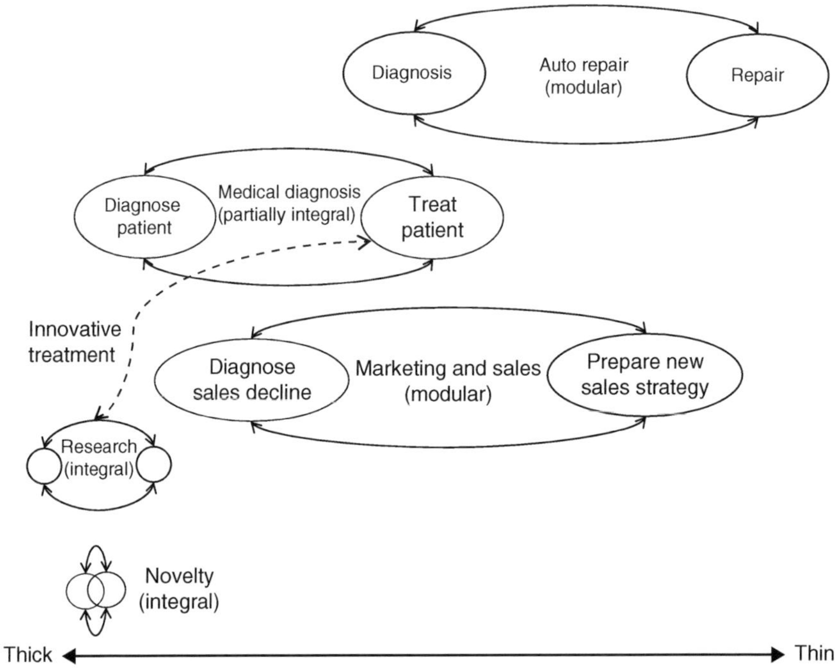

Figure 4.1 Sample range of paired K tasks.

intertwined. Recall from Chapter 2 that we refer to this as knowledge binding—increasingly thick work necessitates late K-binding. For example, in pure research (e.g., medical or astronomical) question and answer are intermingled and recursive—a given answer reveals other related questions. Also, in novel situations the solution to a problem is often constructed on the fly, and there may not be an opportunity for reevaluation, particularly in emergencies. As another example, though we show "diagnose patient" as thicker than "treat patient," in fact, complex treatment is often more thick and innovative than the initial diagnosis and may resemble an integral research task.

4.1.1 Student Exercise

Does repetition of a task make it routine? Is "routine surgery" routine? Does a lecture given to several classes become routine? To enhance understanding of K work, students were given the problem of comparing an assembly line worker, physician, professor, and Broadway actor. As we will see, the comparison among the different types of work, and, indeed, the nature of K work, is best understood using the notions of thin and thick. We initiated the problem by presenting our students with a Venn diagram representing a continuum of thin and thick. We also included Kant's characterization of sensibility and understanding as a way of differentiating, say, a Broadway actor (high in sensibility) with an assembly line worker (low in sensibility). This is shown in Figure 4.2.

We received a variety of answers, several of which are shown in Figures 4.3 and 4.4a–c. In Figure 4.3 the students mapped the primary tasks of each of the workers to the Perrow model.

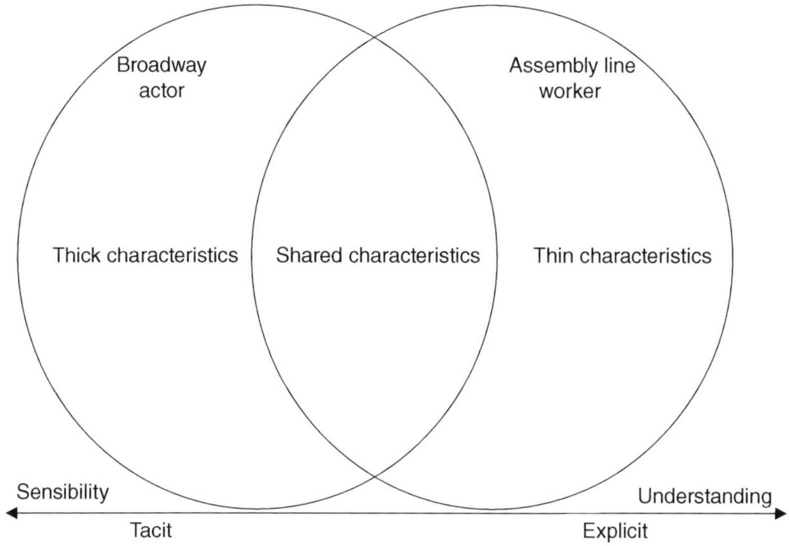

Figure 4.2 Comparing a Broadway actor and an assembly line worker.

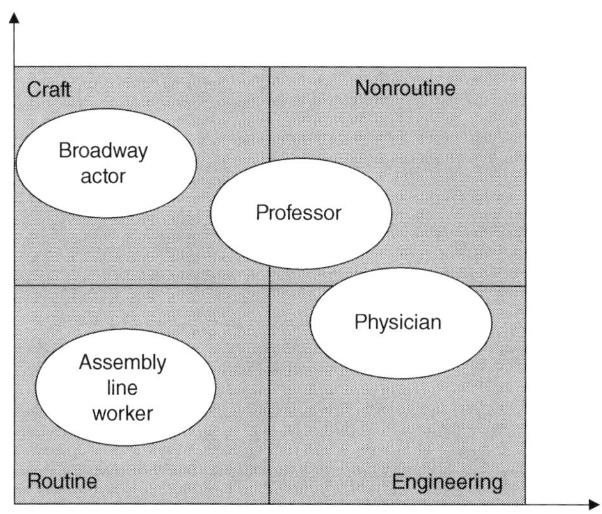

Figure 4.3 Primary task correspondence for Broadway actor, physician, professor, and assembly line worker using the Perrow model.

In Figure 4.4a and b, the students compared an actor and an assembly line worker, whereas in Figure 4.4c the students compared a physician and a professor. In class discussions comparing the actor, physician, and professor, students made the following observations on the three professionals:

- Sensibility and Understanding (Kant 1993). Each (actor, physician, and professor) has sensibility and understanding; it is a question of the proportional amounts of each. The actor is distinguished by a very high level of sensibility; a physician by a very high level of understanding; a professor by moderate levels of both.
- Interactivity with customers and high variability. The level and type of interaction is different for each, but the key is interaction. The interaction of a physician is 1 : 1, the professor 1 : small group, and the actor 1 : very large group. Note too that the interaction shifts along a range from two-way to primarily one-way for the physician, professor, and actor, respectively; however, even the actor receives information through nonverbal signals from the audience.

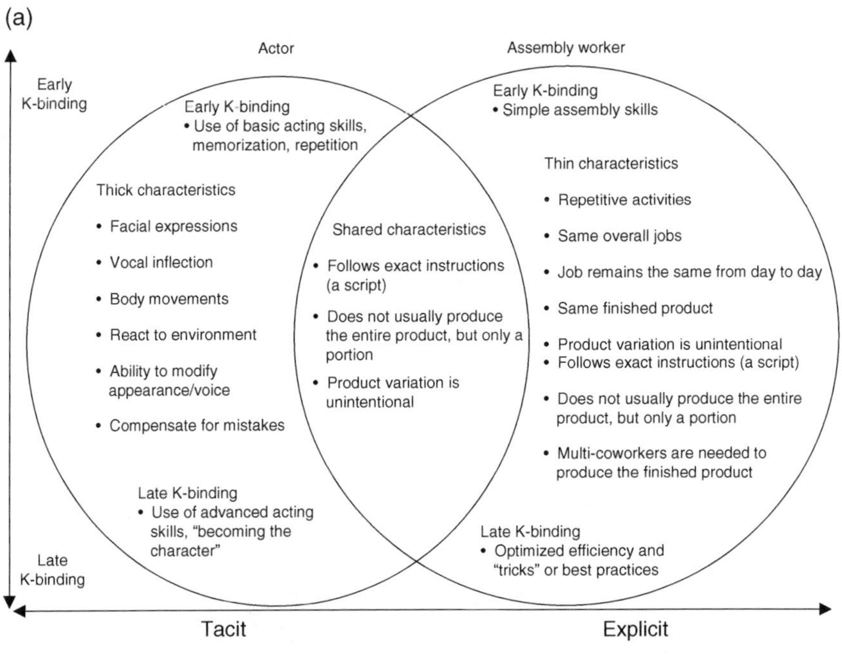

Figure 4.4 (a) Sample solution 1: actor versus assembly worker. (b) Sample solution 2: actor versus assembly worker. (c) Sample solution 3: physician versus professor.

(b)

#	Actors perform following tasks	Thin	Thick	Shared
1	Attend auditions/casting calls and audition for roles		x	
2	Study and memorize dialogue lines and actions from a script	x		
3	Interpret roles and relationships in script		x	
4	Research and develop character from script			x
5	Collaborate with director and other actors in an ensemble			x
6	Rehearse to ensure accurate portrayal of character	x		
7	Perform various interpretations of emotions, actions, and situations		x	
8	Deliver comedic timing, act, sing, and/or dance during performances to portray role to audiences			x

#	Assembly workers perform following tasks	Thin	Thick	Shared
9	Reading of technical schematics to become familiar with the assembly line process	x		
10	They measure parts and use tools to shape and trim them down to appropriate size	x		
11	Connect parts together to assemble required product in given amount of time	x		
12	Use variety of hand and power tools, such as hacksaws, drill presses, and forklifts during assembly process	x		
13	Extensive automation lead assembly workers to work with robots and computer controlled machines		x	
14	Assembler may have to collaborate during the design stage to help improve production efficiency			x
15	Process special orders, assembly line changes			x

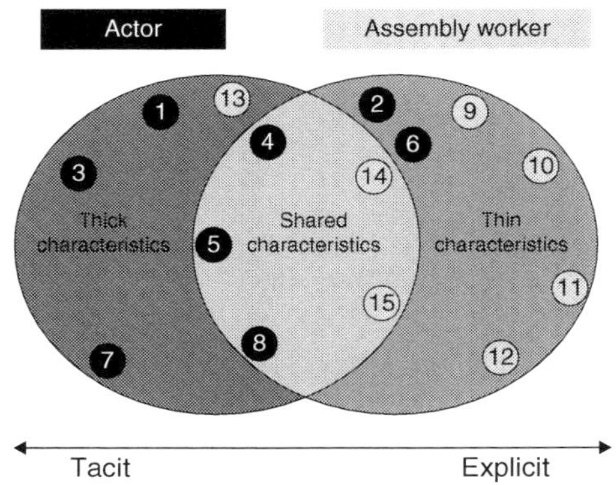

Figure 4.4 (Continued)

(c)

#	Physician are qualified experts who generally perform the following tasks	Thin	Thick	Shared
1	An adequate assessment of the patient's condition, based on the history and physical examination	x		
2	Provide for appropriate investigations		x	
3	Provide for appropriate treatment			x
4	Referring the patient to another practitioner, when indicated		x	
	In providing for care you must			
5	Work within the limits of your professional competence			x
6	Be willing to consult with colleagues	x		
7	Be competent when making a diagnosis or providing treatment			x
8	Keep clear, accurate, medical records which report the relevant clinical findings, the decisions made, and any drugs or treatment given	x		
9	Keep your colleagues well informed when sharing the care of patients	x		
10	Pay due regard to efficacy and the use of resources	x		
11	Prescribe only the drugs, treatments and appliances that will serve the patient's needs		x	
12	Make sure that your prescriptions are written correctly and clearly	x		
13	Abide by the hospital rules, protocols, and standard practice	x		

#	Professor are qualified experts who generally perform the following tasks	Thin	Thick	Shared
14	Manage the teaching, research, and publications in their departments			x
15	Conduct lectures and seminars in their specialties, such as the fields of mathematics, science, humanities, social sciences, education, literature, music or the applied fields of engineering, design, medicine, law, or business		x	
16	Perform advanced research in their fields		x	
17	Provide pro bono community service, including consulting functions (such as advising government)			x
18	Teach campus-based or online courses adopting instructional technology	x		
19	Mentor young aspiring academics (graduate students)		x	
20	Conduct administrative or managerial functions, usually at a high level	x		
21	Assess students in their fields of expertise			x

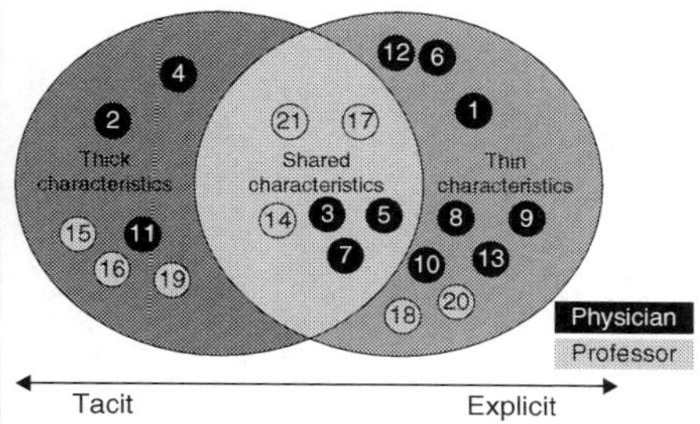

Figure 4.4 (Continued)

- Each programs the task, even the actor who brings something different to every performance. This is in line with Drucker's assertion that K workers program the task, whereas in routine work (e.g., assembly worker) the task programs the worker.
- It follows from the previous bullet that each exhibits a high level of thickness and tacitness in his or her work; that is, a practice character. For the most part, the content of the task is in the head of the actor, professor, or physician. This is in contrast to the assembly line worker whose work is primarily thin—modular, explicit, process oriented, and where the content is usually designed by someone other than the person performing the work. This is discussed in more detail later.
- Each constantly practices, experiments, improves, and researches alternate modes of delivery, particularly the actor and professor.
- Each is fully engaged in the task—his or her heart and mind are in the task.
- There is a level of improvisation in the task, particularly the actor and professor—less so for the physician.
- There is a certain moral imperative in the performance of their respective work, particularly the physician, but the actor and professor too. This raises an interesting question: does a K worker in a large organization, say, a biochemist researching a new drug in a pharmaceutical company, have a moral imperative? The consensus among students is "yes," but "less" than the others; the immediacy and interactivity of the physician, actor, and professor probably account for a portion of the moral imperative. Continuing with that line of thinking, the organization K worker is probably low in sensibility as well; again, this is a function of low client interactivity. Naturally, this depends on the function of the K worker.
- Measurability is an open question. *Outcomes* and *organizational impact* may be a better way of thinking about performance for K work rather than direct measures.
- "Routineness" is mediated by thin and thick. That is, repetition in a thin context (e.g., assembly line worker) is vastly different than repetition in a thick context (e.g., actor, physician, and professor). More precisely, *thick work is never fully repeated.*
- Observe too that the actor and professor share a "theatrical" character in their work, which a physician does not share.

4.2 Classifying Knowledge Workers

We will use two models to classify K work. The first is a two-dimensional classification from Eric Sveiby (1997), as shown in Figure 4.5. In this representation, Sveiby (1997) identifies four "power players." The first is the *Professional* who is strong in professional competence and weak in

Figure 4.5 The four power players—Adapted from Sveiby (1997).

organizational competence. This person comes closest to what most students perceive as a K worker. The professional is a specialist (even an expert), has professional pride, focuses on the job, enjoys problem solving, and dislikes routine. The *Manager* is the opposite of the *Professional,* whereas the professional relies on expertise, the manager relies on organizational competence. The *Support Staff* is relatively low in both professional and organizational competence, though in practice this person may develop organizational competence over time. Note too that this person includes the "clerk" in our example of a K worker from Drucker (1999). Finally, the *Leader* is strong in both professional and organizational competence. The leader does not necessarily correspond to someone high in the organization chart; rather, the leader is "informally appointed" by peers, is typically a former expert, and may be located anywhere in the structure of the organization.

The next classification comes from Tom Davenport and also includes four knowledge roles. The first are *K-oriented Personnel.* The K-oriented person comprises most of the specialized staff in a service organization and includes, for example, mangers, analysts, engineers, marketing professionals, secretaries, and clerks. Observe that this classification is particularly useful since it embraces virtually everyone in an office, even the clerk (see Drucker 1999) mentioned earlier. Each of these people manipulates information and needs to create, share, or discover K, but gets paid to do something other than managing knowledge itself. The line here can easily get blurred: the marketing person who creates a new product plan has created K, but the emphasis (what the person gets paid for) is on the plan itself, not the K content as such—fuzzy indeed.

The next category is the *KM Worker* (or KM Team). This classification includes two roles. The first is the traditional KM worker who creates and uses knowledge, such as a chemist in a pharmaceutical company. The second is a relative new K role, that of the *librarian* (also known as the *synthesizer, integrator,* or *editor*). We have added, extended, and elaborated on the possible KM worker roles to include the following:

- K engineer who extracts K (tacit K capture) in the development of an expert- or other K system.
- K engineer—Nonaka version. Recall from Chapter 2 Nonaka's perspective of a K engineer as a middle-level manager who (1) leads new product development projects and (2) functions as the "middle" in a "middle–up–down" structure, that is, someone who integrates strategic intent (mostly tacit) from the top and market and resource realities (mostly explicit) from the bottom.
- Conducts readouts after projects to extract lessons learned and possible best practices.
- Manages and participates in the K-base mentioned in Chapter 2 (recall this is our proposal that we believe necessary to make the hypertext organization work).
- Participates in K audits and K inventories for the organization.
- Identifies and refines the organization's necessary K based on the organization's strategy, K audit, and K inventory.
- Participates in the development of organization's intellectual capital model of itself.
- Extracts, structures, refines, and maintains K in specific KM systems or the organization's K-base. This includes building K stores and K maps.
- Synthesis of new K from the other activities listed here.
- Manages or is a member of a team in a "center of excellence" or "center of business knowledge."

The third classification is the *Manager of K projects.* This person has two sets of skills: expertise and knowledge in a specific domain as well as that of a traditional project manager. That is to say, the team leader in a development effort should be an expert in that domain. For example, recall from Chapter 2 that the director of a new drug development effort is a microbiologist. Nonaka's K engineer may fit into this category, but it is actually broader than that implied here.

The last classification is the *Chief Knowledge Officer.* We have refined this K role to include a *K officer* who is responsible for the KM activities of a business domain. We have seen the K officer (and not the Chief K Officer) emerge as a new job role in organizations in which we have taught. Here too we have extended the roles of a K officer (or Chief K Officer) to include the following:

- Establishes, coordinates, and oversees the organization's K-base. This includes the establishment of new entities and systems, such as internal libraries, research centers, and networks.
- Builds and refines the organization's intellectual capital model, including its internal relationships and measures.

- Provides guidance into (1) the reengineering of the organization's knowledging processes and (2) reuse strategies for K systems.
- Oversees and participates in the organization's or business area's K audit and K inventory. Note that the K audit and K inventory are periodic activities.
- Establishes linkages and alliances with external suppliers of knowledge, such as universities and vendors.
- Establishes internal and external K communities.
- Leads the development of K strategies, both technological and business.

Enterprise-wide KM requires an umbrella organization or department to establish an enterprise-wide context for coordination and integration of K activities. In Chapter 7, we propose a Chief Knowledge Office as a governance structure to include the librarians (synthesizers, integrators, or editors), the Chief K Officer, and key systems and business managers.

Finally, here we present a working definition of a knowledge worker from Tom Davenport (2005) in his book "Thinking for a Living," which itself is a metaphoric description of a K worker. Note too that this definition comes closest to the traditional KM worker described earlier and is the one with which most of our students feel comfortable.

Defining Knowledge Workers
Knowledge workers have high degrees of expertise, education, or experience, and the primary purpose of their jobs involves the creation, distribution, or application of knowledge

4.2.1 Degrees of Expertise

In Wiig's (1993) model of knowledge, he identifies "degrees of internalization," which are actually levels of expertise. These degrees range from *novice, beginner, competent, expert,* to *master*. The primary value of this classification is to create an image of expertise rather than strict design guidelines. For example, most organizations will assign a new employee with little or no experience (say, a college hire) to an entry-level position in a given area, such as marketing. With experience, the person develops increasing expertise and eventually, perhaps, becomes a recognized expert, even an acknowledged leader (as described earlier). The development of expertise approximately follows the progression shown in Table 4.1. Note that we have suggested examples to each level of expertise. It is important to recognize that these levels of expertise may be relative; an organization may have an expert who is recognized as a leader, but that is only for that organization and time (its particular K-space). In another organization, the same person may not be as highly regarded.

Table 4.1 Degrees of internalization—extension of Wiig (1993).

Type	Description	Examples
Novice	Barely aware or not aware of the K and how it can be used	New hire. New functional assignment
Beginner	Knows that the K exists and where to get it, but cannot reason with it	Trainee. Light experience
Competent	Knows about the K, can use and reason with the K given external K-bases such as documents and people to help	After completion of formal study (e.g., college course)
Expert	Knows the K, holds the K in memory, understands where it applies, reasons with it without any outside help	With extensive experience
Master	Internalizes the K fully, has a deep understanding with full integration, into values, judgments, and consequences of using that K	Leader (Sveiby 1997), elite expert

There need not be a conscious classification as the person develops expertise. Recall from Chapter 2 that deep smarts (Leonard and Swap 2004) involves the development of an "experience repertoire"—in effect, climbing the tacit hierarchy shown in Figure 1.14. Thick expertise is cultivated—initiated, developed, refined—through extensive experience, often more than 10 years. Other researchers have suggested 10,000 h (Gladwell 2007). In either case, the lesson is the same—it takes a long time to cultivate expertise.

4.2.1.1 Implementation: Airline Industry

In the airline industry, expertise and experience are regulated. In 2013, the Federal Aviation Administration (FAA) issued a ruling that commercial airline pilots and first officers (i.e., copilots) hired by US carriers must have at least 1500 h of flight time. This was in response to an airline crash in 2009 where both the pilot and first officer were relatively inexperienced.

Pilot expertise is not based exclusively on experience. Aeronautical decision making (FAA 1991) is a systematic approach to the mental process used by aircraft pilots to consistently determine the best course of action in response to a given set of circumstances. Jensen (1995) defines "pilot judgment" in terms of a pilot's skill repertoire, knowledge, and motivation to choose a "suitable" course of action. Similarly, the FAA's definition of aeronautical decision making includes both process and outcome components. Hunter (2003) argues that pilot judgment and aeronautical decision making can be used interchangeably to represent the multidimensional nature of these concepts in the aviation world, whereas the one-dimensional definitions of judgment and decision making, while applicable, do not express the complexity of the concepts in an aeronautical environment. A pilot's expertise (skills and knowledge) can influence his or her judgments. This is consistent with general research that supports the idea that experts make better judgments.

Pilot judgment and expertise are improved with the development of specific skills. These include perceptual–motor skills (e.g., mentally simulating problems and their solutions), which are enhanced with practice and come very close to the SECI (socialization–externalization–combination–internalization) notion of internalization. Also included are cognitive skills (e.g., improving recall through mnemonic devices) and metacognitive skills (e.g., the ability to monitor one's current level of understanding and to decide when it is and is not adequate for the task at hand). The latter is particularly interesting; among the findings of Leonard and Swap (2004) (Chapter 2) is that the expert can identify when generalizations will not work in an unfamiliar situation and go beyond them to adeptly assemble disparate elements and formulate a unique response to a novel situation. For example, an "expert pilot" is one who knows his or her own limitations and is motivated to keep a safe margin above that level. Also, the expert pilot possesses the discipline and strength of ego to acknowledge his or her limitations in every situation and seek help from crew members to offset the limitations. Note that the term "expert pilot" (FAA Aeronautical Decision Making, Adams, Ute, and Kasarskis) is used to denote a pilot with a variety of in-flight and preflight airline decision-making instances in a variety of settings.

A key learning in our discussion is that pilot judgment and expertise are closely linked and do not arise exclusively from the number of hours logged or the number of years of flying experience. Skills specific to the domain (in this case, piloting an aircraft) are necessary, such as perceptual–motor, cognitive, and metacognitive (essentially, thinking about thinking) skills. Of particular importance, common with expertise in general, is the ability to know when schemas (mental models) based on past experience are not sufficient and an original response must be constructed. *Cognitive flexibility* is thus important in applying expertise and judgment in new or unusual situations.

4.3 Tacit Aspect of Knowledge Work

Recall from Chapter 2 that all work lies on a continuum of thin and thick—process and practice, and (administrative) work and knowledge work, respectively. This too corresponds to Figure 2.19 where thin and thick are layered. That is to say, we have alternating thin and thick ontologies cross-referenced with thin and thick epistemological tasks. This is shown in Figure 4.6, which extends Figure 2.19 to include a thin–thick epistemological range.

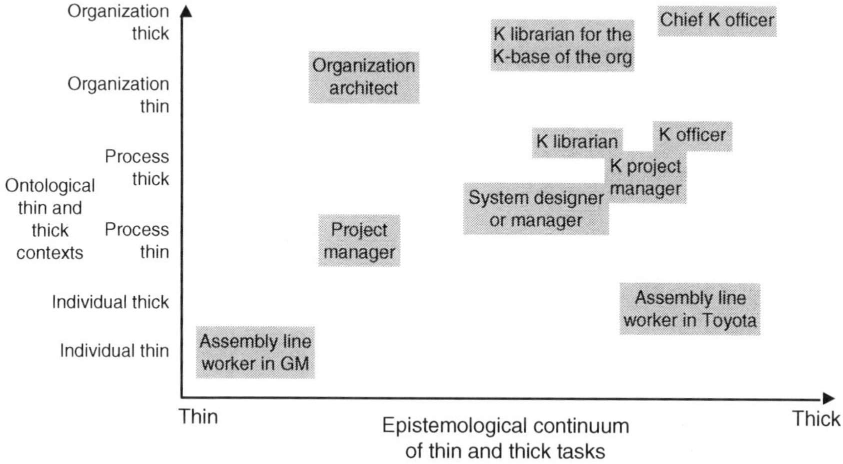

Figure 4.6 Sample K roles in context.

4.3.1 Pencil Metaphor

Let us look more closely at Figure 4.6. Certain comparisons seem obvious: the project manager and K project manager operating in thin and thick process layers, respectively. Other comparisons are more instructive on the role of culture and its impact on thin and thick design: there is a dramatic difference, say, in the job role of an assembly line worker in GM and Toyota, respectively. Each has a primary role of working on the assembly line; however, in Toyota the context is thick. What does this mean since both perform "routine work"? Several years ago Larry Prusak gave a talk at Stevens and suggested a pencil metaphor, based on his visit to Japanese organizations in Japan. In effect, a new worker in a Japanese organization is expected to "use the pencil" and function as a knowledge worker—to surface problems and suggest improvements. This is particularly strong in Toyota where the workers see their role as experimenters (see Spear and Bowen 1999). Moreover, the Japanese experience suggests that *all work—even routine work—contains* a *residue of deep tacit understanding* that may be leveraged for improvement.

The Toyota assembly line worker isn't simply given the assignment of improvement—the organization must play its part too. Organizational and process enablers and characteristics abound, such as constant training, accorded authority and responsibility, trust, autonomy to make decisions (within limits), and engagement with one's work.

In addition, this suggests that the Venn diagram comparing an assembly line worker with a Broadway actor, physician, and professor implies a thin context for the assembly line worker—the students in our school probably conceptualized assembly line work as fully (or mostly) thin. In contrast, if we were to present the same problem to Japanese students, the results may be different; the assembly line worker in Toyota would operate, at least in part, as a K worker and share thick characteristics with, say, a Toyota manager or process designer. Note too that the pencil metaphor is one way of thinking about our idea of a "tacit hook" described in Chapter 2.

4.3.2 Industry Example

This was further shown in a case study of a telecommunications company where a trouble ticket process was reengineered and conversations eliminated (Sachs 1995). In effect, the tacit K of the individuals could not be leveraged through practice communities (as existed in the old process). It was thought that routinization—deskilling coupled with the use of technology—would produce a higher-quality system than previously, when in fact, it produced the opposite effect. The elimination of discourse structures (even informal conversations) meant that problems could not easily be surfaced, and the dual benefits of both promoting and sharing tacit understanding through conversations could not be realized. As importantly, this observation applied to a trouble ticket system where the constituent tasks (we presume) are routine or engineering at best. In effect, a partially integral process was replaced by a modular one.

4.4 Characterizing Thick Knowledge Work

We assert that K work is an example of thickness. The following characteristics apply to both—in fact, we use the terms K work and thick work interchangeably. This is not an exhaustive list, but representative of what distinguishes K work from other forms of work.

1) *Individual space and autonomy.* A key characteristic of K work is autonomy. Recall from Chapter 2 that autonomy is considered an enabler—but it is also an essential characteristic of K work. In the same way that industrial work relies on control and standardization, K work requires the opposite. As Davenport (2005) has stated, "thinking for a living engenders thinking for oneself." More generally, autonomy, ambiguity, authentic passion, creative tension, relational contracting, and cooperative psychological contracts are all representative of the immediate individual environment (organizational and social) and psychological disposition of the K worker. Each of these factors structures, motivates, and characterizes K work.

   Creativity is a topic not commonly covered in KM. However, there are several key overlapping characteristics that we want to briefly mention. Creativity, or, let us call it the creative side of knowledging, depends on an individual's intrinsic motivation and skills. Google calls these people "smart creatives," and they like to think this is whom they hire (Schmidt and Rosenberg 2014). (We will revisit hiring K workers later in the book.) Amabile (1997) has proposed a "3 component model of creativity" that includes *expertise, creativity skills (thinking),* and *intrinsic task motivation.* Creative thinking includes an appropriate cognitive style in adopting new perspectives in problem solving as well as a persistent working style. (We may add imagination too.) Naturally, certain personality traits are more conducive to creativity and include several that characterize K work, such as independence and tolerance for ambiguity. Intrinsic task motivation, in turn, has been discussed here and in KM generally and includes, for example, engagement through deep interest and involvement in challenging work. Again, KM and creativity are interconnected and promoting one promotes the other.

2) *Tacit work structure.* A second characteristic is that the structure of K work cannot be articulated. In contrast to thin work, which is typically structured with workflows and process maps, K work does not lend itself to detailed task sequencing. For example, we (rhetorically) ask our students to draw a workflow of a criminal trial. What would you draw? Nothing meaningful. The legal process is embedded in the practice of the trial itself. On the other hand, we observe the salient features to be the commitment, passion, and expertise of the legal professionals—defense attorney, prosecutor, and judge—as well as their interactions. Expertise embodies the work, not a workflow. *This juncture of traits—commitment, expertise, and human communications—is a distinctive feature of the structure of K work.*

Whereas a criminal trial clearly involves K work, the trouble ticket system and the Toyota cases mentioned earlier also benefitted from the application of K work principles. (Aside: Enablers and characteristics are often embodied in the same principle or aspect of K work.) On the surface, the trouble ticket system is characterized by routine work yet benefits from informal social systems used to locate and solve problems through collaboration. Communications, relationships, and coordination are as important as detailed sequencing of explicit tasks. What makes the Toyota experience mentioned earlier particularly interesting is that Toyota formerly combines both approaches—the use of detailed task analysis and the leveraging of tacit K and experimentation. Observe too that the Toyota approach may be represented with layers of thin and thick activities, as described in Chapter 2.

Finally, whereas both K work and more routine work benefit from deep tacit understanding and practice, the formal structuring of work is typically not very helpful in K work. We thus see two types of K work structures: the relatively isolated K worker (e.g., physician) and the K worker embedded in an organization. The former does not benefit from the formal structuring of work as found in an organization, whereas the latter requires characteristics of thin and thick interconnectivity, mediated by collaborative structures. Finally, *all work contains a residue of tacit knowledge.* Hence, elements of thick work (or practice) may peek into thin (or process work); practice, in turn, is intertwined with a corresponding process.

3) *Engagement and motivation.* A third characteristic is that the K worker should be fully engaged in the work. Returning to the criminal trial example, if you were the defendant, you would want a defense lawyer who is almost fanatical in his or her belief in your innocence. That is to say, a K worker can't just "show up" but must have his or her heart and mind in the work. This commitment is what we refer to as *engagement.* This is associated with another characteristic that we can relate to cooperative psychological contracts: the *passion* and *energy* that a K worker brings to a task. This psychological orientation, in turn, promotes experimentation and inventiveness.
4) *K sharing.* A K worker does not easily share his or her knowledge (Davenport and Prusak 1998). Think of it this way: would an auto mechanic who has spent his entire life accumulating tools simply give his toolkit away? A K worker has a toolkit too—his or her lifetime of accumulated knowledge and experience. This speaks to the failure of performance reviews—an artifact of industrial organizations. Rather than seeking to elevate the role of the K worker to that of a partner or some other form of tenure (which would encourage sharing K), the organization tries to persuade or intimidate workers to share knowledge through promotions or bonuses (or the absence thereof). Furthermore, in several of the organizations in which we teach, the students tell us there is a ranking of performance according to quotas. Even more amazing, a worker who has received an excellent or top ranking cannot again receive that ranking for another 2 or 3 years; it must be distributed among workers and departments over several years. What could be more insulting to K workers? None of this facilitates the transmission of knowledge from one worker to the next. The transformation from industrial to knowledge thinking is a difficult transition indeed!
5) *Social and collaborative aspects of K work.* A key structural element of K work is the context beyond the individual space. This comprises the social aspects of K work, including culture, collaboration, socializing structures, and social networking. We dedicate Chapter 6 to this important aspect of K work.
6) *K Practice.* K work is a form of practice that, as discussed later, embodies the social dimension of knowledge. (Also, see Chapter 6 on the social aspects of KM). In many ways, *practice is the fundamental unit of K work.*

4.5 Architectural Perspective of Knowledge Work

In Figure 4.7, we illustrate work using the process molecule introduced in Chapter 3. The figure shows a historical progression of how work was typically configured in the last 125 years or so. The top figure shows work as *craft.* Prior

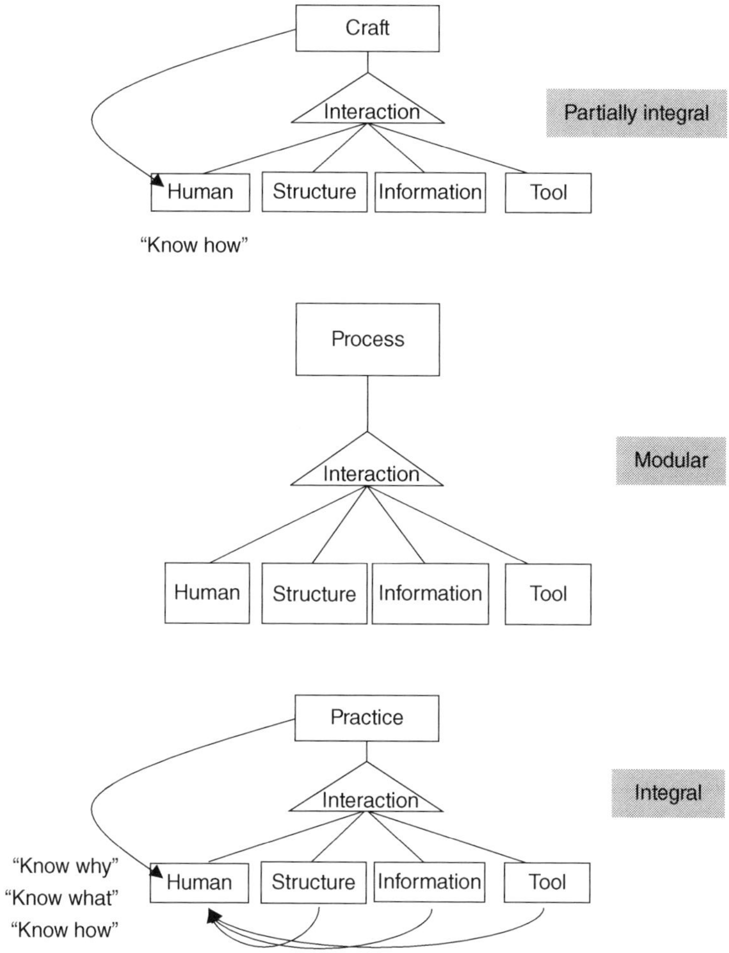

Figure 4.7 Historical progression of work.

to Frederick W. Taylor and Scientific Management, most work was perceived as craftwork. Craft is partially integral and "know-how" embodies the craft person who performs the work.

The center figure shows *process* work. In process, the work is fully thin and modular. That is to say, the work is specified by someone other the person performing the work, typically a specialist or engineer. Task content—the structure and workings of the task—is engineered in a formal sense, outside of task execution. This manifests itself as modularity, which means the absence of K-based interdependencies; hence, know-how, know-what, and know-why all lie outside the work molecule itself. Know-how, for example, is engineered into the specification and workflows, not in the work performer. Furthermore, the specification may be sufficiently rigorous to permit automation—that is, the elimination and substitution of the human component with a robot in the tool component. Unlike craft work, process components are substitutable. Finally, as Drucker (1999) has stated, the primary management question in process is "how do we do the task," an implementation concern. Again, the content of the task has been fully specified well before its implementation (i.e., early K-binding). It is important to note that process is foundational to the design of complex work in the industrial society. (Hence, the rise of the performance review, an artifact of thin work, which matches actual and specified performance, again, in an engineering sense.)

The lower figure illustrates K work, which we are calling *practice*. Practice is fully integral where the description of the work cannot be easily separated from the human component (i.e., the K worker) who does the work. Task content (structure and workings) is inside the head (at least partially) of the person performing the task; hence, know-how, know-what, and know-why embody (or should) performance. There is little separation between specification and implementation (i.e., late K-binding), each residing in the K worker's head. As Drucker (1999) states, in contrast to process work, the key management question in K work is "what is the task?" With thick work, the worker programs the task (paraphrasing Drucker 1999). This is in contrast to thin work where the task (already engineered) programs the worker. Finally, in contrast with thin work and process design, in thick work the person *cannot* be separated from the work and its content. This has all sorts of design implications: the work itself, its social and organizational context (everything from office design to culture), hiring, reward, and compensation practices, and so on. In practice, this means that many organizations may have to make significant shifts in their leadership, people, and design strategies.

4.6 Process and Thin Work

According to Hammer and Champy (1993), "a process is a collection of activities that takes one or more inputs and creates an output that is valuable to a customer." This is often represented as a sequence of interconnected tasks and is called a workflow model.

In addition, Peter Keen (1997) states "a process is any work that is recurrent, requires coordination, affects some aspect of organizational capabilities, can be accomplished in different ways that make a difference in terms of cost, value, service, or quality." This broadens the definition beyond workflows and includes "soft" processes such as mergers and acquisitions, and even culture. Knowledge may come into play if we consider that "processes are a source of 'firm-specific' special competence that makes for competitive difference far more than does strategy, products, structure or industry." The latter quote is very interesting—it raises the competitive importance of process. We can think of this as an aspect of a resource-based approach to strategy. The key words "special competence" suggest processes not easily imitated.

We think of a process as work that may be codified in a formal sense. By formal, we mean the work may be described, even specified, in a logical sense, with precise and transactional contracts. This is one reason for organization molecules; recall in Figure 3.14 a process molecule, which may be used to precisely specify and configure a process or constituent tasks—the process or task itself and constituents (implementation) at the next lower level. Such thin work may include decision making, but this too may be specified with contracts. A little fuzzier, perhaps, are thick tasks embedded in a thin process, as shown in Figures 2.19 and 2.20: here the approach (as shown in the illustrations) is to separate the thin and thick tasks as well as the thin and thick process perspectives. Thinness is designed with engineering precision, using, for example, precise contracts, workflows, and other explicit and modular artifacts. Thickness, on the other hand, is designed with an integral perspective, which combines many K-spaces or system levels into, for example, a practice as described later. Note too that in thin design, emotions and ethics are not necessarily embodied in the process, whereas in thick design they often are necessary constituents.

4.7 Practice and Thick Work

Davenport views practice as a way in which K workers interact with the environment and how they go about solving problems, that is, an understanding of how and why thing get done. This is in keeping with the anthropological analysis associated with KM. A similar idea is structuration theory, which concerns the social interaction of people with their environment, and in the process, how each reproduces and shapes the other. More generally, "structuration means studying the ways in which social systems are produced and reproduced through social interaction" (Giddens 1984).

(Aside: At the organization level, structuration suggests a synthesis between the environment and organization's internal resources. We will revisit this in a future chapter when we discuss K strategy, but briefly, this is a way of conceptualizing the synthesis of the strategy of position (environment) with the resource-based approach to strategy (internal resources).)

In order to explain and understand the essential nature of this social interaction, we must more fully explore the notion of a practice. After all, practice extends well beyond the notion of work and into every discipline and aspect of life, including anthropology, culture, history, philosophy, psychology, science, technology, and so on. Vegetarianism, for example, is a practice, as is adherence to most religions. How does this compare to, say, a medical or legal practice? Let us delve into this idea of practice a little further. Here we review several ideas associated with contemporary practice theory (Schatzki *et al.* 2001):

- Most practice theorists conceive of practices (minimally) as arrays of activity. More specifically, theorists in philosophy and the traditional social sciences conceive of practices as *arrays of human activity*. Additionally, a practice is organized by a "pool of understandings" (i.e., abilities and sayings that pertain to action), a set of rules, and a teleoaffective structure (i.e., the emotions, moods, beliefs, hopes, and expectations that guide action) (Schatzki *et al.* 2001).
- Other theorists define practices as the skills, or tacit knowledge and presuppositions, which underpin activities (e.g., Dreyfus 1991; Turner 1994).
- A significant "posthumanist" minority centered in science and technology studies claim, additionally, that the activities bound into practices also include those of nonhumans, such as machines and the objects of scientific investigation. We consider this to be exceedingly important where it is difficult to conceive of modern practice without technological interventions and support.
- According to some practice theorists, practices are the chief and immediate context within which the preponderance of bodily properties crucial to social life are formed, not just skills and activities, but bodily experiences, surface presentations, and even physical structures as well.
- Perhaps most importantly, all practice theorists acknowledge the dependence of activity on *shared skills or understandings* (which are typically viewed as embodied). However, there is significant disagreement on what, if anything, beyond shared understandings is necessary to explain practices. Yet again, theorists assert, despite such disagreements, the proposition that the maintenance of practices, and thus the persistence and transformation of social life, rests centrally on the successful imbibing of *shared embodied know-how*.
- Social order is established primarily through social practices and the actions of the mind. This is in contrast to other popular approaches: individualist and rational. According to Schatzki *et al.* (2001), individualist conceptualizations attribute the layout of social affairs to features of individuals and their direct interactions, for example, agreements, shared internalized norms, skills, mutual understandings, reciprocal interpretations (ethnomethodology), negotiation and mutual adjustment (symbolic interactionism), and so on. Among advocates of rational choice or game theory, social orders (e.g., markets) are unintended by-products of actions governed by rational calculations.
- Returning to social order as social practice, practice thinkers acknowledge the importance of agreements and other interactions, as well as the

significance of skills and interpretations. However, they treat these phenomena as features of, or as embedded in, practices. That is to say, social order is instituted within practices.

- Practice is more than activity, but serves as the context for the activity. This is because practice "houses the semantic spaces." A semantic space embraces three interwoven *fields of meaning*: (1) a person's state of *being* (e.g., feelings, desires), (2) the *situation* a person confronts; that is, the people, events, and things with which a person interacts, and (3) the *behavior* the person engages in response to the situation.
- "Mastery of a practice cannot be gained from books or other inanimate sources, but can sometimes, though not always, be gained by prolonged social interaction with members of the culture that embeds the practice" (Collins 2001).

4.7.1 Revisiting Tacit Knowledge in a Practice Context

Collins (2001) has described three conceptions of tacit knowledge in the sociology of scientific practice. The first is the *motor-skills* metaphor, which corresponds to Nonaka's technical dimension of tacit knowledge. The second is the *rules-regress* model, which emphasizes cognitive abilities and approximately corresponds to Nonaka's cognitive dimensions of tacit knowledge (i.e., mental models). We have seen earlier, for example, that an expert pilot must excel in both these dimensions—motor skills and cognitive flexibility.

The third form of tacit knowledge is the *forms of life* approach. We start with the observation that people in different social groups take different things to be certain knowledge, and, furthermore, they are not aware of the social basis of their certainties. That is, what we take to be certain has more to do with the social groups in which we are embedded than with rationality. It follows, then, that the only way to obtain scientific knowledge is through immersion in the relevant social group. This way of looking at things not only explains the difficulty in transferring knowledge and its consequences (i.e., its stickiness) but also provides a *strategy for gaining it—more social interaction*. The basis for our beliefs is to be found in the history of the social groups in which we are embedded.

This third aspect of tacit knowledge—social embedding—brings us closer to seeing *practice as a coalescence of tacit knowledge and social consensus within groups*. The interweaving of technical or motor skills, cognitive mental models, and social consensus in a practice group is illustrated in Figure 4.8.

As illustrated, we assert that tacit knowledge as such is not wholly an individual characteristic and, in fact, cannot be separated from the social group in which it arose. The practice groups may be, for example, vegetarian or religious, but may also be medical or any other professional practice group.

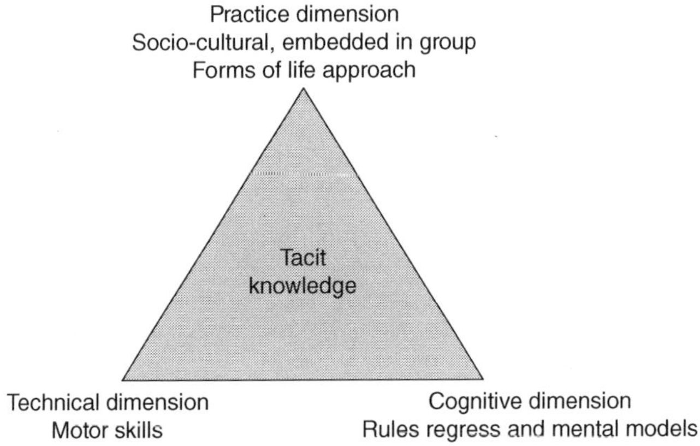

Figure 4.8 Three dimensions of tacit knowledge.

Tacit Knowledge
Tacit knowledge comprises three interrelated dimensions: motor skills, cognitive flexibility and mental models, and socially imbibed practices.

Recall from Figure 2.5 that we illustrated knowledge with a combination of a database (explicit K) and a neural net (tacit K). Whereas it is possible to imagine a neural net replacing a fixed or final body of tacit knowledge based on the first two conceptions described earlier, the third conception makes it difficult to conceive of such nets having formed these traditions in the same way as humans formed them (Collins 2001). Furthermore, social practice is constantly changing within a given group; that is, databases and neural nets capture a snapshot of K, not its social evolution.

4.7.2 Unraveling Knowledge and Practice

Recall that in Plato's Justified True Belief justification bridges belief and truth. We assert that the principal orientation for the K worker is in justification—this is where most of the work is located. In one sense, both beliefs and truths are essentially independent and do not arise from the conscious effort of individuals; it is the workings of justification that consciously surfaces and connects those truths and beliefs. It follows, then, that while science establishes truths and beliefs (i.e., theories), K workers practice justification. Justification, in turn, establishes or solidifies truths and beliefs. However, there is an important difference here between theory and practice—*it is science that builds theoretical knowledge; practice builds practical knowledge.*

What does this mean? Every rational person understands the difference between what is true in a theoretical sense and acting on that truth. *We tend to make judgments based on theoretical K and decisions based on practical K, though each is intertwined with the other.* Understanding the nature of this difference is less clear and is at the heart of practice. Let us more fully explore this idea by revisiting Kant (1993).

First, recall that truth and belief are interwoven through justification. Similarly, practical applications of knowledge—in the form of practical actions (recall from earlier our first description of practice as arrays of activity)—are always related to reasoning, which is the faculty of the human mind that enables logical thinking. Kant (1993) wrote the trilogy of critiques that dealt with reasoning and practice. His *Critique of Practical Reasoning* published in 1788 is the second of his three critiques, the others being the *Critique of Pure Reason* (1781) and the *Critique of the Faculty of Judgment* (1790). *Pure reason, practical reason, and judgment answer, respectively, what we must think, what we must do, and how we hold something to be true or false.*

For each of the first two critiques; that is, *Pure Reason* and *Practical Reason*, Kant (1993) wrote a briefer, less scholarly work on the same topics. These shorter works follow the *analytical* or *regressive* method: they begin with experience and regress upon their a priori presuppositions or principles without which it would not be possible to have that kind of experience. Evidently, it is important to note that in these shorter works Kant (1993) has clearly developed the connection between *knowledge* and *experience*.

Another point worth noting is that in these shorter works the starting points are found in mathematical and scientific knowledge as well as the "common knowledge of morality." This brings together the *two basic kinds of knowledge*; that is, the (1) scientific and mathematical knowledge and (2) the knowledge of commonly understood ethics which is embedded within the customary concepts of morality. In each of these, everything is based "upon something already known as trustworthy (in other words what we call *Knowledge*) from which we can set out with confidence and ascend to sources as yet unknown." Regarding "sources as yet unknown," we can interpret these as "a kind of hidden knowledge." For these, Kant (1993) has developed a philosophy of "forms of intuition" and "categories of judgment," dealing with "moral law and freedom."

Apart from the earlier-mentioned shorter works, the method used by Kant (1993) in the writing of the actual *Critiques* is synthetic. That is, the actual *Critiques* begin with principles and thence proceed to experiences, which they organize, conceptualize, and render intelligible. According to Kant (1993), only by this method can philosophical knowledge "present all its articulations, as the structure of a peculiar cognitive faculty, in their natural combination." Hence, the *Critique of Practical Reason* begins with definitions, and proceeds in a quasi-deductive manner to the formula of the *moral law* and the theory of *freedom*.

Wow! This may seem rather complex, but it is foundational to our conception of practice. Let us briefly review. Knowledge and experience—pure reason and practical reason, respectively—are intimately related and arise from logical thinking. Note too that practical K and experience are referred to as *practical reason* and not merely practice or experience, underling the rational basis of practical knowledge and decision making. Judgment, in turn, may be applied to theoretical or practical knowledge. There are thus two types of K: scientific that yields Truth in a pure sense and practical K embedded in moral and ethics. That is, the factors that influence practical reason are based (or should be) on moral law and freedom, the latter being a precondition for the former.

Back to K work. Justification—judgment and action—is at the heart of building new knowledge, whether theoretical or practical. The K worker rarely creates the truths and beliefs that arise from social practice, but must constantly build justification. Justification is the principal work of K workers; for the most part, they create neither beliefs nor truths (except for Einstein). That is to say, *we believe justification and the development of practical knowledge within a social network is at the heart of K work. The artifacts of a practice (e.g., protocols, precedence) represent the corresponding social group's justification, linking in turn the group's beliefs and truths, but in the context of the group's practical knowledge.*

4.7.3 The Social Continuum

Figure 4.9a illustrates a social continuum between theoretical and practical knowledge as conceived by Kant (1993) and extended by the authors. At each pole—theory and practice—we have illustrated Justified True Belief as a function of reason. Theoretical knowledge demands a search for certain K or Truth. However, with practical knowledge, the goal is not Truth as such, but some other combination of factors representative of the social or practice group. While Aristotle suggests one such factor—happiness—Kant (1993) has developed a more complex and sophisticated approach centering on *duty* and *moral vision. We believe it imperative that the goal of practical knowledge be noble,* such as the Kantian advocacy of duty and moral behavior. Without a higher goal, the mitigating factors may take on an expediency that undermines the inherent truth and reasoning upon which practical reason should be based. For example, we have all seen instances of the legal system ignoring exculpatory evidence that results in convicting an innocent person for the political gain of a prosecutor or judge. Furthermore, the offending parties are almost never held accountable, thereby reinforcing the amoral goal of the corresponding practice group.

The more cloistered one's study, the more refined one's conception of the truth—this has led to the image of a scholar in an "ivory tower." On the other hand, this refinement is necessary to develop complex theories and moves our

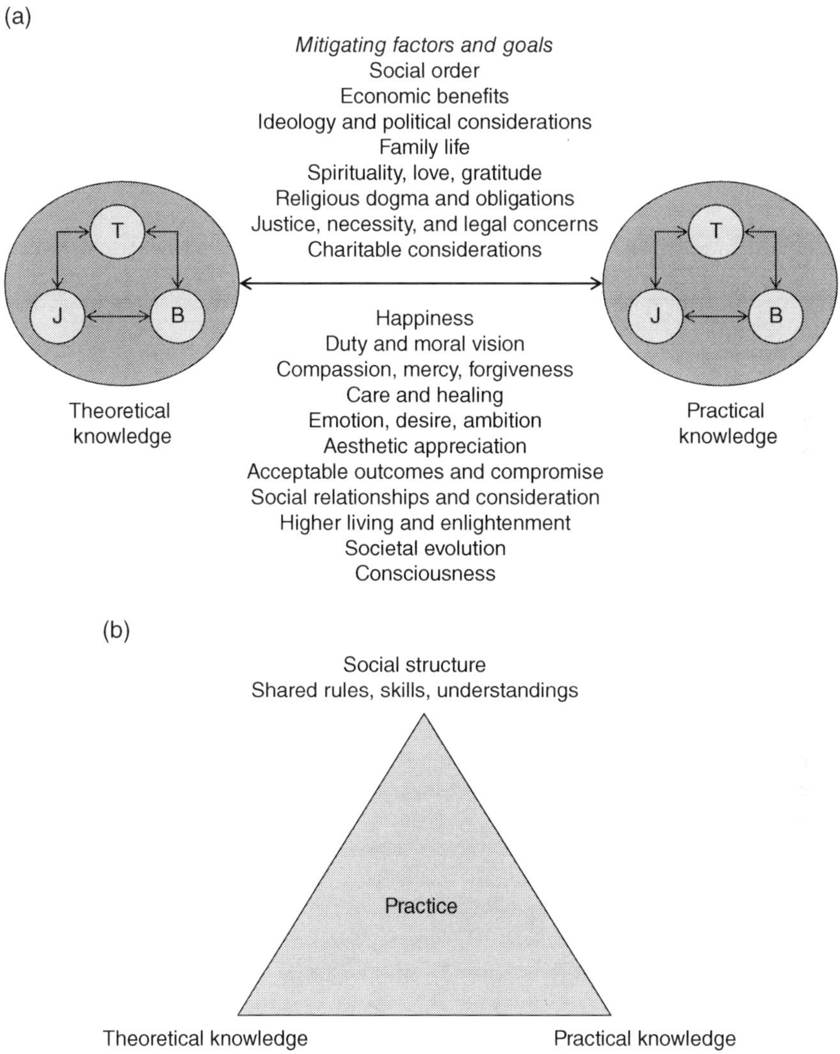

Figure 4.9 (a) Social practice continuum: theoretical–practical knowledge interaction. (b) Three dimensions of practice.

conception of reality to the left in Figure 4.9a. In contrast, with less seclusion and more social interaction, refinement goes in the other direction toward *practical living*; thus, practical knowledge makes sense only within a particular social context of life. Practical K is not necessarily a compromise or something less than theoretical K, but an accommodation and synthesis demanded by life and living.

4.7.4 Structure of Practice: Social Knowledge

Figure 4.9b illustrates the structure of practice along three dimensions. In the same way that Figure 4.9a shows the application of theoretical K (leading to the development of practical K), Figure 4.9b shows the application of theoretical and practical K in a social setting (leading to practice).

The key idea is that *personal judgment is not sufficient* in the application of K. There must be a structure that serves as a *clearinghouse* where understandings and skills are tested and shared, rules and methods developed, and norms of behavior established. Collectively, these areas represent the best practices of the profession or practice group, for example, the "standards of care" for medical practice. The structures may range from professional groups to organizational problem solving and decision-making groups.

Naturally, a given K worker may belong to any number of practice groups that must be resolved before socially acceptable behavior can take place. It is worth noting that the standards of the practice group often have precedence over those of the organization, for example, a physician working for an organization must first abide by the standards of medical care before considering the objectives or goals of the organization.

Finally, we can say that both the individual K worker and the social group mediates the interaction of theoretical and practical K. Using Kant's factors, moral vision should guide the social relationship of the practice group, whereas duty should govern the behavior of the practice. *It is through practice that social K takes shape and evolves. By definition, practice always produces a change in the real world.*

4.7.5 Enabling Practice in an Organization

The following are the key factors in building practice and developing engaged K workers—*practice and K work are virtually synonymous.* Naturally, the organizational and process enablers discussed in Chapter 2 apply, but here we focus on practice in an organizational setting:

1. *Freedom.* If we focus on the Kantian factors of duty and moral vision, practical knowledge implies that one must be free to make a moral decision in the first place; in Kant's words, "ought implies can." In an organizational context, freedom suggests few if any organizational constraints, at least on ethical behavior. This may sound obvious, but our experience in some organizations bears witness to an enormous pressure to do, first, what is best for the company, and, second, doing the right thing.

2. *Duty and moral vision.* Duty implies something more than business goals or performance. Rather, it implies adherence to a higher motivation: a moral vision. Moral vision in turn means that the practice group or organization must stand for something more than strategies and that something must be *ennobling* in nature. Moral vision should be at the centerpiece of an

organization's purpose. For example, Johnson & Johnson has a credo that states the company's first obligation is to doctors, nurses, patients, mothers, and fathers. Organizational efficiency and stockholder considerations are listed too but are secondary to its first duty. Similarly, Toyota's first principle (of 14) is to "base your management decisions on a long-term philosophy, even at the expense of short-term financial goals" (Liker 2004), that is, a philosophical sense of purpose that supersedes immediate gain.

As another example, let us consider Whole Foods. In *Conscious Capitalism*, cofounder and co-CEO John Mackey (2013) describes an essentially moral vision for Whole Foods. His personal awakening led him to higher levels of consciousness, which, in turn, has penetrated into all aspects of Whole Foods from culture and leadership to "higher purpose" core values. Interestingly, Mackey bases the company's higher purpose on four transcendent ideals, the first three of which were articulated by Plato: The Good (service to others), The True (discovering knowledge), The Beautiful (creation of beauty), and The Heroic (courage to do what is right to change the world). The higher purpose of Whole Foods has progressed more to the heroic category and includes (1) to help evolve the world's agricultural system to be both efficient and sustainable, (2) to raise the public's collective awareness about principles of healthy eating, (3) to help end poverty around the world through its foundation, and (4) to make conscious capitalism the dominant economic and business paradigm to promote human flourishing. This is representative of what we mean by ennobling goals and moral vision.

As yet another example, Deloitte assessed their best teams and discovered that high performance is driven by three factors, one of which is a company mission that *inspires* the individual. We will revisit the Deloitte case later in the book.

In Chapter 1, we concluded with the statement that the ancient lines of thinking (Indian, Chinese, Greek) "elevate life and leads ... to knowledge, righteous behavior, and spiritual wisdom." Similarly, Mackay sees individuals everywhere raising their consciousness, meaning (1) to be fully awake and mindful, (2) to see reality more clearly, and (3) to more fully understand our actions. What makes the Whole Foods experience meaningful is that it illustrates that consciousness can by elevated from individuals to organizations. *Consciousness* is one way of encapsulating moral ideals and to make them practical to every human and organization. Further, consciousness captures the very best of the mitigating factors shown in Figure 4.9a. Finally, all such statements of moral vision must extend beyond the slogans found in a town hall meeting. They should be made manifest in policies, rewards, systems and structures, and other activities, and imbibed into the cultural assumptions of the organization.

3. *Coffee Metaphor*. Recall that tacit knowledge has three dimensions: technical or motor skill, cognitive or mental model, and a social-group dimension that

embodies practice. We often use a cup of coffee as a metaphor in describing practice and knowledge. The coffee beans represent an individual's intrinsic expertise, cognitive knowledge, and motivation. However, to bring forth the flavor of the coffee, it is necessary to apply heat—this heat is the social interaction that all K workers and practice demands. Without social interaction or heat, our knowledge would not fully develop and flower.

4. *Social Interaction.* KM and practice perspectives demand strong social interaction of several types. We conceive of practice interaction as three concentric circles. The inner most circle is the social interaction within the practice group. This may take place within or outside an organization, as in, for example, a professional group. Additionally, a practice group may be a local problem-solving group within an organization. The second circle is the larger working group within which the member is part. This may take several forms, particularly for K work: a formal functional structure or a secondary informal structure such as a community of practice. Finally, a tertiary level of interaction may exist that represents the wider community; in an organization this may be a product or market segment group. Note that these concentric interactions may be overlapping, as in, for example, a professional practice group that spans organizations and is simultaneously local and global.

5. *Collaborative problem solving.* We believe the key to enabling practice and knowledging is to organize K workers (or all workers) around *collaborative problem solving.* Toyota considers problem solving to be the "silver bullet" around which it simultaneously builds competence, develops people, builds engagement, and realizes organizational benefit—in fact, all the positive attributes associated with a K culture and organization. There is perhaps no better way to engage and demonstrate trust to a K worker than to present him or her with a challenging problem. Problem solving brings forth and, as importantly, develops most of the desirable characteristics of K work and spills over into the practice or culture of the organization. Learning and growth, respect for people, building leader skills, promoting trust, and so on, arise from collaborative problem solving.

6. *Concentrated study.* In addition to the organization- and work-related activities discussed earlier, we believe concentrated study—education—is necessary to continually build expertise. This takes us to the left side of Figure 4.9a as we seek to cultivate certain K in a given domain. We will discuss specific people and culture-related strategies later in the book, but in terms of practice, we think it essential that all the trappings of an expert be cultivated. This includes continual teaching as well as improvement (Drucker 1999). In particular, we illustrate (metaphorically) the importance of corporate-level education to our students with a quote from Liker (2004): "Toyota trains people like they are all surgeons."

4.7.6 Toward a Fuller Understanding of K in Action: The K-Molecule

Recall from Chapter 3 that we introduced the concept of an organization molecule as a domain-specific design construct that configures, specifies, and describes an implementation. For example, Figure 4.7 illustrates the historical progression of work using the process molecule. Recall too that Figure 3.14 illustrates the information as well as the process molecule.

We have often been asked by students to create a knowledge molecule. The idea of an organization molecule is based on systems and lends itself to hierarchical analysis. In contrast, knowledge and knowledging (K&K) is more complex and integral, and more evolutionary and historical than other organizational domains; hence, we have resisted force-fitting one idea into another. However, in writing this book we have come to realize the importance of taking a stand and advancing our conception of the essential character of K&K. We have distilled the essence of four figures to collectively represent an *interacting K-molecule*: the level of analysis or K-space local to the work (Figure 2.17), the theoretical–practical K interaction and corresponding moral vision of the practice group (Figure 4.9a and b), the practice or K-molecule itself (Figure 4.10), and its organizational or work manifestation (Table 4.2).

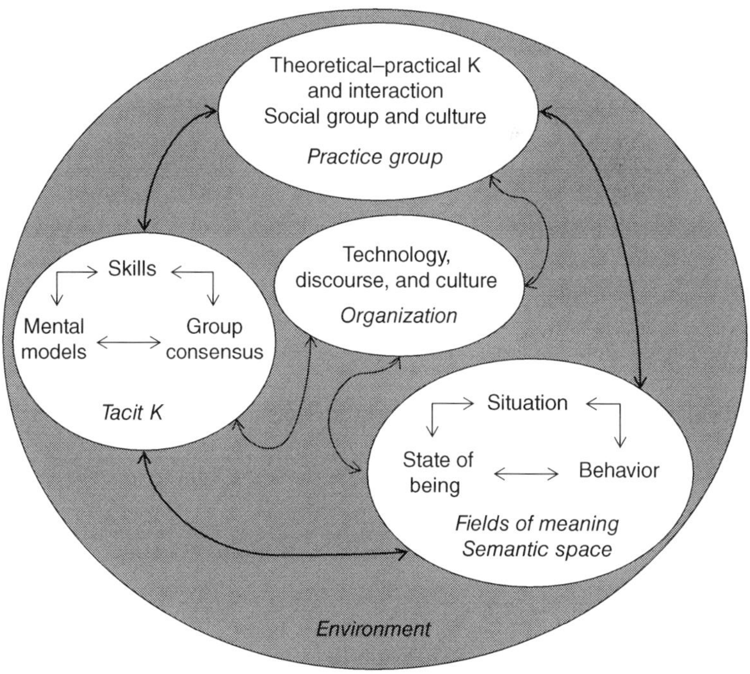

Figure 4.10 Workings of practice—core K-molecule.

Table 4.2 Characteristics of thin and thick work.

Work characteristic	Thin work	Thick work
K-orientation	Process	Practice with embedded process
K principle	Theoretical K—pure reason	Practical K—practical reason
Systems analysis perspective	Simple; separate analysis of levels is possible	Complex; simultaneous analysis of several levels and perspectives is necessary
Information continuum	Towards data pole	Towards knowledge pole
Structural design	Emphasis on asymmetrical structures and hierarchies	Emphasis on symmetrical structures and interconnection of communities
Design approach	Engineering-like; task, modular, rule-based, and procedural	Anthropological-like; observational, integral, social, and practice oriented
Knowledge construction	Early K-binding; templates	Late K-binding
Knowledge exchange	Precise and transactional	Relational and psychological
Knowledge dimension	Primarily explicit	Primarily tacit, but also explicit
Organizational manifestation	Process, workflows; high formalization; templates with decision making	Practice and expertise; judgment, experience, activity, and unstructured decision making
Semantic understanding	External to the worker	Embodied in fields of meaning
Operational and management perspective	Routinization; reduce human intervention; reduce uncertainty	Encourage and design problem-solving; experimentation; build understanding
Image	Formulated organization	Self organization; cultural and K practice

Here we briefly describe the key elements of each figure.

1) Figure 2.17 illustrates K-spaces. In effect, this is representative of fractal spaces in which K&K takes place. In this illustration, integral design extends from the psychological to the global community.
2) Figure 4.9a is our representation of the interaction of theoretical and practical knowledge and the mitigating factors that influence its manifestation. Similarly, Figure 4.9b shows the structure of practice.

3) Figure 4.10 is our representation of a K practice in action. Observe the complexity of ideas and their interrelationships, and how each is a composite intertwined with other such composite ideas. Whereas practice is partially based on pure reason, it emphasizes the practical and arose within the history of human social communities, from the immediate family to global societies. In many ways, practice (Figure 4.10) comes closest to a stand-alone representation of a K-molecule.
4) Table 4.2 summarizes the key differences between thin work or process and thick work or practice. While there is some overlap with Table 2.1 (i.e., thin and thick characteristics at the organizational level), Table 4.2 focuses on K at the work level and on distinguishing process and practice.

4.7.7 Implications of Practice

Figure 4.11 shows the interactions of the K-molecule; that is, the interaction of the ideas represented by the figures mentioned earlier. Observe that the core K-molecule is based on our representation of a practice (Figure 4.10) and the interaction of theoretical and practical K.

Let us summarize several key elements.

- Practice often embodies a process, taken from the social group in which the practice arose. This includes constituent standards, procedures, precedents, and protocols.
- In contrast, process as such is unconnected from a practice (from which it may have originally risen). That is, process is a distilled piece of information—distilled from a practice or an engineering design. The human constituent (see the process molecule) is unaware of its knowledge and social context. In this case, even tacit "know-how" is devoid of its social dimension.
- We separate organizational and practice communities and resist the popular trend of referring to most organizational groups or teams as practice communities. We reserve the term *practice community* to mean a social group that embodies shared skills and understanding (i.e., technical, cognitive, and social consensus dimensions of tacit K), group history and evolution, and the social group's interpretation of its corresponding duties and (moral) goals—that is, ascertaining true K and acting on that K in a social context. Further, semantic meaning is established by the practice, which thus serves as its own context. The practice is additionally mitigated by the psychological state of the practitioner and available technology. In an organizational setting, practice is further constrained by the organization's culture, standards, and group decisions.
- Semantic understanding embraces the internal emotional state of a practitioner, the immediate situation, and the consequential activities. This is referred to as the fields of meaning.

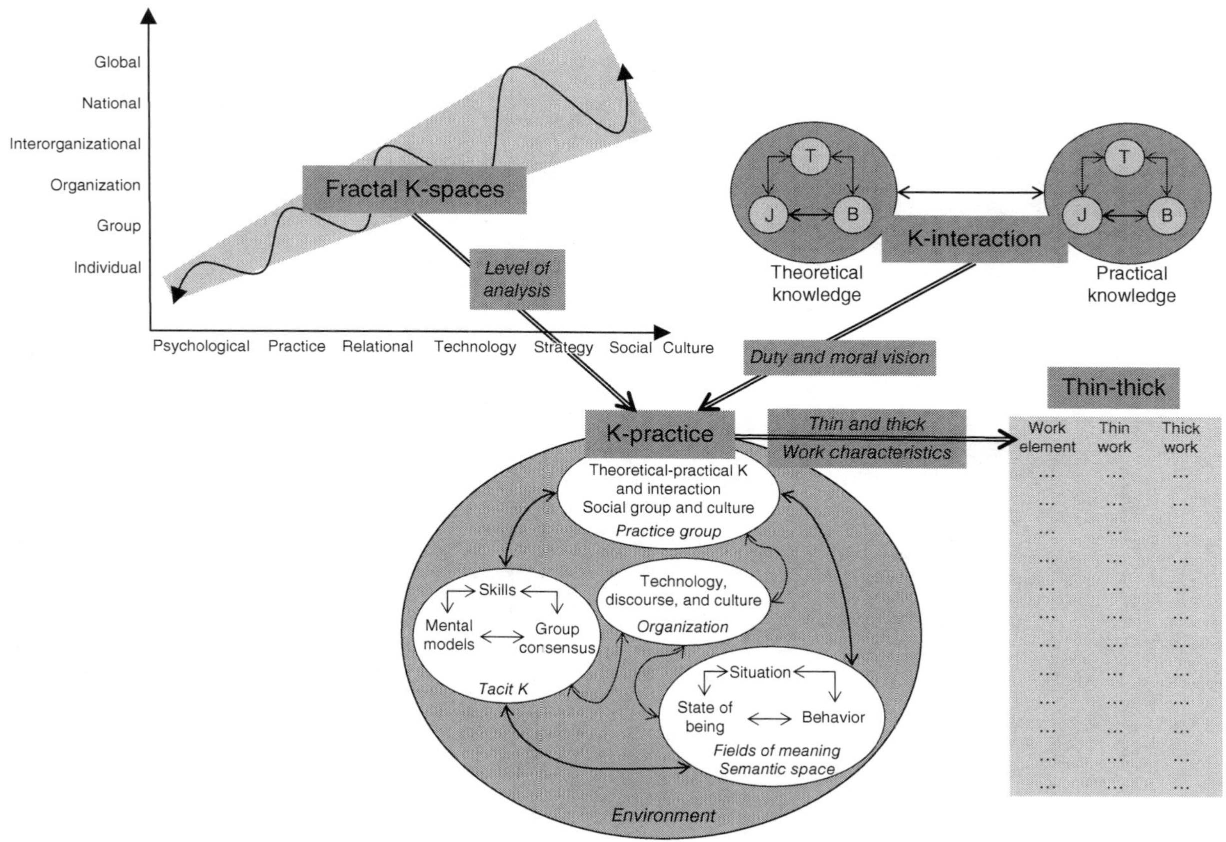

Figure 4.11 K-molecule interactions.

- We recognize that there may be overlap between organizational and practice communities. Small, self-organizing groups often arise in an organizational context to solve problems and may in time evolve into a practice group in the classical sense; that is, an organization-level community of practice may qualify as a true practice group. Note too that the design artifact "discourse structure" discussed in Chapter 2 refers to both organizational and practice groups.
- Practice groups may have a representation, in whole or part, in a given organization. For example, a K organization such as the film company discussed in Chapter 2 includes practice and thick work throughout the organization from actors to directors and producers. Each of these interacting practices is simultaneously individual and organizational in nature.
- Process work often involves social interaction, but this is drawn primarily from an organization and not necessarily a practice community.
- Thin–thick interaction is a continuum that extends to each of the work characteristics in Table 4.2. As an example, let us consider decision making. In a management sense, decision making is associated with informating and reducing uncertainty (see Earl—Figure 1.6). James March (1994) has described this category of decisions in terms of rational theories of choice or rule-based theories. Such decisions are often embodied in the organization's craftwork templates (see Chapter 2). As the situations become more ambiguous (as opposed to uncertain), decisions take on a social character as organizations leverage decentralized structures and the background of culture and other organizational institutions, in effect, drawing on larger communities to make decisions. That is, decision making shifts from a thin to a thicker activity. March (1994) also comments on the social basis of meaning, and this in turn brings us closer to a practice model of decisions. Finally, decisions in a knowledging sense are embodied within the K activities associated with expert decision making discussed earlier; that is, decisions are intertwined with activities, often constructed from pieces of mental models and experiences assembled in real time. The latter characterizes a fully thick decision activity.

4.8 Knowledge Work and Supporting Technology

The top part of Figure 4.12 is Davenport's classification of knowledge-intensive processes. From our perspective, this is a variation of Perrow's classic task model shown in Figure 3.2. In this case, both the unanalyzable and variability dimensions are collapsed into a single dimension called complexity of work on the X-axis, whereas the Y-axis captures the (simplified) social nature of the work (individual or group). The bottom part of Figure 4.12 illustrates the information system support for each classification of work.

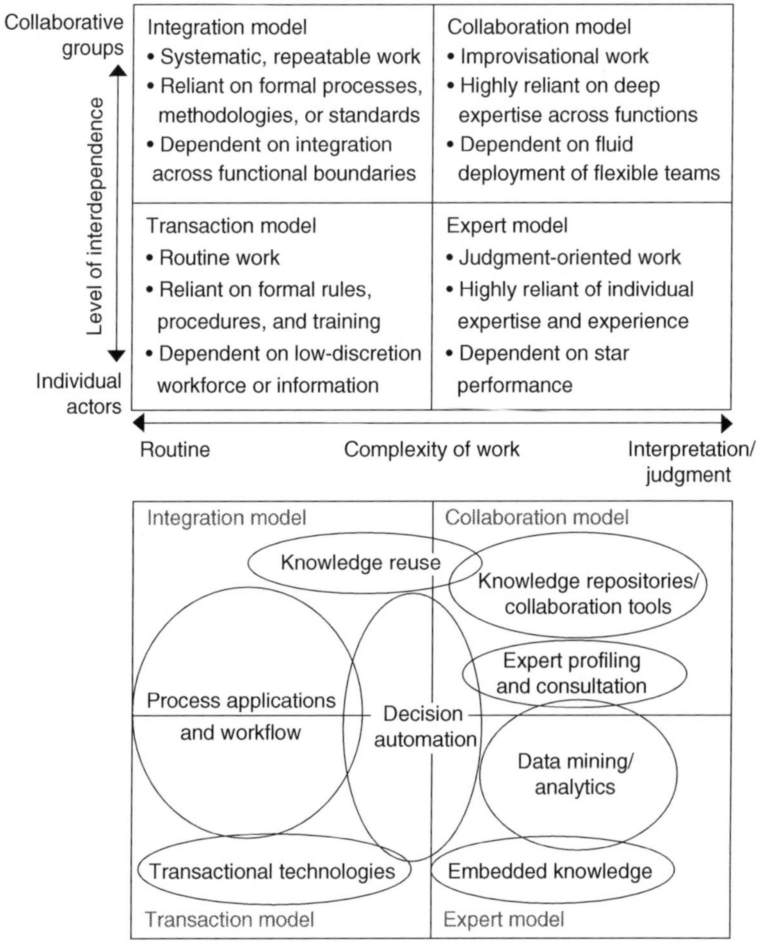

Figure 4.12 Classification for K-intensive processes and technologies—Adapted from Davenport (2005).

In this model the notion of a *script* is introduced as "structured reuse." Reuse is necessary if we are to promote improvement through structural or technology interventions. This idea of a script is not very different from our idea of an explicit thread of information shown in Figure 1.13. We have updated this representation model to include a *K-nugget* (or simply a *nugget*) as a thread of explicit knowledge, also shown in Figure 1.13. This is necessary if we are to properly understand and design technology interventions. The left side of the representation is an explicit orientation of work, whereas the right side is primarily tacit. Explicit and tacit orientations, in turn, dictate the nature of the work and the type of technology interventions. Clearly, explicit or, shall we say, thin work benefits from scripts embedded in the work processes and information systems. On the other hand, tacit or thick forms of work benefit from the K-nuggets embedded in practices, social interactions, and KM systems. We have included these notions in Table 4.3, which extend the original Davenport model with other elements we believe necessary to understand K work and its technology interventions.

Let us be clear on the difference between a script and a K-nugget. A script is typically developed by an expert for use by a lesser-skilled or knowledgeable worker; that is, a script is a structured piece of reusable information. Thus, a script does not necessarily carry deeper meaning to the lesser-skilled worker. In contrast, a nugget is designed by an expert for reuse by him- or herself or other experts; that is, a nugget embodies some piece of expertise or understanding. Hence, a nugget represents an explicit thread of knowing (supported by tacit knowing) and is thus meaningful to a K worker.

Naturally, in addition to the systems shown in Figure 4.12, there are many other technologies that support K work—expert systems, for example.

Table 4.3 Characterization of K-intensive processes and technologies—extension of Davenport.

	Transaction	Integration	Collaboration	Expert
Character of work	Process; routine	Process; systematic and repeatable; elements of practice	Process and practice; improvisational look and feel	Practice with embedded process; judgment oriented
Contingency	Formal rules, procedures, and training	Formal processes, standards, and methodologies	Cross-functional expertise	Individual expertise and judgment
K and arch orientation	Explicit and modular	Explicit and modular	Tacit and integral	Tacit and integral
Formality	High	Moderate–high	Low–moderate	Low, except constraints imposed by practice
Technology intervention	Script reuse embedded in procedures	Script reuse embedded in architecture and methods; functional integration possible thru architecture	K-nugget reuse embedded in technology; teams and other discourse structures	Embed K-nugget in practice/ process and technology
Example	Call center	Software development	Investment banking	Physician

Of particular importance are role-specific portals, the most popular of which is a *business intelligence portal* (BI portal). The BI portal organizes semantic access to underlying data structures and applications based on job or organizational role (e.g., executive, analyst, operational worker). By semantic access, we mean the user need not know the location of data. Rather, the user simply requests information necessary to accomplish some task and the BI system will navigate the user to the appropriate source. In the BI portal, navigation is structured and designed into the system based on three criteria: job roles, underlying data, job tasks. (This is in contrast to a general portal such as Google, where there are no constraints.)

4.8.1 Student Exercise: Integrated Job and Technology Analysis

Let us illustrate the analysis associated with K work and corresponding technology support. Our student teams were given the assignment of analyzing a K job (e.g., professor, chemist) and creating a set of support systems in a K workstation. The assignment included the following phases (and is shown in Figure 4.13).

Also, this represents a design problem for an *integrated K-space* discussed in Chapter 2 and shown in Figure 2.17. In this case, we are designing the lowest-level integration space, identified in Figure 2.17 as the *desktop space*:

- Job analysis—select a K job, identifying job requirements, skills, education, and so on. This is followed by a classification according to both Sveiby (1997) and Davenport.
- Task analysis—this is the key element in the assignment. The students were asked to decompose the given K job into discrete roles or tasks (some students did both) and then to characterize each task using several concepts or artifacts, such as explicit–tacit, modular–integral, script–nugget, and so on.
- Cognitive analysis—perform a cognitive analysis, according to the job, guided by the task analysis. The most well-known reference model is the Bloom taxonomy.
- Design of technology support—based on the earlier analysis, the students are to design a set of integrated systems (with portal or workstation access), including the contents of each system.
- Mapping to an organizational KM cycle—the students may map to any of several models discussed in class (and discussed in Chapter 5). The most popular KM cycle chosen was the Wiig (1993) cycle.

Several K jobs were particularly popular, perhaps reflecting the students' experiences. These included chief information officer (CIO), application architect, software engineer, vice-president–IS, financial risk manager, chief marketing officer, and knowledge management officer (KMO), among others. We have chosen to illustrate the KMO (business level)—we will make use of this information in Chapter 5, when we discuss the design of an organization's

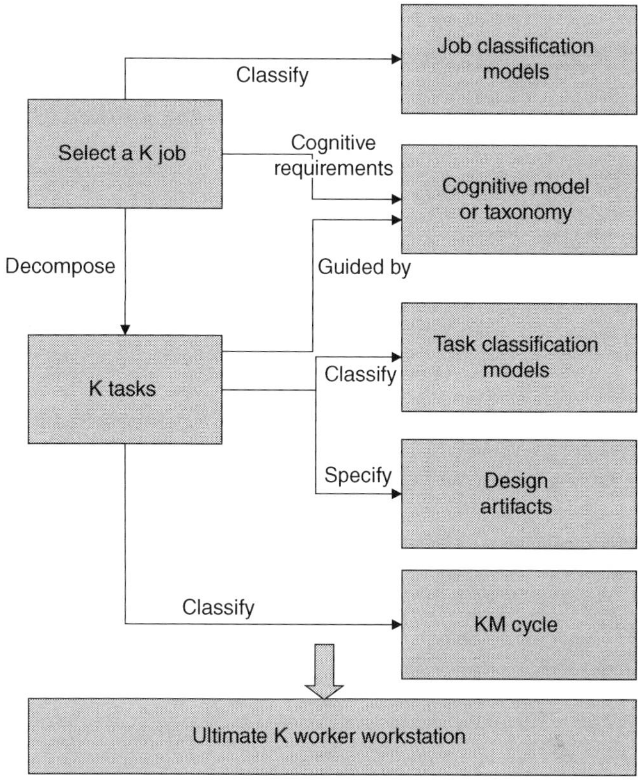

Figure 4.13 Student assignment—analysis of K job.

K-base. See Figure 4.14a–g. Note too that we are presenting only a small portion of the total solution; for example, while Figure 4.14g shows the system layout of the KMO workstation, the actual project presentation also includes a detailed sampling of the content of each constituent system.

4.8.2 Decision Support Systems

Decision support systems (DSS) are those systems that support decisions in unstructured situations where automation is neither possible nor desirable. By unstructured situation we mean one where cause-and-effect relationships are not known, there is uncertainty rather than risk, variables may not be measurable, or there are multiple conflicting goals.

Figure 4.15 illustrates the evolution of DSS along the information continuum. On the left side of the continuum, we see the operational systems of the organization that are generally known as online transaction processing systems (OLTP). The context for OLTP is transactional; hence, the key design constraint

(a)

KMO key responsibilities

- Promote knowledge sharing through the organization's operational business processes and systems by, among others, strengthening links between knowledge sharing and the information systems, and improving integration among information systems in the organization, to facilitate seamless exchange of information across systems
- Promote collaborative tools such as activity rooms to facilitate sharing of ideas and work among internal teams and external partners
- Provide support for the establishment and nurturing of communities of practice, including workshops, one-on-one guidance, and troubleshooting
- Share experiences across communities of practice, business units, and networks on innovative approaches in knowledge sharing, including preparation of case studies
- Help monitor and evaluate the knowledge sharing program, including external benchmarking and evaluation programs/opportunities
- Help disseminate information about the organization's knowledge sharing program to internal and external audiences, including organizing knowledge sharing events (such as knowledge fairs, site visits, interviews), maintaining communications on knowledge sharing across the organization, participation in orientation and training sessions, and preparation of brochures/presentations.

KMO hiring requirements

- Advanced degree in business and/or social sciences. Preferably, a Master's degree in Knowledge Management with experience in enterprise business and Professional Certification in Knowledge Management
- Several years of experience with the organization's operations and business tools
- Familiarity with developing and delivering knowledge sharing programs and with the company's information infrastructure
- Experience in establishing effective partnerships within and outside the organization

(b)

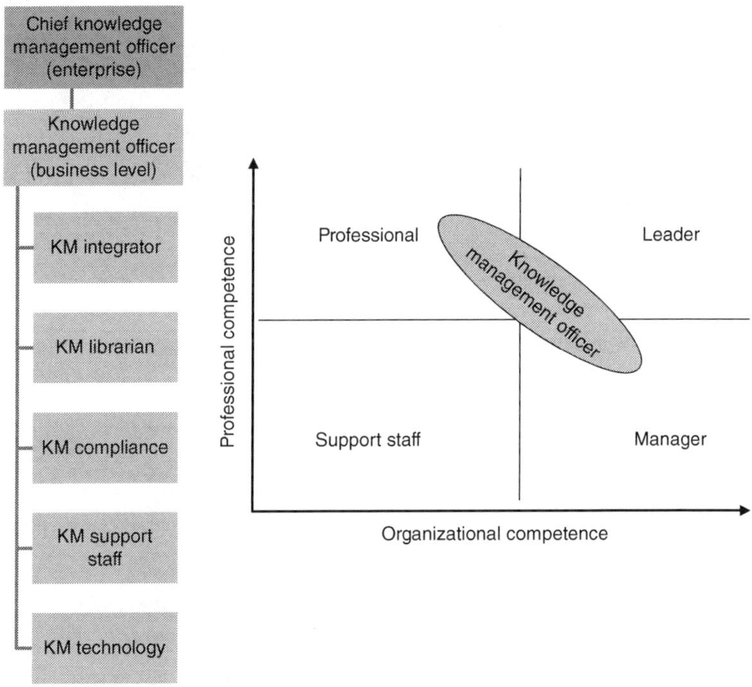

Figure 4.14 (a) Student assignment—KMO responsibilities and requirements. (b) Student assignment—KMO job classifications. (c) Student assignment—KMO key tasks. (d) Student assignment—KMO task classifications. (e) Student assignment—KMO Bloom cognitive analysis (partial). (f) Student assignment—KMO task artifact analysis. (g) Student assignment—KMO workstation arrangement.

(c)

KMO key tasks

1. Promotes adoption and use of knowledge management in the business through the development and support of communities of practice, including consulting and training/capacity building.
2. Ensures effectiveness of KM programs and activities by analyzing KM activities, monitoring their use, and measuring their impact
3. Tests KM products and services for effectiveness, quality, and responsiveness.
4. Serves as key contact for data requests by responding to queries and by assisting clients in accessing KM tools and resources
5. Utilizes knowledge bases, networks, communities of practice, and other sources to synthesize responses to queries
6. Conducts outreach programs and oversees training for staff and clients in use of KM products.
7. Provides oral and written responses and briefings for queries by staff and managers and by external clients
8. Contributes to KM resources by preparing best practices and capturing KM practices through written stories and video debriefings
9. Reviews the artifacts for content accuracy, details, standards, and usefulness as submitted by the review team managed by the Knowledge Management Librarian
10. Approves the artifacts and releases to Knowledge Librarian for publishing into the knowledge base
11. Interface with Governance process for KMS changes as a result of service request–related changes to KMS content
12. Conduct Continuous Improvement audits on KMS process.
13. The Knowledge Management Officer is ultimately responsible for the content integrity of the knowledge base and is therefore the final approver of artifacts before they are released for publishing.
14. Supervises Knowledge Analysts.

(d)

Craft	**Nonroutine**
Promotes adoption and use of knowledge management in the business through the development and support of communities of practice, including consulting and training/ capacity building (1)	Ensures effectiveness of KM programs and activities by analyzing KM activities, monitoring their use, and measuring their impact (2) Serves as key contact for data requests by responding to queries and by assisting clients in accessing KM tools and resources (4) Interface with Governance process for KMS changes as a result of service request–related changes to KMS content (11) Conduct Continuous Improvement audits on KMS process (12)
Routine	**Engineering**
Tests KM products and services for effectiveness, quality, and responsiveness (3) Utilizes knowledge bases, networks, communities of practice, and other sources to synthesize responses to queries (5) Conducts outreach programs and oversees training for staff and clients in use of KM products (6) Provides oral and written responses and briefings for queries by staff and managers and by external clients (7) Contributes to KM resources by preparing best practices and capturing KM practices through written stories and video debriefings (8) Supervises Knowledge Analysts (14)	The Knowledge Management Officer is ultimately responsible for the content integrity of the knowledge base and is therefore the final approver of artifacts before they are released for publishing (9) Approves the artifacts and releases to Knowledge Librarian for publishing into the knowledge base (10) Reviews the artifacts for content accuracy, details, standards, and usefulness as submitted by the review team managed by the Knowledge Management Librarian (13)

Integration model	**Collaboration model**
Utilizes knowledge bases, networks, communities of practice, and other sources to synthesize responses to queries (5)	Promotes adoption and use of knowledge management in the business through the development and support of communities of practice, including consulting and training/capacity building (1) Serves as key contact for data requests by responding to queries and by assisting clients in accessing KM tools and resources (4) Conducts outreach programs and oversees training for staff and clients in use of KM products (6) Contributes to KM resources by preparing best practices and capturing KM practices through written stories and video debriefings (8)
Transaction model	**Expert model**
The Knowledge Management Officer is ultimately responsible for the content integrity of the knowledge base and is therefore the final approver of artifacts before they are released for publishing (9) Interface with Governance process for KMS changes as a result of service request–related changes to KMS content (11) Reviews the artifacts for content accuracy, details, standards, and usefulness as submitted by the review team managed by the Knowledge Management Librarian (13)	Utilizes knowledge bases, networks, communities of practice, and other sources to synthesize responses to queries (5) Ensures effectiveness of KM programs and activities by analyzing KM activities, monitoring their use, and measuring their impact (2) Supervises Knowledge Analysts (14)

Figure 4.14 (Continued)

(e)

Level	Description	Action verbs
1. Knowledge	Oversee update and archival process (utilizing subject matter experts) aimed at ensuring knowledge capital is current and relevant Work with engagement teams to harvest knowledge in alignment with the approved content strategy/knowledge plan Perform proactive knowledge harvesting for reusable/best practice knowledge capital from existing engagements, subject matter experts, and other sources, and so on	Contributes to KM resources by preparing best practices and capturing KM practices through written stories and video debriefings (8) Reviews the artifacts for content accuracy, details, standards, and usefulness as submitted by the review team managed by the Knowledge Management Librarian (9) Approves the artifacts and releases to Knowledge Librarian for publishing into the knowledge base (10)
2. Comprehension	Regularly monitor the communication vehicles, which enable easy access to knowledge capital by our global practitioners (home pages, newsletters, practice aides, etc.) Develop a good understanding of the subject area capabilities in subject area	Promotes adoption and use of knowledge management in the business through the development and support of communities of practice, including consulting and training/capacity building (1) Conducts outreach programs and oversees training for staff and clients in use of KM products (6) Provides oral and written responses and briefings for queries by staff and managers and by external clients (7)
3. Application	Work with Senior Knowledge Integrator and/or Knowledge Manager to collect, synthesize, package, and deploy knowledge capital that supports assigned content area(s), and ensure Knowledge Analysts correctly post materials to the relevant libraries/repositories and home pages Work with Portal Manager to ensure that online content for each service line is managed and maintained effectively in line with KM governance framework`s and processes	Reviews the artifacts for content accuracy, details, standards, and usefulness as submitted by the review team managed by the Knowledge Management Librarian (9) Interfaces with Governance process for KMS changes as a result of service–request related changes to KMS content (11)

(f)

Modular job tasks (thin)

- Working with requestors (1)
- Reporting (7)
- Best practices (8)
- Testing for accuracy, etc. (9)
- Compliance (11)
- Quality improvements (12)

Integral job tasks (thick)

- Ensures effectiveness (2)
- Conducts analysis, metrics (3)
- Promoting adoption and use (4)
- Build knowledge base, networks, community of practice (5)
- Develop programs and training (6)
- Approves and releases artifacts (10)
- Develop strategy (13)
- Supervising, mentoring and coaching (14)

Figure 4.14 (Continued)

is consistency of update. In practice, this leads to the removal of redundancy with techniques such as normalization, the result of which is a large number of relatively small tables (in relational database management systems (RDBMS)). Reporting instances—OLTP snapshots—are an early form of DSS that support tactical decisions. As we move to the right, we have operational data stores

(g)

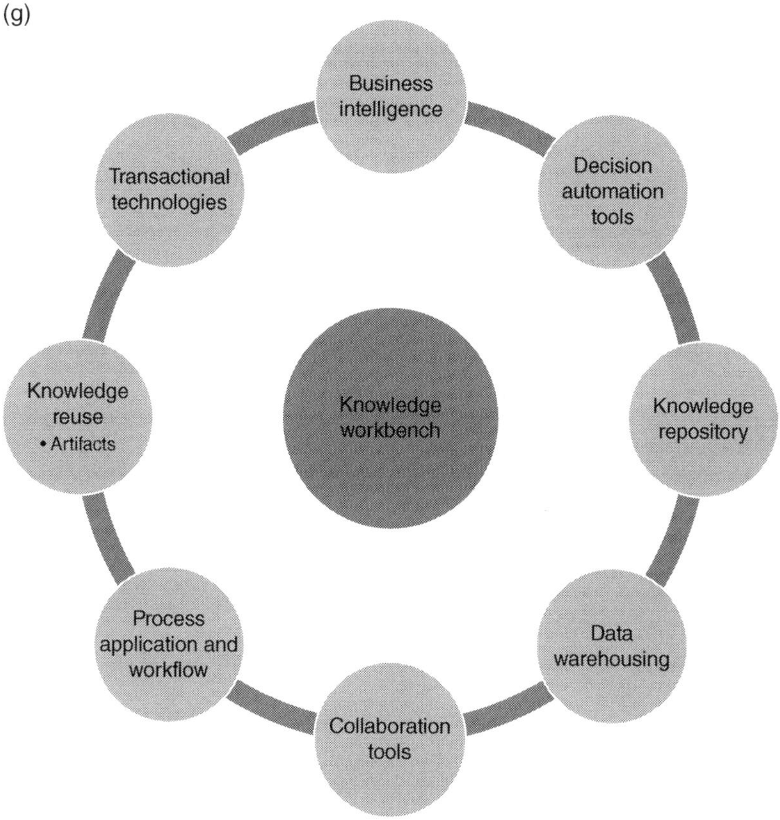

Figure 4.14 (Continued)

(ODS), which are integrated operational systems that support, for example, the location of a lost item in an order fulfillment system. In each of these systems—OLTP and ODS—the design primarily supports operations and secondarily supports decision making. On the right we have online analytical processing systems (OLAP). The context of OLAP is specifically decision support; hence, the key design constraints are speed of retrieval, visualization, and value-added (Kimball *et al.* 2008). In contrast to OLTP, OLAP systems are characterized by a relatively small number of very large tables that may alternatively be implemented in proprietary multidimensional databases (MDDB) or, metaphorically, "cubes." In general, OLAP is multidimensional and is the core data design concept in BI systems.

Finally, there is a certain amount of correspondence between DSS and those shown in Figure 4.12. The latter systems support decisions through scripts and

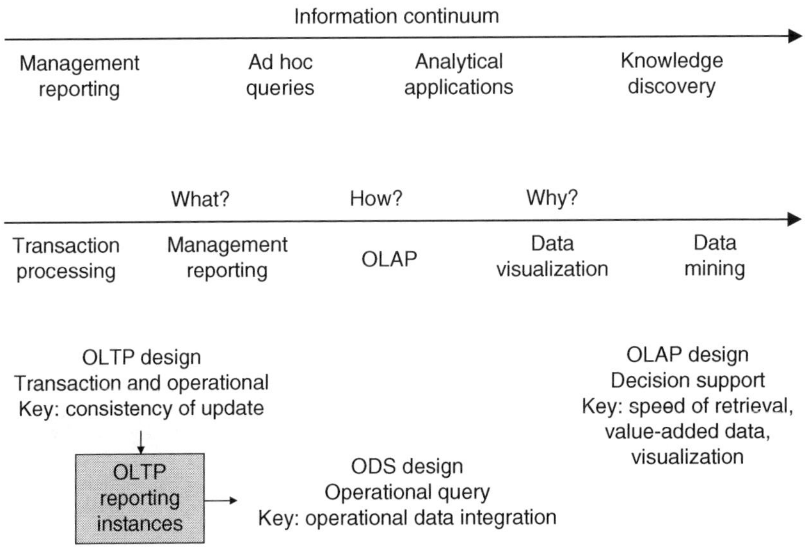

Figure 4.15 Information continuum and the evolution of DSS.

K-nuggets; data mining (also shown in Figure 4.12) attempts to uncover patterns that become K-nuggets and is part of the BI suite of systems.

4.8.3 Business Intelligence

In the same way that business processes are one of the last remaining sources of advantage (see Keen (1997) earlier), process–data integration represents a new generation of organization-specific advantages and decision making. We say organization-specific because both details of process implementation and data are specific to an organization. Business intelligence (BI) is a term commonly used to describe the collection, refinement, storage, and delivery of data. Applications include everything from simple reporting to analytics. Whereas the IT department is generally responsible for the processing of data including reporting, the business units largely perform their own analytics. Hence, both the analytical talent and the understanding of issues reside in the business units.

Architecturally, BI encompasses data warehousing, cleansing, access, reporting, analytics (e.g., mining), visualization, and portal development. An enterprise-wide perspective of BI may be used as an integrating construct for all the data and DSS of the organization. Furthermore, the data warehouse environment (the back-end of BI) increasingly contains real-time data, reducing the need for reports from operational systems, referred to as OLTP reporting instances in Figure 4.15. When we consider that the modern organization is characterized by K work and decisions at all levels, the BI concept may support the entire suite of applications—operational and strategic—in an integrated manner that in turn promotes business process integration.

There are two dimensions in BI that are particularly important in KM. The most obvious is data mining—that is, *knowledge discovery in databases* (KDD). The analyst seeks to reveal and evaluate patterns in data (i.e., K-nuggets) that will lead to improved decision making, especially in marketing and sales but increasingly throughout the organization. Furthermore, analytics has expanded to include, for example, social network analytics, web analytics, and business process optimization. Applications include forecasting, predictive modeling, optimization, and so on.

The second dimension is the *interconnectivity of BI systems with business processes*. If organizations are to progress to "fact-based" decisions and a K culture, the very character of work must shift and include a seamless integration with data. As an example, Continental Airlines developed a real-time BI system that, in essence, drives all operational and analytical decisions though its BI system. Applications include flight management, customer value analysis, marketing insight, fare design, and recovering lost airline reservations (Anderson-Lehman *et al.* 2004). Similarly, Cardinal Health coupled BI with dense, robust social networks for quick identification of expertise and the collaboration necessary to solve problems and exploit opportunities (Carte *et al.* 2005). In general, the most successful BI implementations appear to be enterprise-wide BI interconnected with the processes of the organization; that is, BI is embedded in the work of the organization. In effect, the organization is explicitly or implicitly reengineering its business processes through BI and changing the way work is done and thereby the culture of the organization.

4.8.4 Big Data

Organizations are increasingly driven by the interweaving of information and decision-making processes. The latter requires an ever increasing quantity of data in a wide variety of areas. This is all within the realm of BI and process integration; however, the advent of big data extends the data to sources outside the organization on a scale that typically changes BI, analytics, and data management. All this, in turn, is impacting both job roles and organizational design.

"Big data" refers to "datasets whose size is beyond the ability of typical database software tools to capture, store, manage, and analyze" (McKinsey Global Institute 2011) and is characterized by the "three Vs": *volume, velocity,* and *variability*. In sequence, the three Vs refer to data's size, constant streaming, and different data types (text, numerical, image, video, audio, etc.). All three Vs need not be present at the same time; sufficient quantities of any one V, data volume for example, qualify as big data. Also, Dr. Divesh Srivastava (2014) suggests the need for a fourth V, that of *veracity*. With data streaming into the company 24 × 7, the data cannot be made to conform to existing data models, business rules, and constraints. Srivastava suggests that we let the "data speak for itself" and derive the rules and models from the content of big data. This is the opposite of traditional data management where the constraints (e.g., functional dependencies), rules, and models are developed first and the data content made to conform to those models.

The point is that the scale of big data (some combination of the three Vs, but not necessarily all three Vs at the same time) imposes information processing and organizational design constraints. Big data is typically sourced from web sites and social networks outside the organization, creating a somewhat new and uncontrolled environment. The following is a brief summary of the design implications of big data, taken from Belanger (2014).

- The scale of big data mitigates against the common practice of separating data processing (in the IT department) from data analytics (in the business units). For the present, big data must be coresident with the analysts in the business units, at least partially disintermediating the IT department.
- The scale of big data requires new and extensive programming skills, with an emphasis on web programming and parallelization of algorithms. This has led to the emergence of a new job role, the *data scientist.*
- The velocity of data means that organizations will have to adopt data streaming practices in addition to database management. This extends to data streaming analytics as well. Big data is primarily real-time data.

4.8.4.1 Big Data and Organization Design

With the advent of big data, Galbraith (2014) suggests new strategies for the organization with regard to job roles and design considerations. In the same way that organizations require a CIO, managing big data requires a *chief digital officer* (CDO) role. Depending on the company, the CIO may fulfill both roles (as in P&G), or a separate CDO job may be deemed necessary (Starbucks). One driving force for a CDO is that data and analytics is being matrixed throughout the organization, similar to HR and finance, and thus requires a separate champion and leader. It is hard to imagine a 21st century organization with neither a chief K officer nor a CDO—data is an integral constituent of knowledge.

A second driving force associated with big data is the increased speed of decision making. Galbraith (2014) discusses the need to increase the "clock speed" of the organization. In the same way that computer designers constantly increase the clock speed of computers, organizational designers must increase the clock speed of decision making, moving toward a goal of real-time decision making. Certain business units would benefit from increased clock speed, in particular advertising, customer relationship management, new product development, and supply chain management.

There are two organization design strategies emerging for increased clock speed in decision making. The first is to distribute the data and talent (e.g., data scientists, hardware and software engineers, web and social media experts) into the business units. The second is to colocate the talent into a semi-independent functional unit. In this book, we have been making the argument for the second alternative in most, if not all, knowledge domains: knowledge is best cultivated in dedicated functional units. As we mentioned in Chapter 3, functional design is best for the K organization. If data truly becomes a commodity, then the possibility of cross-distributing the talent, activities, and projects in a federal-like structure is feasible—but this is not likely for the foreseeable future.

4.8.5 The Internet of Things

Recall from Chapter 1 that one driving force behind the knowledge revolution is miniaturization. The Internet of Things (IoT) represents a fusion of miniaturization and big data. Life in the supermarket would be difficult without product barcodes. Microelectronics, man-made materials, and product design make possible sensory devices everywhere in the environment, from inside the human body, on a wrist, automobile, appliances, cities, forests, in fact, everywhere in the world.

We illustrate IoT in Figure 4.16 as concentric circles. The circles are shown as dotted lines because part of the strength of IoT is the interconnectivity of

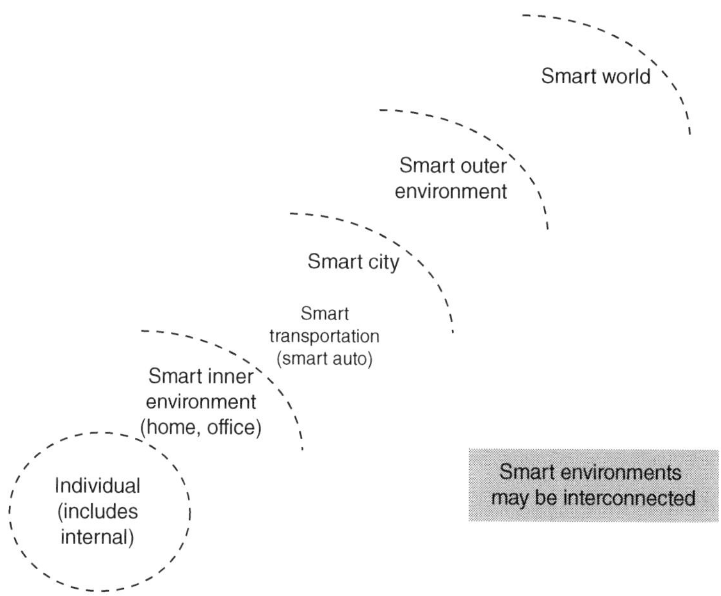

Figure 4.16 Smart environments.

things at all levels in our environment. We think of IoT as yet another fusion: from the organizational perspective, a fusion of big data, decision making, and business processes, with applications such as supply chain management. But the promise of IoT extends the fusion idea in all directions. Health monitoring, smart robots (Chapter 1), smart appliances and homes are increasingly common, as are smart cities. Governments can now sense forest fires as they start. On the other hand, surveillance is another application, which may raise privacy issues.

Recall from earlier that the posthumanist perspective on practice includes technology—IoT is a good example of practice technology that is limited only by our imagination. For example, the smart watch is a relatively new device, which may be used to fuse health, exercise, schedules, meetings, flight schedules, weather, public transportation, and whatever else is our fancy or desire.

As another example, Google (Barr 2014) has announced its "Nanoparticle Platform" project, which is part of a wider effort inside Google to develop new technologies capable of improving healthcare. The idea is quite simple but profound: the development of a pill that will send microscopic nanoparticles into the bloodstream in an effort to identify cancers, imminent heart attacks, and other diseases, long before symptoms may appear. The particles could then report their findings to a wearable device, typically a bracelet or a wristband that uses the magnetic cores to gather the nanoparticles back together and to read what they have found. The device could be configured to send information to a physician over the Internet. (Announcement made by Andrew Conrad, head of the Google X research lab, at the Wall Street Journal's WSJD Live conference in Laguna Beach, California.)

4.8.6 Cognitive Systems

A cognitive system is one that enables collaboration between people and machines. Its distinguishing features are as follows (Hurwitz *et al.* 2015).

- *Learning.* A cognitive system leverages data to make inferences about a domain, issue, topic, or person based on training and observations from data. Typically, the data in a domain qualifies as big data and includes structured data in databases and unstructured data (three Vs). The techniques used to acquire, manage, and learn from data vary but the central approach is to apply the algorithms associated with *machine learning.* The essence of machine leaning is the detection of patterns in data followed by some action or recommended action that improves performance. Thus, a cognitive system learns from data (i.e., discovering patterns with or without input from a user regarding the pattern itself) and improves its own K and performance without reprogramming.
- *Modeling.* To learn, a system needs to create a model or representation of a domain—a nugget of K in a given domain—and assumptions that guide

which algorithms are to be used. That is, a context is created between the data and the model through the development of a corpus or a machine-readable representation of the complete record of a given domain or topic. The two most common forms of K representations that are employed are taxonomies and ontologies.

  - Taxonomy. A taxonomy is a hierarchical representation of elements in a given area of interest. For example, we have all seen a hierarchical representation of life on earth that covers every known animal, plant, and bacteria to any level of depth.
  - Ontology. An ontology is a taxonomy with embedded definitions, rules, and a vocabulary that collectively captures a common understanding of a community in a given area of interest. Thus, the development of an ontology often uncovers the shared assumptions, beliefs, and practices among members of a community.

- *Generating hypotheses.* A cognitive system is probabilistic and generates the most appropriate answer (a candidate explanation or hypothesis) based on the data at hand. This includes the evaluation of conflicting hypotheses based on the current state of K in the system.

Cognitive systems may be considered an emerging technology with a potentially wide set of applications. One growing KM-oriented application is the transfer of K and best practices—for example, teaching new physicians medical best practices. In these types of applications, a cognitive system builds a dialogue between human and machine where K and best practices are learned by the system as opposed to being programmed through rules.

4.9 KM Tools and Technologies

There can be no doubt—tools and technologies (T&T) change the way people think and behave. The K-oriented concepts we have been discussing are foundational; however, T&T are often necessary to promote and optimize knowledging. In fact, we may state that T&T animate knowledging and in effect make ideas and K both tangible and easily transferrable.

In general, KM tools fall into three categories: data-oriented, process-oriented, and social networking tools. Further, we may visualize different levels of abstraction, from (atomic) tools to (composite) technologies and systems. Also, tools tend to be explicit (database) or tacit (social) oriented. In our earlier discussion, we focused primarily on explicit K-oriented technologies and systems, such as BI systems and K repositories.

As an example, a database may be seen as a basic data-oriented tool that leads to higher levels of abstraction and virtually all of the technologies we have been discussing. This includes data warehouses, business intelligence suite of tools, big data, data analytics (and its derivatives including forecasting, predictive modeling, and optimization), K repositories, K reuse, and so on. Furthermore, these data-technologies have been *extended* into AI with products such as IBM's Watson.

Examples of process technologies are workflow management systems and embedded K systems. The former embodies architectures, methodologies, standards, and best work practices. Such systems keep track of work instances within the flow, sign-offs, approvals, and so on. Embedded K systems integrate K-nuggets into processes through technology.

The third category of T&T are those associated with social networking. Starting with the Internet, we may expand (really, an explosion) into all sorts of communication, cooperation, and collaborative T&T. This includes, for example, e-mail, research forums, wikis, blogs, and so on. Such tools are often intertwined with database and process technologies. As an example, the Alzheimer Forum (i.e., Alzforum.com) is an online forum for both researchers and patients that includes research databases, research papers, and other resources. We will discuss social networking and collaboration in Chapter 6.

The three types of T&T often converge to produce higher-level technologies. For example, simulation tools or expert systems are composites that have application as educational platforms. As another example, consider robotics—a complex system that integrates several lower-level T&T. KM T&T typically support high level functions, such as K creation and K sharing (see Dalkir (2011) for sample KM tools). Finally, observe that it is sometimes difficult to ascertain the demarcation between a tool, a technology, and a system.

4.10 Robot Economy

It is apparent that we are rapidly moving to a robot economy. We think of robotics as the next level of automation and, more broadly, routinization. Recall that Scientific Management includes a large measure of routinization where craft or nonroutine tasks are simplified. That is to say, complex tasks are analyzed and redesigned into routine tasks (see Figure 3.3). In practice, partially integral work (top of Figure 4.7) is transformed into modular work (middle of Figure 4.7). This transformation requires that human know-how be removed from the head of the (craft) worker, resulting in a kind of "deskilling" of that worker. In the redesigned work, the structure and workings of the task (i.e., the know-how) are now specified by engineers. There is thus a formal separation between design and implementation. If sufficiently routine, the "tool" component (Figure 4.7) can be a machine or robot that replaces the "human" component, resulting in an automated task.

Until recently, most automation was confined to the factory floor. One limiting factor has been the availability of robots with sufficient motor skills and navigation capabilities to perform the necessary physical manipulation required for manual work. A second limiting factor has been the "know-how" capabilities of the robots that limited the extension of automation beyond only the most routine tasks. However, the combination of AI and advanced robotics has changed all that. As a result, today we ask if you would trust yourself to a driverless taxi?

As interestingly, is it possible for robots to replace K workers; that is, can we replace a fully integral practice (bottom of Figure 4.7) with a modular design (middle of Figure 4.7)? Can "know-why" and "know-what" be instilled in robots? Where would the next great symphony come from, or, for that matter, the next great baseball player?

4.10.1 Impact of Automation

According to Chui *et al.* of McKinsey (2015, 2016), the automation of entire jobs or occupations is not likely in the foreseeable future. Rather, certain activities will be automated, not unlike, for example, the redefinition of a bank teller's job caused by the introduction of ATMs. Also, the highest-paid occupations, including physicians and CEOs, have a significant amount of activity that can be automated.

The McKinsey research studied about 2000 individual work activities from more than 800 occupations and made the following observations:

- Automation of activities. The key finding is that automation is about activities within occupations and not individual jobs. The researchers assessed 18 different capabilities that have the potential for being automated. These include social (e.g., social and emotional sensing), cognitive (e.g., understanding natural language, creativity), and physical (e.g., fine motor skills, mobility) capabilities.

  Perhaps as compelling, the researchers assert that automation is no longer confined to routine, codifiable activities. The speed of advances in AI and machine learning suggest that activities requiring tacit K or experience are no longer immune from automation. "Automation, now going beyond routine manufacturing activities, has the potential, as least with regard to its technical feasibility, to transform sectors such as healthcare and finance, which involve a substantial share of knowledge work."
- Job and business process redefinition. Fewer than 5% of occupations can be entirely automated. On the other hand, about 60% of occupations can have 30% of their constituent activities automated. Organizations will have to redefine jobs and reengineer business processes.
- Impact on high-wage occupations. A significant percentage of activities performed by those in the highest paid occupations can be automated. For example, it is estimated that 20% of a CEO's working time can be automated.
- Future of creativity and meaning. It is acknowledged that creativity and sensing emotions are difficult to automate. On the other hand, the amount of

time that workers spend on these capabilities is surprisingly low. Only about 4% of the activities across the US economy require creativity while only 29% of activities require sensing emotion, each at a median human level of performance. It is anticipated (or hoped for) that automation will free workers to spend more time on that which is truly creative or emotion-based. We would add that this is a function of enlightened leadership that looks beyond short-term gain.

- Trade-offs. The fact that automation of a given human activity is *technically feasible* does not mean that it should or will be automated. The trade-offs are many: (1) technical feasibility (really a precondition). (2) cost of hardware and software, (3) cost of labor and its supply-and-demand quotient, (4) top-line benefits (increased output, higher quality), which typically outweigh reduced labor costs, and (5) regulatory and social-acceptance issues. Therefore, a business case will have to be made for automation-related decisions. For example, consider the third trade-off discussed earlier: the activities in the accommodation and food service sector have the highest automation potential, but among the lowest human costs where workers are in abundant supply.
- Susceptibility and activity groups. A given occupation consists of multiple types of activities, each with varying degrees of automation potential (i.e., technical feasibility). Chui *et al.* (2015, 2016) have identified seven high-level groupings ranging from the least, less, and highly susceptible candidates for automation. The activity groups range, in increasing potential for automation, from (1) managing others (9% of activities have automation potential), (2) applying expertise to decision-making, planning, and creative tasks (18%), (3) stakeholder interactions (20%), (4) unpredictable physical work (25%), (5) data collection (64%), (6) data processing (69%), and (7) predictable physical work (78%). *It is the first two groups—managing others and applying expertise—that the researchers identify as K work.* Together, managing others and applying expertise occupy 21% of the time in all US occupations.
- Sectors and susceptibility. The researchers examined 19 sectors such as construction, manufacturing, finance and insurance, and so on. As we would expect, there is an uneven distribution of activity groups when cross-referenced with sectors.
  - The technical feasibility for automation is highest in the accommodation and food service sector.
  - The technical feasibility for automation is lowest in the education sector. The activities with the least susceptibility to automation (i.e., managing others and applying expertise) account for one-half of the activities in the education sector.
  - Within a given sector, there is variation based on specific occupation. For example, finance relies on professional expertise (e.g., stock traders and investment bankers), yet 50% of the overall time is spent in collecting and

processing data, where the technical feasibility of automation is high. Similarly, healthcare has an overall automation potential of 36%, but is lowest for healthcare professionals whose daily activities involve the application of expertise and direct contact with patients.

4.10.2 Polanyi Paradox

Recall from Chapter 1 the contributions and models of Michael Polanyi. It has been observed that there exists an interesting situation, known as the Polanyi Paradox. The essence of the Polanyi Paradox is that "we know more than we can tell." That is to say, a form of self-ignorance characterizes tacit-oriented activities (e.g., driving a car) where K of a task (often common, everyday activities) is beyond explication. See Chapter 1 for our description of tacit K.

McAfee and Brynjolfsson (2016) describe the phenomenon by comparing two AI systems used to play chess and Go, respectively. The first is IBM's Deep Blue, which defeated a human champion in chess, and the second is Google's AlphaGo, which defeated a Go champion. First, let us compare the two games. Each is a complex strategy game where chance and luck play no role. However, unlike chess, no human can explain how to play Go at the highest levels. Top players cannot assess their knowledge about how they are able to perform so well—a good illustration of the Polanyi Paradox where top Go players cannot articulate their moves.

Deep Blue made optimal chess moves through brute force by sifting through millions of possibilities that it created. In contrast, there are many more moves and complex positions in Go than in chess, preventing computers from simulating moves. In response, AlphaGo learned winning strategies almost entirely on its own by assessing instances of successes and failures from examples and experience. The examples came from huge libraries amassed over the 2500-year history of Go. The system does not rely on human knowledge about the task at hand, in effect, doing an end-run around the Polanyi Paradox.

4.11 Closing Remarks

We have seen how K work—manipulating information—characterizes virtually all office work and most 21st century work in general. We have also commented on certain artifacts of the Industrial Age that are hampering the full flowering of organizations in the knowledge era. On the other hand, organizations *are* shifting, slowly, even generationally, from thin to thick assumptions, designs, and manifestations.

We have argued in previous chapters that a key characteristic of a K organization is the interweaving of thin and thick elements, both from within and from without, across K-spaces. We have seen that this may be operationalized by (1) interweaving symmetrical and asymmetrical structures; (2) the interleaving of thin and thick organizational contexts; and (3) the superposition of discourse and practice structures over thin and thick layers, even beyond the boundaries of the organization. It is thus essential that we create interactions across the entire sphere of K-spaces. Similarly, in this chapter we have extended this line of thinking into K work—most work has a mix of thin and thick attributes and is not very different from the design of knowledge discussed in previous chapters.

K work is no longer confined to the laboratory but exists throughout the organization in various forms and in various degrees. In our conception of work, we use the terms thin and thick to represent this range of K work. This raises several important issues. In the same way that operations has a different set of rules and expectations than R&D, we must figure out how to maintain the efficiency expected in operations and yet promote the experimentation and knowledging necessary for R&D. That is to say, *we have to bring the culture of R&D into the front office,* blurring the line between the two areas. This is one reason we have advanced new ideas and artifacts over those of the Industrial Age. Hence, we have argued in favor of a thin–thick design paradigm, functional structures, practice communities, discourse structures, extensive education, and against performance reviews, at least as commonly implemented. Equally important, we advocate ennobling goals for organizations, not merely because it is just and moral, but because it is necessary for the flowering of knowledge and society.

As we have argued in this book, knowledge is intricately intertwined with moral behavior and social responsibility. This is not an Eastern view, but one shared by all cultures, including that of the Greeks and their Western progeny. It is only recently, and one with a technical or organizational bent, that we have narrowly defined knowledge with anecdotes or with technical definitions. This extreme simplification may be convenient for managers, but it unduly disconnects knowledge from its roots and essential character—and opportunities and challenges in the modern world.

Practice

The continuum between theoretical and practical knowledge gives rise to the nobility of knowledge and its offshoots—religion or philosophy, for example. Determining the mitigating factors of an organization or area is an opportunity to raise the status of an organization above that of a mere business to an important social institution. We have advocated the Kantian mantra of duty and moral vision—consciousness is one such noble implementation, made manifest in Whole Foods, for example, with goals such as eliminating world hunger.

Justification—judgment and action—is at the heart of building new knowledge, whether theoretical or practical. Basic research puts knowledge creation at the theoretical pole, whereas practice would be toward the practical K pole, mitigated by human and social factors. The mitigating factors, or categorical imperatives, must be tied back to theoretical K (i.e., certain K) or the practice would be corrupted. This intermediate point (Figure 4.9a) is where most K workers are located—balancing the needs of truth and practice (knowingly or unknowingly) with the obligations imposed by the social group (Figure 4.9b).

Practice is the basic unit of K work. We conceive of a K practice as a complex set of behaviors that both emanate from and construct K-spaces, and include, for example, the history of the practice group, social responsibility, ethics, individual emotions, and everything in between. A practice establishes its own meaning and context. Clearly, it follows from earlier that a *practice is a social knowledge construction* that manifests itself in a collection of behaviors. These behaviors include structured rituals as well as innovative social constructions within a community. In this regard, a community includes everything from a profession to a religion to a national culture. The knowledge content within a practice has a large tacit component but is sufficiently broad to encompass all forms of knowledge—this is particularly important in professional practice of every type.

We have been arguing that knowledge is complex and cannot be neatly fitted into a single model or framework. Throughout this book, we have maintained that knowledge is inherently noble and elevates life, for the individual and for society. Accordingly, K work is, at once, individual, social, organization (ontological), as well as intellectual, moral, and aesthetic (epistemological). Tacit K, for example, is *not* fully individual but brings with it a social history that underpins its assumptions and workings. Hence, our core model of knowledge—a *K-molecule*—is centered around a K practice that, in our conception, brings balance to science, moral law, and the aesthetic, without sacrificing immediate situational factors or organizational obligations.

Technology

Technology and data appear to be "taking over the world." We must recognize that both practice and KM will be strengthened with close technology integration—this is as much a function of our imagination as the technology itself. Google's nanoparticle project is instructive: we see a fusion of knowledge and technology with noble goals.

Data and knowledge are increasingly matrixed throughout the organization. As the organization develops new techniques and technologies to leverage and manage big data, new organizational designs are emerging. These designs are very nearly the same as in K organizations—functional structures and practice communities, for example. An additional requirement imposed by big data is task–technology integration that together with KM represents an opportunity for coordinated design. The ideas of scripts and K-nuggets, for example, help sort out the types of systems and their content appropriate for a given K task. As we have seen in our student exercise on integrated job–technology design, new technologies such as portals and mobile devices have made access and integration even easier.

Automation through AI and robotics is challenging the stability of K work and the perceived safety of K workers. Robots will magnify the importance of thin and thick design, as modern work will consist of interleaving elements of both. We must accommodate the social, cognitive, and physical capabilities that increasingly are performed by robots and incorporate these factors into our designs of K work and K processes. Enlightened design suggests that organizations will embrace AI and robotics along with a corresponding redefinition of work that enables creativity, social interaction, and collaboration. Moreover, the robot economy requires integrated design of business processes and tasks, new technology, and jobs.

4.12 Class Exercises

1. Will artificial intelligence and robots replace knowledge workers? Justify your answer giving examples of K-related activities, contexts, costs, and other factors that may come into play.

2. Analysis project. Using Figure 4.9a as a guide, assess the knowledge activities in your organization or school, identifying the mitigating factors that may explain such behavior. Are these factors (goals) noble or compromised by other interests? What would you change? How would you institute a more enlightened and socially conscious goal on an organization-wide basis?

3. Using Table 4.2 as a guide, develop a table for your organization or school comparing its processes or thin work with its practices or thick work.

4. Give an example where K workers are being used as "tools" without a noble or categorizing imperative. What would you change to bring the imperative more in line with the ennobling goals we have discussed?

5. Expertise project. Assess and characterize your expertise, identifying its basis, sources, and state of development. How do you account for the evolution of your expertise? Do social or cultural factors play a role?

6. Practice project. Describe the application of your K in the exercise of your job. Discuss the mitigating factors that influence how you carry out your duties. Include the organizational or professional structures that exist and how they guide your behavior.

7 Kant (1993) has extensively explored "a priori synthetic knowledge"; that is, how we come to know things that we have never experienced. Provide your own thoughts on this problem and how this relates to K practice.

8 Integrated project. Select and analyze a K job from your organization or school and design its supporting technology using Figure 4.13 and the sample solution as a guide. Also, extend the assignment to include (1) specific organizational and practice communities that may be appropriate and (2) big data or IoT task integration that would promote K.

9 Discuss the most common tools and technologies used in the exercise of your job? What additional tools would you like to have access to?

10 Discuss how your organization accounts for big data. Include in your answer the issues and strategies associated with data ownership (e.g., IT, business units, combined or hybrid unit), departmentalization, and decision making.

References

Amabile, T.M. (1997), "Motivating Creativity in Organizations." *California Management Review*, 40(1): Fall 1997.

Anderson-Lehman, R., Watson, H.J., Wixom, B.H., and Hoffer, J.A. (2004), "Continental Airlines Flies High with Real-Time Business Intelligence." *MIS Quarterly Executive*, 3(4): December 2004.

Barr, A. (2014), "Google's New Moonshot Project: The Human Body." *Wall Street Journal*. July 27, 2014.

Belanger, D. (2014), "Seminar Notes on Big Data." Stevens Institute of Technology.

Carte, T.A., Schwarzkopf, A.B., Shaft, T.M., and Zmud, R.W. (2005), "Advanced Business Intelligence at Cardinal Health." *MIS Quarterly Executive*, 4(4): December 2005.

Chui, M., Manyika, J., and Miremadi M. (2015), "Four Fundamentals of Workplace Automation." *McKinsey Quarterly*. November 2015.

Chui, M., Manyika, J., and Miremadi M. (2016), "Where Machines Could Replace Humans: And Where They Can't (Yet)." *McKinsey Quarterly*. July 2016.

Collins, H.M. (2001), "What Is Tacit Knowledge?" In *The Practice Turn in Contemporary Theory*. Eds. Schatzki, T.R., Knorr-Cetina, K., and von Savigny, E., Routledge.

Dalkir, K. (2011), *Knowledge Management in Theory and Practice*. Second Edition. The MIT Press.

Davenport, T.H. (2005), *Thinking for a Living: How to Get Better Performance and Results from Knowledge Workers*. Harvard Business School Press.

Davenport, T.H., and Prusak, L. (1998), *Working Knowledge: How Organizations Manage What They Know*. Harvard Business School Press.

Dreyfus, H. (1991), *Being-in-the-World: A Commentary on Heidegger's Being and Time, Division One*. MIT Press.

Drucker, P.F. (1999), *Management Challenges for the 21st Century*. HarperBusiness.

Galbraith, J. (2014), "Organization Design Challenges Resulting from Big Data." *Journal of Organization Design*, 3(1): 2–13.

Giddens, A. (1984), *The Constitution of Society*. University of California Press.

Gladwell, M. (2007), *Outliers: The Story of Success*. Little, Brown and Company.

Hammer, M., and Champy, J. (1993), *Reengineering the Corporation: A Manifesto for Business Process Revolution*. Harper Collins.

Hunter, D.R. (2003), *Measuring General Aviation Pilot Judgment Using a Situational Judgment Technique*. Federal Aviation Administration.

Hurwitz, J.S., Kaufman, M., and Bowles, A. (2015), *Cognitive Computing and Big Data Analytics*. John Wiley & Sons, Inc.

Jensen, R.S. (1995), *Pilot Judgment and Crew Resource Management*. Avebury Aviation: Ashgate Publishing Co.

Kant, I. (1993), *Critique of Practical Reason*. Third Edition, Trans. Beck, L.W., Macmillan.

Keen, P. (1997), *The Process Edge*. Harvard Business School Press.

Kimball, R., Ross, M., Thornthwaite, W., Mundy, J., and Becker, B.J. (2008), *The Data Warehouse Lifecycle Toolkit: Practical Techniques for Building Data Warehouse and Business Intelligence Systems*. Second Edition. John Wiley & Sons, Inc.

Leonard, D., and Swap, W. (2004), "Deep Smarts." *Harvard Business Review*. September 2004.

Liker, J.K. (2004), *The Toyota Way*. McGraw-Hill.

Mackey, J., and Sisodia, R. (2013), *Conscious Capitalism*. Harvard Business Review Press.

March, J.G. (1994), *A Primer on Decision Making: How Decisions Happen*. The Free Press.

McAfee, A., and Brynjolfsson, E. (2016), New York Times Newsletters Opinion Today. March 16, 2016.

McKinsey Global Institute. (2011), *Big Data: The Next Frontier for Innovation, Competition, and Productivity*. McKinsey.

Sachs, P. (1995), "Transforming Work: Collaboration, Learning, and Design." *Communications of the ACM*, 38(9): September 1995.

Schatzki, T.R., Knorr-Cetina, K., and von Savigny, E. (2001), *The Practice Turn in Contemporary Theory*. Routledge.

Schmidt, E., and Rosenberg, J. (2014), *Google: How Google Works*. Grand Central Publishing.

Spear, S., and Bowen, H.K. (1999), "Decoding the DNA of the Toyota Production System." *Harvard Business Review*. September–October 1999.

Srivastava, D. (2014), "Data Quality: The Other Face of Big Data." Stevens Institute of Technology Research Colloquium. November 5, 2014.

Sveiby, K.E. (1997), *The New Organizational Wealth: Managing and Measuring Knowledge-Based Assets*. Berret-Koehler Publishers.

Turner, S. (1994), *The Social Theory of Practices: Tradition, Tacit Knowledge, and Presuppositions*. University of Chicago Press.

U.S. Federal Aviation Administration (1991), "Aeronautical Decision Making." (Advisory Circular 60-22). Washington, DC.

Wiig, K. (1993), *Knowledge Management Foundations*. Schema Press.

5

Organizations and Knowledge

Chapter Preview

Introduction

The central theme of this chapter concerns organizations and knowledge. We argue that organizations are in transition from the first to the second divide in knowledge management (KM). This chapter reviews first divide thinking and models, followed by our contribution to second divide representations.

The standard material (KM models and activities) is supplemented with several of our design representations that extend our discussion from previous chapters. We include our proposed K-base and its governance structure. In addition, we selectively review the literature on organizational learning (OL) and its relationship to KM. We illustrate the application of KM concepts with a case study conducted by the authors and one of our graduate students on the Peace Corps (PC).

Food for Thought

1) Imagine the flow of knowledge through your organization or school. How does this take place and what is it that actually flows?
2) Describe how K is formed in your organization or school.
3) Does your organization or school reuse K? How does this take place?
4) How is "social K" related to both individual K and organizational K in your organization or school?
5) How easy is it to find the "right expert" that can assist you at your place of work or school?
6) Think about recent "learnings" in your school or organization. What did you learn, and how did you learn it?

Designing Knowledge Organizations: A Pathway to Innovation Leadership, First Edition.
Joseph Morabito, Ira Sack, and Anilkumar Bhate.

Topic Layout

The figure below illustrates the topic layout of Chapter 5. As shown on the left side of the figure, we continue our discussion of K and KM and introduce an integrated K model that ties together the models we have been discussing. The left section informs the organization on the issues associated with K and organizational design. The center of the figure represents the organization in practice. We review several organization-level KM cycles and models and introduce an integrated K cycle. We further discuss several organizational design representations and include new design concepts. Also reviewed is an illustrative case study on the PC. The right side of the figure shows organization-level K-related activities that inform and are informed by the organization in practice. This includes K exchange and transfer activities as well as the activities associated with OL. We also introduce the design and management of the organization K-base, including its governance structure.

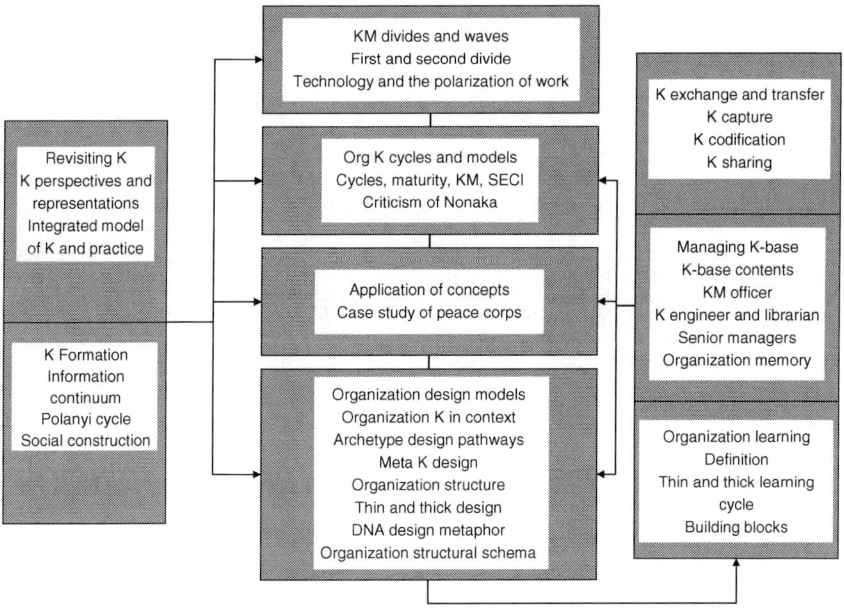

Concept roadmap to Chapter 5.

5.1 Organizations and KM

A colleague of ours asks, “Knowledge management had been growing in popularity until recently; since around 2000, it seems to me relatively little has taken place to change business practice. Why is this so?”

5.1.1 First Divide in KM

Recall the two divides in KM. The first divide represents KM as an evolutionary progression of management ideas, starting with management by objectives and continuing with organizational development, total quality management (TQM), business process reengineering (BPR), and finally KM (Despres and Chauvel 2000). We may supplement this management progression with a corresponding evolution of "traditional" information technology (IT), starting with data processing and continuing with management information systems (MIS), decision support systems, expert systems, TQM, BPR, business intelligence (BI), and similarly culminating with KM. By traditional IT, we mean that which is commonly found in large organizations, including the waves listed earlier. Note too the significant crossover between the two waves, as each reinforces the other—that is, several management paradigms depend on or are enhanced through IT. *Collectively, the two waves of management and traditional IT represent the first divide.*

Knowledge is often conceived either as an *object* or as a *flow*, the former implying an explicit and the latter a tacit orientation. It is the tacit perspective that makes KM so special—in particular, the extensible nature of K assets is due to their action in the imagination (recall the Introduction)—and brings us closer to a *second divide* perspective of KM. Unlike data or information, K is not consumed when used but grows and is refined—made manifest and enlarged—with application. In principle, organizations recognize the importance of tacit K and seek to develop those areas in which tacit K flourishes (e.g., communities) and to optimize tacit–explicit interaction. Thus, KM in a corporate sense may be defined as "the identification, optimization, and active management of explicit or tangible informational assets (such as data physically stored in a computer or on a piece of paper) and tacit knowledge (information and insights residing largely in people's heads)" (Cortada 2011).

Explicit Approaches. Many organizations manage K as an object, with little to distinguish it from data or information. As Cortada's definition of KM implies, organizations recognize the value of both their hard and soft assets, but in practice most emphasize the former. In fact, this is the locus of most KM initiatives, with success coming more by accident than by design (Cortada 2011). Accordingly, the strategy of both consulting firms and large organizations is to gather and store the tangible assets (data, information, K) of the organization into a database and make them generally available. In effect, most success in managing explicit KM lies in the realm of BI and data analytics, along with those technologies discussed in Chapter 4 (e.g., K repositories).

Tacit Approaches. Organizations are increasing their attention to tacit K, largely through the development of communities. Communities vary in type and purpose and may include, for example, project teams (work-centered), lateral processes (coordination-centered), and communities of practice (CoP) (K-centered). A formalized outgrowth of a given CoP is a center of excellence or competence. In large organizations, communities may be networked or virtual. A research-based community—collaboration across organizational boundaries—is an important new development for organizations. We discuss this more fully below and in Chapters 6 and 7.

5.1.2 Transition to the Second Divide in KM

The progression of *communities*, starting with a work group and leading to a CoP, is representative of a shift from first to second divide KM thinking. Note that we list communities under tacit-oriented approaches because the development of tacit K requires human interaction and sharing; however, there are large measures of explicit K involved in the activities of any community. Furthermore, Drucker (1988) states that K workers, in addition to continuously improving themselves, should engage in continuous teaching. We may add that such experts—the elite among K workers—in addition to participating in communities, should cultivate and share their knowledge through internal consulting and publications as well as teaching. We may add too that the cultivation of expertise involves participation in social and global communities, spanning the boundaries of organizations and even nations. Finally, organizations are increasingly looking to increase both casual and work-related interactions through the development of open office spaces, discussed subsequently.

Tacit–Explicit Interaction Approaches. This is a relatively new and difficult paradigm to implement. Tangible integration typically includes a *best practice* or a *project experience readout*, which commonly contains unequal elements of both tacit and explicit K. A given best practice typically involves elements of management and technology waves, such as software, architecture, or BPR approaches. These often evolve into standards of a sort, and may include a methodology or procedure, particularly if explicit or technology-based, such as a software development lifecycle or a BPR framework.

The most significant manifestations of tacit–explicit interactions are the *K-conversion modes* described by Nonaka in the SECI model: socialization, externalization, combination, and internalization. We have described this model earlier in the book and continue our discussion here.

K Culture Development. Most managers rightly see the development of a K culture as a slippery concept. In essence, it involves the long-term and continuous promotion of values (hopefully enlightened), K-oriented leadership skills, designs (all levels), policies, activities, and systems that operationalize a given KM cycle (discussed later). We believe the ideas relating to the second divide are necessary elements in the development of such a K culture.

5.1.3 Second Divide KM

Second divide KM is the great challenge of our time: the K-based transformation of human interactions, organizations, societies and, indeed, the entire globe. In this book, we have presented ideas and trends, also with links to management and technology waves, all necessary, in our opinion, to take us into the second divide. These additional waves do not replace but supplement and reinforce management and technology waves.

Figure 5.1 illustrates the impact that KM divides and constituent waves have had on organizations. All new ideas are met with a buildup that hypes their importance, but they eventually settle down into a steady state. Such ideas include first divide waves such as BPR and BI. Further, there is a cumulative effect that magnifies their impact, more so than isolated waves.

One answer to the question posed at the start of the chapter is that most organizations leverage KM as they do other management or technology waves. Failure to appreciate the revolutionary nature of KM prevents a breakout into second divide thinking where the full benefits of KM may be realized. A second divide breakout requires embracing notions from a wide variety of disciplines. Our discussion of philosophy, for example, describes not only the ancient roots of knowledge but also its critical role on moral and social development, and its corresponding effect on K practice and the ennobling goals of organizations and institutions. Similarly, our discussion of thin and thick is essentially a discussion of culture and ethnography.

Second Divide KM
The essence of second divide KM is the introduction of waves and patterns of thinking, sourced from a variety of disciplines outside of traditional management or IT, all necessary to design and manage knowledge in the 21st century.

As the definition of "corporate KM" (our term) suggests, there are attempts to facilitate the development and transfer of tacit K. Teams exist everywhere, but these are designed as work-focused groups. Communities (formal and informal) are increasingly common but are not necessarily designed into the fabric of the organization with the intent of consciously promoting K. However, the second wave is slowly making its way into organizations. For example, until recently many of the organizations in which we teach have spent a great deal of energy disengaging employees from each other—most being *required* to spend a significant amount of their time working from home. Today, there is a growing recognition that this approach is *exactly opposite* as that required for a K organization. In fact, as we will subsequently discuss, organizations are now spending their efforts bringing employees back together through new office

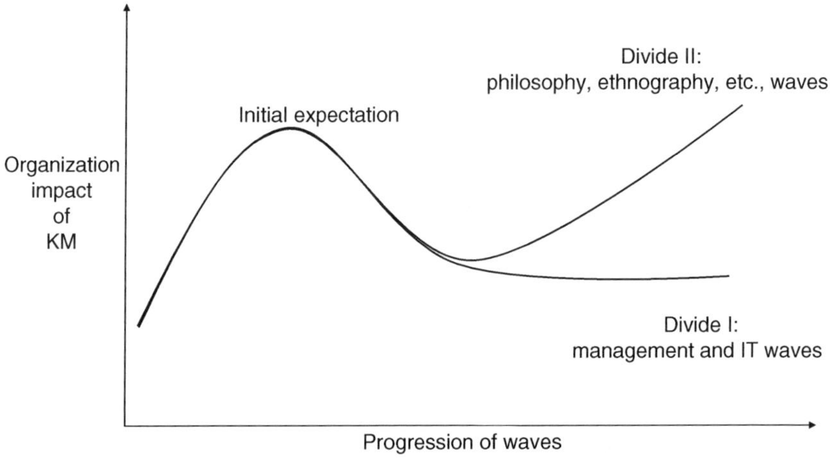

Figure 5.1 KM divides and waves.

configurations to promote interaction and, hence, the development and transfer of tacit K. The second wave *is* slowly but decidedly creeping into organizations, with a new and younger generation leading the way!

5.1.3.1 Importance of the Second Divide in KM

Our answer to the question posed at the beginning of this chapter is that first divide thinking (in and of itself) leaves little room for the evolution of KM. Traditional management and corporate IT do not carry the rich artifacts necessary to expand and bring forth entirely new modes of thinking. While living in the first divide, we will continue to do what we always have done—build detailed workflows (not good); design, manage, and evaluate all work the same way (not good); create more communities (good); and so on.

Figure 5.2 lists several of our ideas that we have discussed previously and that we believe are necessary to move an organization from first to second divide thinking. We have given examples in our discussion of these ideas when we introduced them in Chapters 1–3 and in detail in Chapter 4 in our discussion of K work. In this chapter, we continue our discussion of these ideas in additional contexts.

5.1.3.2 The Changing Landscape

In a 2015 article in *Entrepreneur*, several leading futurists and cultural anthropologists were asked to forecast revolutionary trends in business over the next 5 years (Ankeny 2015). Among the more interesting forecasts were a (1) shift in business from product to service to experiences and (2) a more connected and informed customer, particularly in terms of a "digital customer experience." This shift corresponds to movement along the information continuum—recall that

- Integrated understanding of K models
- K-triangle and the fusion of Indian, Chinese, Greek, and Western thought
- Information continuum: the interaction of {data-information-K}, {explicit K-tacit K}, and {social K-individual K}
- K-worlds and the representation of art, science, and craftwork
- Social continuum–theoretical-practical K interaction with mitigating factors
- Consciousness and mindfulness: ennobling goals
- Three dimensions of tacit K: technical, cognitive, and social practice
- Three dimensions of practice: theoretical, practical, and structural
- K-molecule and the balance between theoretical and social knowledge
- Creative process of knowledging
- Thin and thick design spaces–ethnographic and cultural considerations
- Knowledge spaces
- Interaction and balance between asymmetrical and symmetrical structures
- Communications modes and the contract continuum
- Collaborative technologies and systems
- Knowledge construction and K-binding
- Anthropological analysis
- Integral and modular design
- Big data and the Internet of Things
- Role of social K in the development of individual and practical knowledge
- The role of robots and automation in the design of work

Figure 5.2 Sampling of second divide KM waves.

tacit K contains a large experiential component. We see a corresponding shift in strategy that consciously or unconsciously raises business strategy to a tacit-oriented KM strategy. For example, as part of its brand development, Sony leveraged the idea of "ba" (recall from Chapter 2 that ba refers to the space where knowledge creation takes place) and developed an environment (i.e., METREON) where their products may be experienced firsthand (Akutsu and Nonaka 2004). This is not very different than what we see in Apple stores as customers experience and handle products. Further, customer "experience" goes well beyond product sampling. Indeed, it is part of an ongoing strategy that experiments with ways to draw in customers and improve their experience. For example, Starbucks has announced specialized store formats that are designed for specific places and cover everything from express to luxury "Reserve" stores, all with the goal of transforming customer experience (Foroohar 2015).

As another example, recall from Chapter 4 that Google calls its top employees "smart creatives." According to Schmidt and Rosenberg (2014), this follows from the shift in business models mentioned earlier, a shift that is mirrored by a corresponding shift in power within companies. It is *individuals* and *small teams* that can have "massive impact." Hence, the smart creatives are those employees that can have such an impact. *Smart creatives combine technical knowledge, business expertise, and creativity* to produce new ideas, experiment, sometimes fail, and try again. When you "give them lots of freedom they can do amazing things, amazingly fast" (i.e., recall the key enabler of autonomy). Naturally, this needs to be nurtured by a culture that cares about those things that attract and engage smart creatives (e.g., ennobling goals and purpose), acts positively to realize those goals, and promotes human interactions, even "serendipitous connections."

5.1.3.3 Revisiting Polanyi: The Polarization of Work

At a 2014 meeting of the Federal Reserve Bank, David H. Autor (2014) described the Polanyi Paradox: "our tacit knowledge of how the world works often exceeds our explicit understanding." While technology improves our lives, it also threatens economic stability and may bring social and political disruption. A contemporary manifestation of Polanyi's paradox is labor market polarization: the simultaneous growth of high-education, high-wage and low-education, low-wage jobs.

One central issue in the 21st century is the relationship between computer technology and skilled labor. At a practical level, Polanyi's paradox means that many familiar tasks, ranging from the quotidian to the sublime, cannot currently be computerized because we don't know "the rules." At an economic level, Polanyi's paradox means something more. The fact that a task cannot be computerized does *not* imply that computerization has no effect on that task. On the contrary, *tasks that cannot be substituted by computerization are generally complemented by it.* This point is as fundamental as automation and is often overlooked.

Furthermore, recall from Chapter 4 that automation typically occurs at the activity level, meaning that a given worker, particularly a K worker, will have a mix of automated, computed-enhanced, and tacit activities. If properly designed, these activities will be complementary and mutually reinforcing rather than working against each other or undermining the motivation of the K worker.

Most work processes draw upon a multifaceted set of inputs: labor and capital; brains and brawn; creativity and rote repetition; technical mastery and intuitive judgment; perspiration and inspiration; adherence to rules and judicious application of discretion. Typically, these inputs *each* play essential roles; that is, improvements in one do not obviate the need for the other. If so, productivity improvements in one set of tasks almost necessarily increase the economic value of the remaining tasks. For example, construction workers have had their work augmented by machines; the result has been a significant improvement in their productivity over the years, and, as importantly, an improved standard of living as the workers have acquired new complex skills.

Polanyi's observation also explains *why* high-level reasoning is straightforward to computerize and sensorimotor skills are not. High-level reasoning uses a set of formal logical tools. They were developed specifically to address formal problems (e.g., counting, mathematics, logical deduction, precise language, encoding quantitative relationships). Sensorimotor skills, physical flexibility, common sense, judgment, intuition, creativity, spoken language, and so on, are "built-in" capabilities that the species evolved rather than developed. Formalizing these skills requires reverse engineering a set of activities that we normally accomplish using only tacit understanding. "Contemporary computer science seeks to overcome Polanyi's paradox by building machines that learn from human examples, thus inferring the rules that we tacitly apply but do not explicitly understand." Additionally, we would add that this reasoning also explains why it is difficult to automate a K practice—*a K practice is multifaceted, embodying and intermingling explicit reasoning and tacit understanding.*

Automation Example. The Toyota principle of stopping where there is a quality problem (i.e., "Jidoka"), sometimes known as mistake proofing, also connotes "automation with a human touch." The central idea is to design processes and equipment to free your workers to perform more value-added work, not to replace them (Liker 2004). That is to say, automation may be considered an opportunity for an organization to raise and leverage the knowledge of their workers—but this may depend on the philosophy and culture of the organization.

5.1.3.4 The Internet and the Second Divide

It is hard to overstate the importance of what we all know: the Internet is a *revolutionary* technology that is *disrupting* the fabric of life, as did the industrial revolution before it. The Internet is a technology that gains its strength—worldwide access, robustness, popularity, and so on—at least in part from the interrelated principles of social K and late K-binding. Recall that K-binding refers to the process of K construction: early knowledge binding is characterized by the creation of knowledge by relatively few people, typically senior managers or domain experts. That knowledge is subsequently operationalized down the hierarchy through the division of labor. In contrast, late knowledge binding requires the accumulation of knowledge from the division of labor, with rapid or even unity of implementation.

From a societal perspective, early K-binding is the tool of the elite: large companies, major media, institutions, governments, and so on. Each of these organizations resists competition, for example, through "barriers to entry"—control over the flow of information, costs, and expertise. The principle of late K-binding in combination with the Internet breaks down these barriers through K transfer and the creation of social K. Recall that the drivers of the K economy include democratization and globalization—attributes of late K-binding, the creation of social K, and the Internet. From this perspective, the Internet is more than a tool of implementation; indeed, it is a driver of social interaction and change.

Perhaps the most dramatic effect of the Internet is as a vehicle for building social knowledge beyond the boundaries of a single company or government. The Internet has spurred revolutions around the globe. We only need to look around us to see the restrictions or threats imposed by governments, even democratic ones. Taxes, control over access, control over search, and censorship over content are the most common threats to knowledging through the Internet. The same can be said of large media and corporations—all are threatened by their heretofore customers. In spite of such obstacles, a traumatic second divide is making its way into global life.

5.1.3.5 Collaborative Systems

Perhaps the most important application of the Internet is the development of collaborative applications and products. Until recently, collaboration technologies centered on document management systems, e-mail, and groupware. The rise of the Internet allows organizations to collaborate across boundaries on complex problems. This form of activity is increasingly the norm for academic, medical, and scientific research. As a social technology, collaborative systems may be considered a critical element in second divide thinking. We will discuss collaboration in more detail in Chapter 6.

5.1.3.6 Mindful KM

Recall from Chapter 4 that the personal awakening of John Mackey led him to higher levels of consciousness, which, in turn, led to the ennobling goals and moral vision of Whole Foods. Mindfulness, a term usually reserved for personal awareness, can thus be elevated and made manifest in an entire organization. Mindfulness—arousing awareness and experience of the present—can lead to future organizational possibilities. Full mindfulness embraces past, current, and likely future trends and knowledge. This is particularly true for organizations and is a useful principle for individual K workers as they grapple with and balance organizational and moral imperatives. Mindfulness is embedded in the ideas of a practice and the K-molecule.

For example, in 1958, Volvo developed the 3-point seatbelt. Volvo made the new safety technology generally available and did not collect royalties. As another example, consider the ennobling behavior of Jonas Salk. In 1955, Salk developed a vaccine for polio using a dead virus strain and *refused* to patent the vaccine. He had no desire to profit from the discovery but wished to see the vaccine disseminated as widely as possible. Similarly, Albert Sabin, who developed a live virus polio vaccine, donated his vaccine strains to the World Health Organization (WHO), guaranteeing wide access to the vaccine at the lowest possible cost. Under the stewardship of the WHO, Sabin's oral polio vaccine became among the most accessible and commonly used vaccine in the developing world and contributed to a dramatic worldwide reduction in polio.

5.2 Knowledge Revisited

5.2.1 KM Versus Knowledge As Such

The distinction between KM and knowledge is critical to understanding our perspective. First divide thinking embodies an organizational perspective of knowledge, and this is what we call knowledge management. KM (unlike K itself) is more a strategy than a discipline. In this perspective, KM means (1) K can be managed and (2) K is an organizational asset. On the surface, this sounds reasonable, but let us look at the implications more closely. In practice, KM gives greater weight to the explicit side of things—static objects are easier to manage than dynamic flows. It follows very quickly that our argument on the interaction of {data–information–K} along the information continuum rapidly devolves into an argument for managing K as data or information. It is easier to manage explicit outcomes than either tacit K or the process of K creation.

K as an asset means that K is an economic resource, perhaps an intangible one, but an asset nonetheless. It exists to improve the economic position of an organization. Organizations typically employ several overlapping approaches to KM, in particular the following actions: (1) the planned generation of K (e.g., implementing the SECI model of K creation, hiring new K employees), (2) the capture and codification of K, (3) building communities and sharing K, and (4) embedding K and the principles of KM into the organization, in particular its business processes and culture. In practice, at the heart of KM is (5) the elicitation of individual K from employees for the benefit of the organization. In this respect, KM is a collection of ideas and techniques intended to maximize economic value of employees' K by converting their tacit K into explicit, actionable K (Kilov and Sack 2005).

KM is not a discipline as such. KM embraces model making and real-world decisions. It has the potential of revising our idea on what a 21st century organization is, and is not—the theory of the firm and our strategic business concepts. We will never get more than marginal improvement if we remain in the first divide and rely primarily on management and traditional IT thinking. The second divide is revolutionary in both its means and its ends. It requires an embrace of many disciplines, models, and perspectives. The ideas listed in Figure 5.2 are our contribution to second divide thinking, but this should be only the beginning of the journey into KM.

As we shift to the tacit side, KM tends to focus on the problems created by knowledge—its intangible character, K creation, social development, stickiness, and so on. Tacit K forces us to deal with the ambiguity and "bounded rationality" inherent in tacit K and its corresponding organizational design and leadership issues (Spender 1993). A key design issue, for example, is tacit–explicit K interaction. Furthermore, "managing" tacit K implies we are managing, or at least influencing, the motivation, thinking, and imagination of people;

however, many present-day organizations strongly rely on first divide artifacts such as benchmarking and performance reviews to manage those who think for a living. Thus, first divide thinking inhibits not only the evolution of KM but it may also handicap and undermine the progression toward second divide thinking.

5.2.2 Knowledge Perspectives and Representations

In this book, we have presented several characterizations of knowledge. Each makes sense from a particular perspective—recall that in Figure 1.1, we suggest that knowledge and knowledging have emerged from virtually every human endeavor. There is no characterization without its unique problems; hence, we have concentrated on those that are sufficiently robust to be useful in several contexts. Ultimately, we need to study knowledge from a transdisciplinary perspective—K in the liberal arts, management, and science are often different but needs to be studied holistically. All knowledge and learning must be respected. This is one reason we have extensively reviewed the ancients and their philosophies. In many respects, their philosophies embrace the dimensions of truth, morality, social responsibility, practice, aesthetics, and science that are the essence of knowledge.

5.2.2.1 Knowledge and the Emergence of Wisdom

Figure 5.3 illustrates the interaction of the key K representations we have discussed thus far. Recall that we have argued that knowledge and wisdom are inextricable.

Figure 5.3 shows K and wisdom in the context of the key K models we have been discussing throughout this book. Observe that wisdom emerges as a derivative and interaction of many different types or perspectives of knowledge. In particular, *wisdom emerges as scientific and social truths collide.* Rigor and relevance are thus brought into harmony.

Wisdom
Wisdom emerges as scientific truths collide with social truths. Wisdom thus brings rigor and relevance into harmony.

Here we provide a very brief refresher (selected) of key K concepts with additional commentary.

5.2.2.2 The Genesis: Ancient Philosophy and Knowledge

As the reader of this book should know by now, our conception of knowledge is largely driven by the civilizations of India, China, and Greece (and

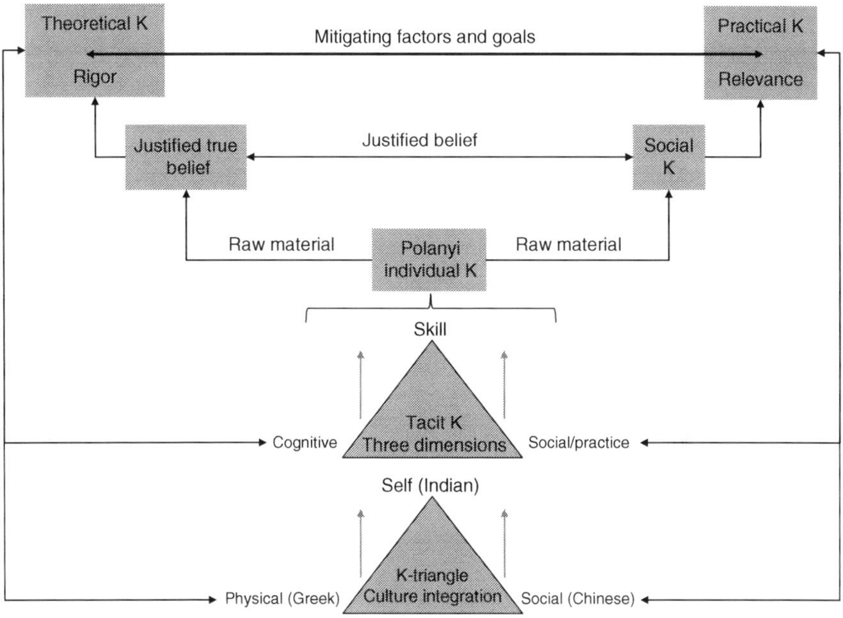

Figure 5.3 Knowledge and the emergence of wisdom.

subsequently Europe), all of which view knowledge in the contexts of right conduct, moral behavior, ethics, science, and politics. In a cultural sense, K is associated with the most noble of characterizations: moral preparation and social action (Chinese), the four Indian aspects of ethical behavior (Dharma, Artha, Kaama, Moksha), the Plato ideal of justice and goodness, and the Confucian thread of conscientiousness and altruism. Indeed, the great traditions intertwine knowledge and wisdom, the practice of which elevates life and leads to righteous behavior and further wisdom. Furthermore, this conception of knowledge underpins a moral-based theory of K in organizations.

We have further introduced the K-triangle (Figure 1.10) as a metaphor for understanding the historical development of knowledge and wisdom: the *Self* (Indian), *Social* (Chinese), and *Physical* (Greek) aspects of life and reality. This gives rise to the idea that knowledge is an *integral* construct and cannot be separated from our imagination, our flesh and blood, the social structures of our family and society, our history, laws and customs, our art and music, and science.

5.2.2.3 Tacit K Dimensions

Tacit K is *not* fully individual but carries with it a social history that underpins its assumptions and workings. Hence, while tacit K is typically characterized as some combination of mental models (cognitive dimension) and sensorimotor skills (technical dimension), an equally important dimension is the social dimension—the history and practice of a particular community in which an individual lives and thinks.

5.2.2.4 Justified True Belief (Plato)

Justified true belief takes us closer to objective knowledge and truth in a pure sense. Figure 5.4 extends Figure 1.11 and further characterizes the visible or explicit and the not visible or tacit elements with Western and Eastern phenomena, respectively. We can make an argument that in a "flat world" East–West distinctions are not as important as they once were, but Nonaka gives many examples of their importance, particularly in World War II. Also, the interaction of tacit and explicit elements gives rise to our working definition of K by Nonaka and Takeuchi (1995), as shown in Figure 5.4. *This particular definition appears to combine both the Plato and Polanyi perspectives.*

The definition of knowledge as justified true belief is not without numerous critics. It is almost always employed to define "knowledge by description" (i.e., propositional knowledge) and is considered third person knowledge (i.e., objective knowledge). "Justification" is not without its issues too—justification is not an objective exercise of inquiry but rather a function of belief and social context. Finally, as illustrated in Figure 5.3, justified true belief interacts with social K—beliefs too must be socially acceptable. Thus, justification and belief are embedded in the social context in which they operate.

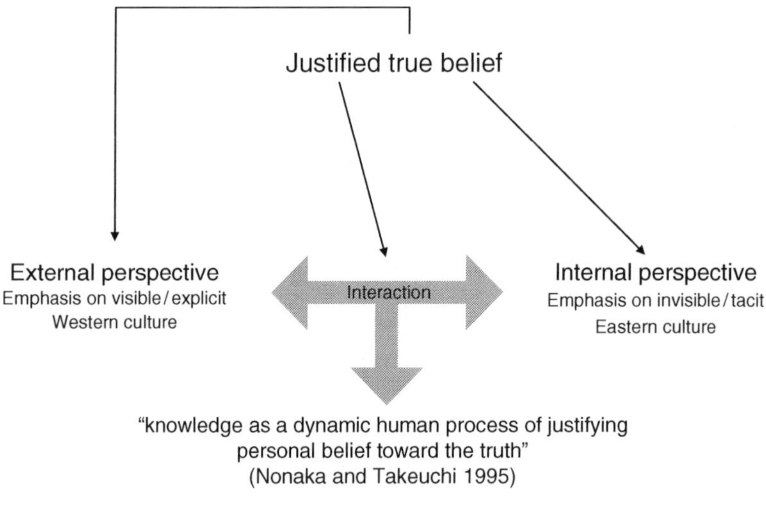

Figure 5.4 Plato's justified true belief revisited.

5.2.2.5 Polanyi's Dynamic Cognitive Experience

Polanyi's perspective is different from that of Plato. Polanyi takes us closer to art and tacit thinking; in fact, it is *Polanyi who coined the term "tacit knowledge."* Polanyi's representation is focused on human experience where knowledge arises (i.e., is created) from the interaction of *tacit knowing* and *explicit knowing* through the awareness and activity dimensions. This interaction, in turn, gives rise to the third dimension of cognitivity, where all knowledge is generated from individual tacit knowing. It is the dynamic cognitive experience of individuals that provides the raw materials for both justified true belief (reinforcing the rigorous pursuit of truth) and social knowledge (that which makes K relevant and useful). Finally, the individual K of Polanyi is more than subjective K for it also includes a universal element. That is, personal K and universal K are mutually dependent. Knowing continually evolves and represents the complex interplay of individual and group, truth and belief, the objective and subjective, and the scientific and social.

5.2.2.6 Social Knowledge

Social K is whatever a society considers K—its "social facts," assumptions, and beliefs, and the "social construction of realty" (Bernstein 2014). The development of social K is not a logical exercise as such but is somehow associated with the social artifacts of organizations, religions, gender, politics, and so on. That is to say, *social K is closer to collective beliefs than it is to universal truths.* This is not without historical consequence—we only need to see how politicians "rewrite history" or are accused of such. The strange combination of power and archeology often determines cultural truths—"culture wars" abound.

Figure 5.5 (a variation of Figure 2.17) illustrates a series of socially oriented K-spaces. Observe that the spaces may overlap and intertwine with each other. For example, a professional practice (e.g., medicine, law, accounting) supersedes an organizational practice; on the other hand, a nonprofessional practice (e.g., database administration) is represented as a best technical practice accepted as an organizational practice. Similarly, organizational and cultural spaces are intermingled, as are cultural and global spaces. It follows that no social space is isolated from inner and outer social influences: every space has its "facts," beliefs, and justifications that must be rationalized with every other space. Today, social media is the driving force behind social K.

5.2.2.7 Social Continuum Revisited

We developed this theoretical–practical K continuum based on the work of Immanuel Kant (see Figure 4.9a). Understanding theoretical K—what is accepted as true in a scientific sense—is different from acting on that knowledge (practical K). We recognize that there is a constant tension between scientific facts and social facts (i.e., group or crowd), with corresponding beliefs and justifications intertwined with both. Yet, while a social community

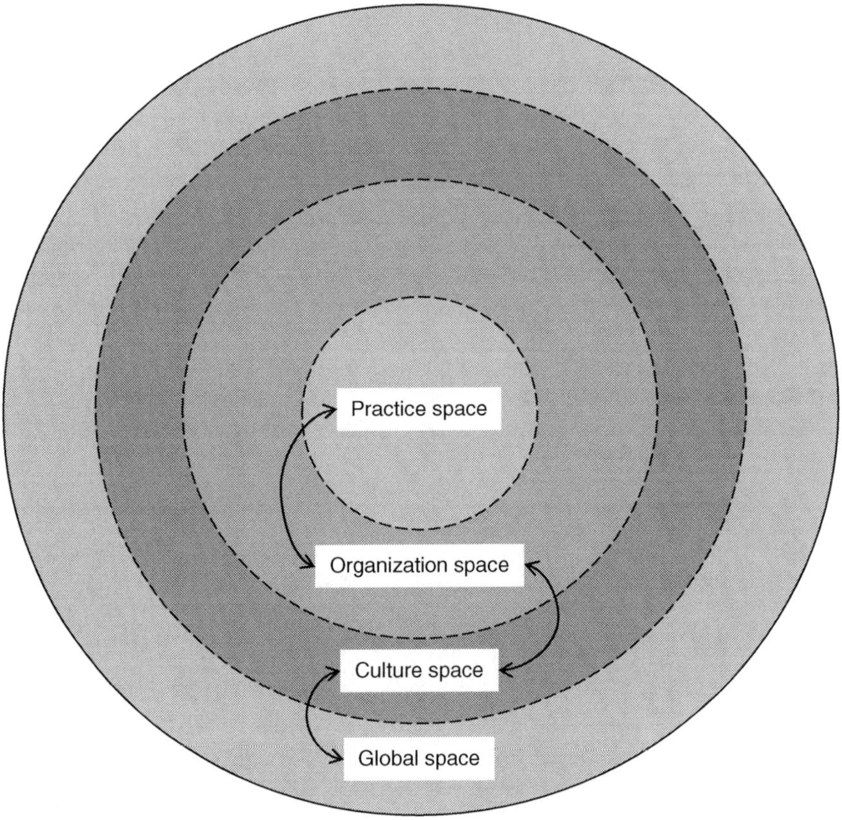

Figure 5.5 Social knowledge spaces.

cultivates practical K, its assumptions and workings should be rational and supported by a corresponding underpinning of theoretical K. In this way, there is a rational check that prevents social K from deviating too far from its theoretical justification and into superstition. This interplay of theoretical and practical is mitigated by a wide assortment of sometimes unrelated factors that characterize human life. They include, for example, religious dogma, compassion, charity, politics, and so on. This is the way K is practiced and manifests itself—along a K continuum that seeks to balance truth and rigor with the demands of the social community for relevance.

5.2.2.8 K-Molecule Revisited

The K-molecule (Figures 4.10 and 4.11) represents the everyday workings of knowledging in an organization. Hence, our core model of knowledge—a *K-molecule*—is centered around a K practice that, in our conception, brings balance to science, moral law, and the aesthetic, without sacrificing immediate situational factors or organizational obligations. Note too that we separate organizational and practice groups, the latter concentrating on K within a domain or discipline, with (hopefully) organizational support.

5.2.2.9 Dialectic Synthesis Revisited

Recall from Chapter 2 our extensive discussion of dialectic and the synthesis of knowledge. The key idea is that an interaction of outwardly opposite positions is an opportunity to synthesize and create a new level of knowledge. Table 5.1 is a brief summary of selected dialectic elements.

Table 5.2 is an example of dialectic elements used by Toyota in their strategy for the development of the first Lexus, LS-400. As can be seen in the figure, the development of the Lexus is characterized by a synthesis of the attributes of a high-performance vehicle with that of a family luxury automobile (Osono 2004).

Table 5.1 Sample dialectic perspectives.

Perspective	Elements	Synthesis
Technical	Tacit–explicit	Higher-level development of K threads
Cultural and philosophical	East–west; body–mind; understanding–sensibility (Kant)	Synthesis is holistic (e.g., both-and) and whole body (e.g., cognitive–sensorimotor)
Management	Hierarchy–task force; top-down-bottom-up; individual–group–organization; global–local; long–term–short–term	Hypertext organization; middle–up–down; K-spiral as K emerges; strategy as a synthesis; strategy as a synthesis
Organization K creation	Primarily tacit (East culture): socialization and internalization; primarily explicit (West culture): externalization and combination	Knowledge emerges as a spiral or synthesis of sequenced K-conversion modes through tacit and explicit K

Table 5.2 Example: dialectic strategy for Lexus LS-400.

Performance vehicle	Family vehicle
Outstanding high speed control and stability	Excellent riding comfort
Fast and smooth ride	Outstanding fuel economy
Superb noise reduction	Light weight
Elegant styling	Outstanding aerodynamic performance
Warm ambience	Functional cabin

Adapted from Osono (2004).

5.2.2.10 Tacit–Explicit Knowledge Revisited

Figure 5.6 is a brief summary of the key comparative properties of tacit and explicit knowledge, sourced from Polanyi, Nonaka, and Spender (1993, 1996). Observe that tacit knowledge is individualized; that is, each of us may share the same outward experience and yet (tacitly) each has a unique experience. This, in part, explains why witnesses to the same crime see things differently.

5.2.2.11 Individual–Social Knowledge

The cognitive and skill dimensions of tacit K are not shared, but its social assumptions and outward manifestations may be shared. In this way, the cognitive dimension, though specific to an individual, is at least in part shaped by the social dimension in which it develops. We can say, therefore, that individual tacit K is an extension of a given social community—a continuum where the individual further refines social K with unique and individualized behaviors, imagination, and sensorimotor skills. A representation of individual-social K comes from Spender and is shown in the next section.

5.2.2.12 Additional Representations of Knowledge

Public–Industry–Organization Knowledge. Figure 5.7 is a representation of three types of knowledge described by Dorothy Leonard. Public knowledge is widely available; industry K contains pieces of proprietary and generally available,

Explicit	Tacit
• Conscious reasoning	• Unconscious reasoning, intertwined with mental models, skills, and social K
• Formalizable and objectifiable, often in a scientific sense	• Embedded in the cognitive–sensorimotor apparatus of a human
• Articulated in a well-structured language	• Manifestation of the social context in which it formed
• Tangible in physical media; for example, computers, paper	• Difficult to articulate, and includes intuitions, hunches, emotions, etc.
• K as an object	• Demonstrated through activity
• Open and generally available	• K as a flow
• Declarative K	• Characterized by the activity in which the K is embedded
• Deductive K	• Individualized experiences
• Procedural K	• Inductive K
• Public, unless kept private	• Personal K

Figure 5.6 Tacit–explicit properties.

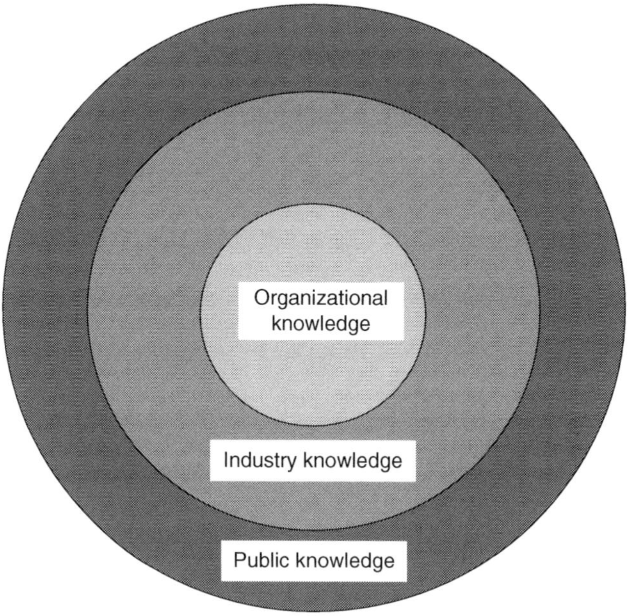

Figure 5.7 Public–industry–organizational K—Adapted from Leonard (1995).

visible K; organizational K is largely proprietary and invisible. That is to say, as we move from the outer to the inner circle, K becomes proportionately more difficult to replicate. Think of it this way: if an organization did not have large measures of proprietary K (e.g., know-how, internal practices, cultural assumptions, patents), it would be a generic entity with little distinctive competence.

Individual-Social Knowledge. Figure 5.8 is a representation of individual and social knowledge advanced by Spender (1993, 1996). Note that both tacit K (taken-for-granted) and explicit K (aware) are shown. Social K of which we are aware (explicit) is called scientific K and comes closest to Leonard's public K as described earlier. Social K taken-for-granted (tacit) is communal K and underlies a group's distinctiveness—this is one source of Leonard's organizational K as described previously. Communal K represents the essence of a group's culture—a practice, organization, or nation—and gives the group its distinctiveness and competitive advantage.

5.3 Organizational Knowledge Cycles and Models

There appear to be endless representations of knowledge, K cycles, maturity models, and models of KM in organizations. Here we present a sampling with selected class exercises.

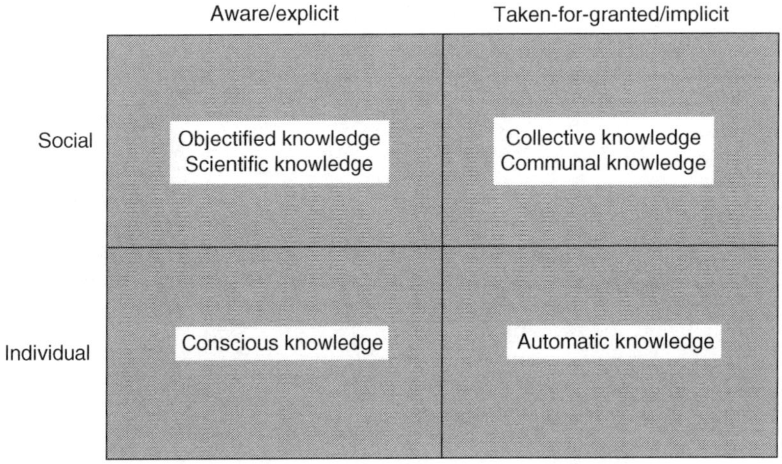

Figure 5.8 Individual–social K—Adapted from Spender (1993).

5.3.1 Knowledge Cycles

We may describe a K cycle as the flow of knowledge through an organization. There are several such cycles, but most have common elements, such as identify, build or generate, capture, distribute, integrate, and apply.

5.3.1.1 Mayer–Zack K Cycle

The Meyer–Zack cycle views a K cycle as an information factory where a repository is built containing modularized and indexed information elements, the contents of which may be assembled and packaged as specific products, depending on the requirements of its users. This is largely an explicit-oriented model, with K essentially treated as information. The top part of Figure 5.9 (light shade) is the model (See Dalkir 2011).

The bottom part of Figure 5.9 (dark shade) is a class exercise illustrating the workings of Meyer–Zack for an enterprise data office. In this exercise, the contents of the information model repository may be seen as containing both scripts and K-nuggets. The components may be subsequently assembled into a variety of data models at differing levels of abstraction, with corresponding software specifications.

5.3.1.2 Wiig K Cycle

Wiig K cycle. The Wiig (1993) K cycle includes both tacit and explicit threads of K, and, in that sense, is broader and more extensible than the Meyer–Zack cycle. The top part of Figure 5.10 is a representation with the corresponding key activities for each component of the cycle. For example, "Hold K" includes people (tacit K) and documents (explicit K). Also, we may extend the activities of "Build K" to include, for example, the SECI-related activities.

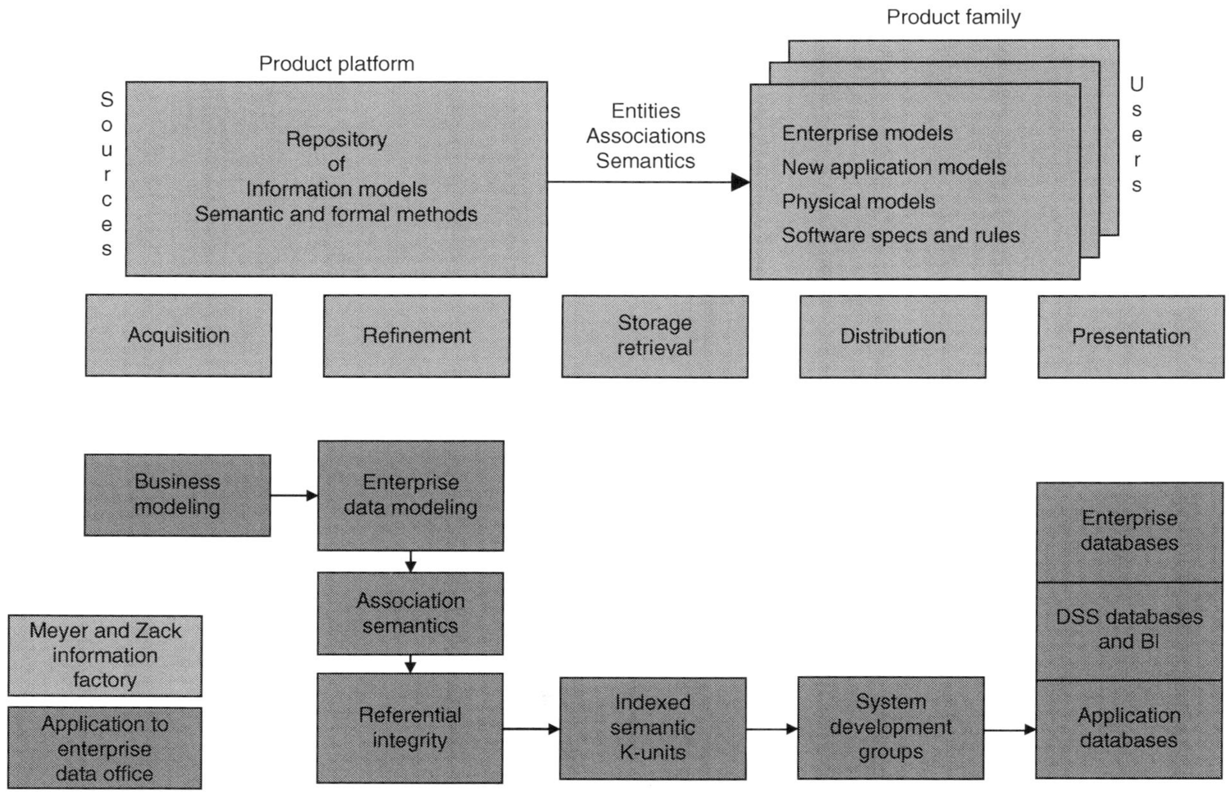

Figure 5.9 Information factory Adapted from Meyer and Zack—with sample application for the enterprise data office.

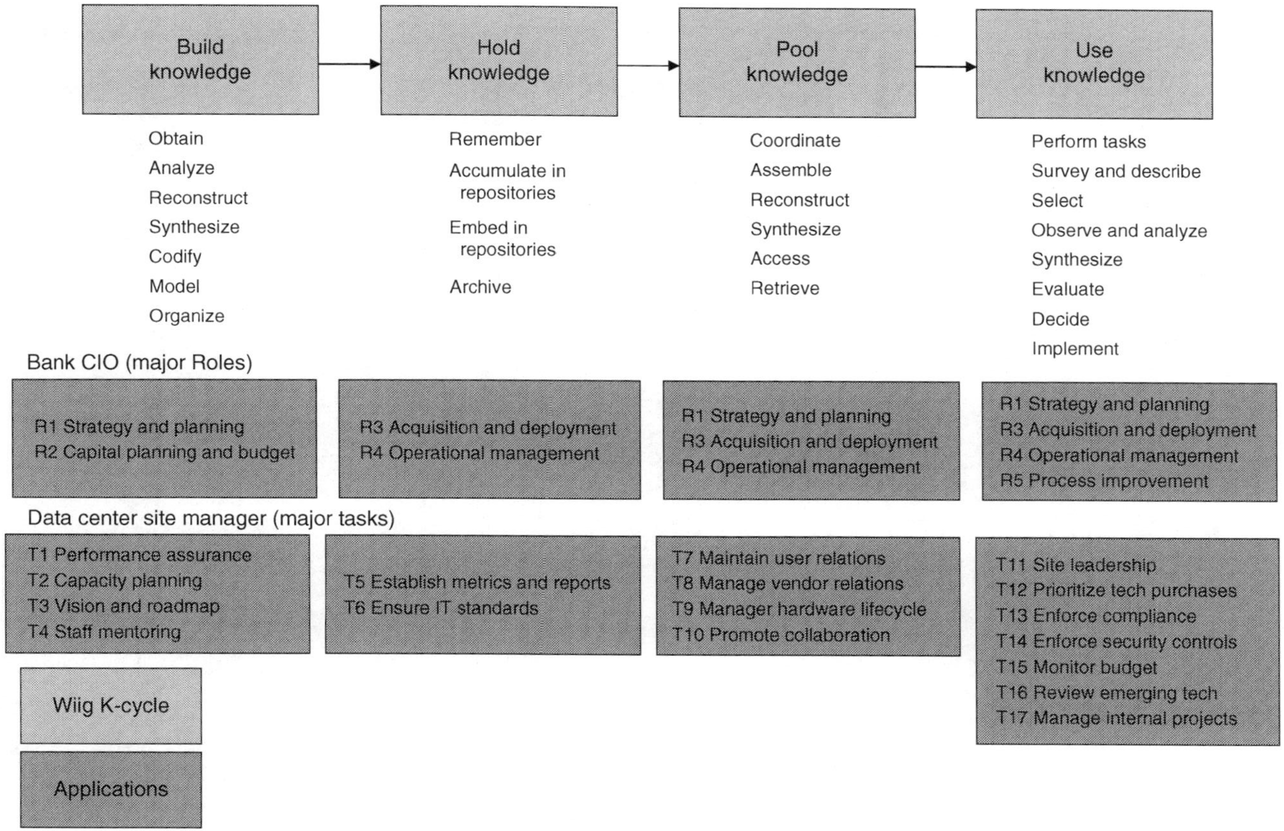

Figure 5.10 KM cycle with key activities—Adapted from Wiig (1993)—with sample application for a CIO and a data center manager.

The bottom part of Figure 5.10 is a summary showing the distribution of tasks in each of the four components of the cycle, taken from the class exercise described in Chapter 4. Recall that the student teams were asked to analyze a K job, decompose the job into corresponding tasks, build a K workstation, and categorize the tasks in the Wiig (1993) cycle. In this way, we can see how and where a given K job impacts an organization's K cycle. Note that certain roles may be repeated; for example, the "strategy & planning" role for a bank chief information officer (CIO) is located in three of the Wiig (1993) steps, but its constituent tasks are different: strategy & planning includes "defining corporate plans, policies, and standards," "establishing IT department goals and objectives," and "direct development and execution of an enterprise-wide disaster recovery and business continuity plan" in the Build, Pool, and Use stages, respectively. Observe too that the distribution of work (roles or tasks) tells us something about the overall nature of a given job. For example, the work of a data center site manager is primarily in K use and application.

5.3.1.3 Authors' Integrated K Cycle

Figure 5.11 is a representation of an integrated K cycle with key activities. We have identified three primary phases: building K, formalizing and refining K, and sharing and applying K. Embedded in the three phases is a feedback phase where value is realized and continually assessed. Also included is a social development phase where social K is created and accessed, including social K outside the boundaries of the organization (See Dalkir 2011).

Note that we have extended the common K cycles into both upstream and downstream activities—social networking is a new and critical step in a modern K cycle. As another example, another key step in our cycle is the determination of value, taken from a wide variety of strategic and tactical activities, including development of the organization's intellectual capital (IC) model, periodic audit and inventory activities, and ongoing socializing and experiencing activities.

5.3.2 KM Maturity Models

The Capability Maturity Model was established by the Software Engineering Institute in 1993. It consists of five organizational maturity levels for the process of software development: initial, repeatable, defined, managed, and optimized. The sequence of steps represents, in turn, increasing levels of formalization and may be used to identify the state of an organization's software development capability and to provide a roadmap for improvement.

Since its initial adoption for software development, maturity models have sprang up for several organizational processes. For example, a process maturity model (PMM) follows the same five-step maturity level and includes, for example, define processes, define key value chains and relationships, develop

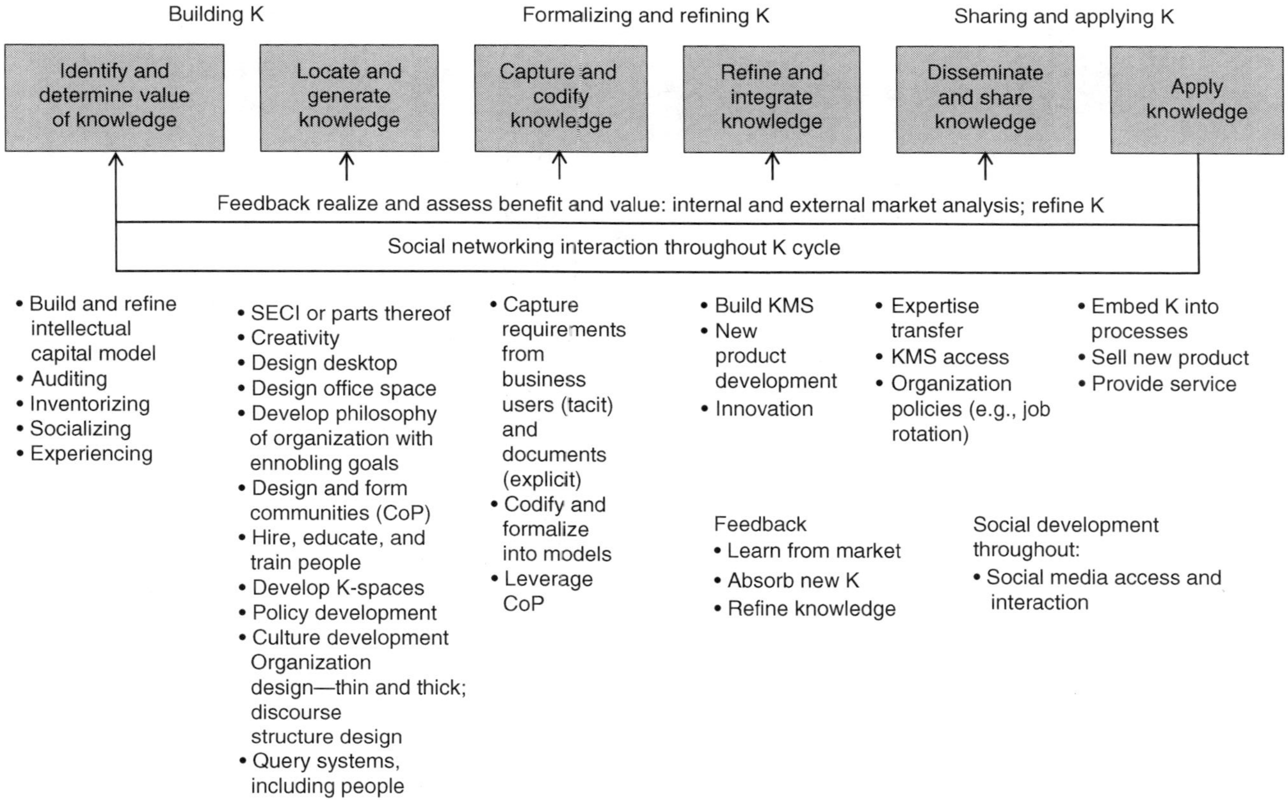

Figure 5.11 Authors' integrated K cycle with key activities.

management reporting systems, develop comprehensive monitoring and control systems, and create continuous teams (Curtis 2004). As another example, data warehouse (DW) development is a process and therefore lends itself to a DW PMM (Sen *et al.* 2006). Recall from Chapter 4 that successful implementations of BI systems were characterized by enterprise-wide BI interconnected with the processes of the organization, that is, embedding BI in the work of the organization. This is in contrast to departmental implementations (e.g., data marts), which are isolated islands and representative of low levels of DW maturity. That is to say, a high level of DW maturity suggests enterprise integration and business process interconnectivity.

As with other organizational processes, there are several KM maturity models. For example, KPMG defines four key areas of KM as people, process, content, and technology; each of these areas, in turn, is characterized by specific activities that may be assessed in a five-level model called "knowledge journey." The five-level model includes K chaotic, K aware, K focused, K managed, and K centric. As with KPMG, most KM maturity models assess KM activities and the level of K sharing in an organization and highlight particular areas such as technology, culture, people, K content, and so on (Khalili *et al.* 2012). This results in a very large number of models, each with a different emphasis. For example, Kuriakose *et al.* (2011) have identified 15 such KM maturity models containing four or five and as many as eight levels.

We suspect the variability in KM maturity models follows from the fact that KM includes virtually every aspect of an organization. A model focused exclusively on process (or any other area) does not provide sufficient coverage. Accordingly, we have shifted our focus from process refinement and maturity to organizational leadership and strategy. In particular, we prefer the intangible or IC models discussed in Chapter 1. The IC models are more illustrative of the many aspects of KM, the differing areas of emphasis, and the different K strategies that exist in organizations. The combination of IC models and K strategies are better indicators and planning tools of KM than is a maturity model.

5.3.3 KM Models

As with the representations in our discussion thus far, there are a large number of KM models. We have sampled the following based on their varying themes.

- Nonaka and Takeuchi's classic SECI model of *knowledge creation*. See below.
- Earl's (1998) model of *knowing and knowledge*. This model includes the characterizations of data, information, and knowledge shown in Figure 1.6. Also included is a representation of knowing and knowledge shown in Figure 5.12. Certain terms may need explanation: "Planned ignorance" implies the organization has chosen to developed little expertise in certain K domains; this is somewhat analogous to a student choosing to major in one subject and thereby ignoring the others. A periodic K audit is necessary to determine if

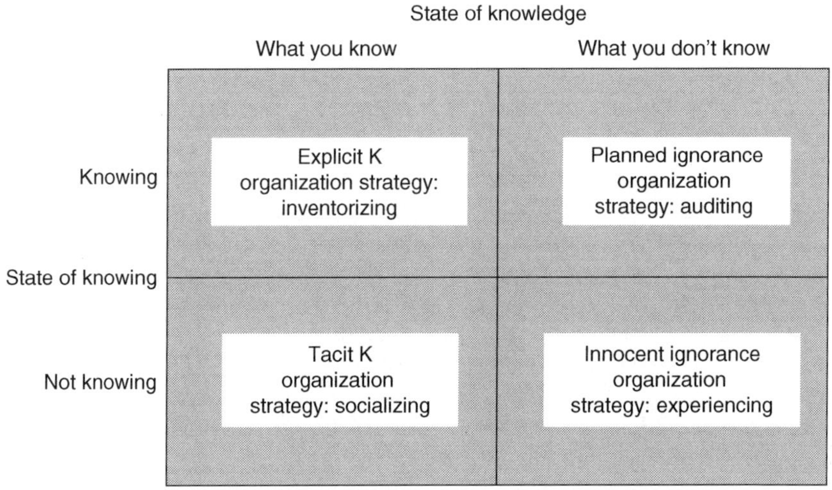

Figure 5.12 States of knowing and knowledge and organizational strategies—Adapted from Earl (1998).

the organization should cultivate expertise in other selected K domains. For example, the development of new program at a university implies that the school has consciously chosen to develop expertise (e.g., hiring new faculty) in new K areas. Similarly, "innocent ignorance" implies K domains that are new or unknown to the organization; here, experience (e.g., through an emerging technology department) will expose the organization to unknown K domains. We will continue this discussion on K strategy in Chapter 7.

- Snowden's (1998a, b) approach centers on *decision making* with a particular emphasis on the level of certainty with respect to objectives and the management of K assets. Figure 5.13 illustrates two aspects to Snowden's framework. The central theme is *KM ecology* and the management of both tacit and explicit K assets. See Dalkir (2011) for additional discussion.
- Boisot's I-Space model is a three-sided cube of K categories: codified versus uncodified; abstract versus concrete; and diffused versus undiffused. The underlying concept is that data may be understood and transformed into information through codification—that is, the creation of content categories (1998). Information, in turn, is shared, which subsequently is internalized into individual K. That is, knowledge itself is not shared, as is the preinternalized information component (2002). The importance of this model is that it brings us closer to the notion that *K is embodied in social action*; in the organization perspective, this highlights the importance of *CoP as organization-level social groups that generate K*. See Dalkir (2011).
- A variety of *IC* models. Recall from Chapter 1 that Andriessen reviewed 25 models and that we selected 4 illustrative models, each representing a

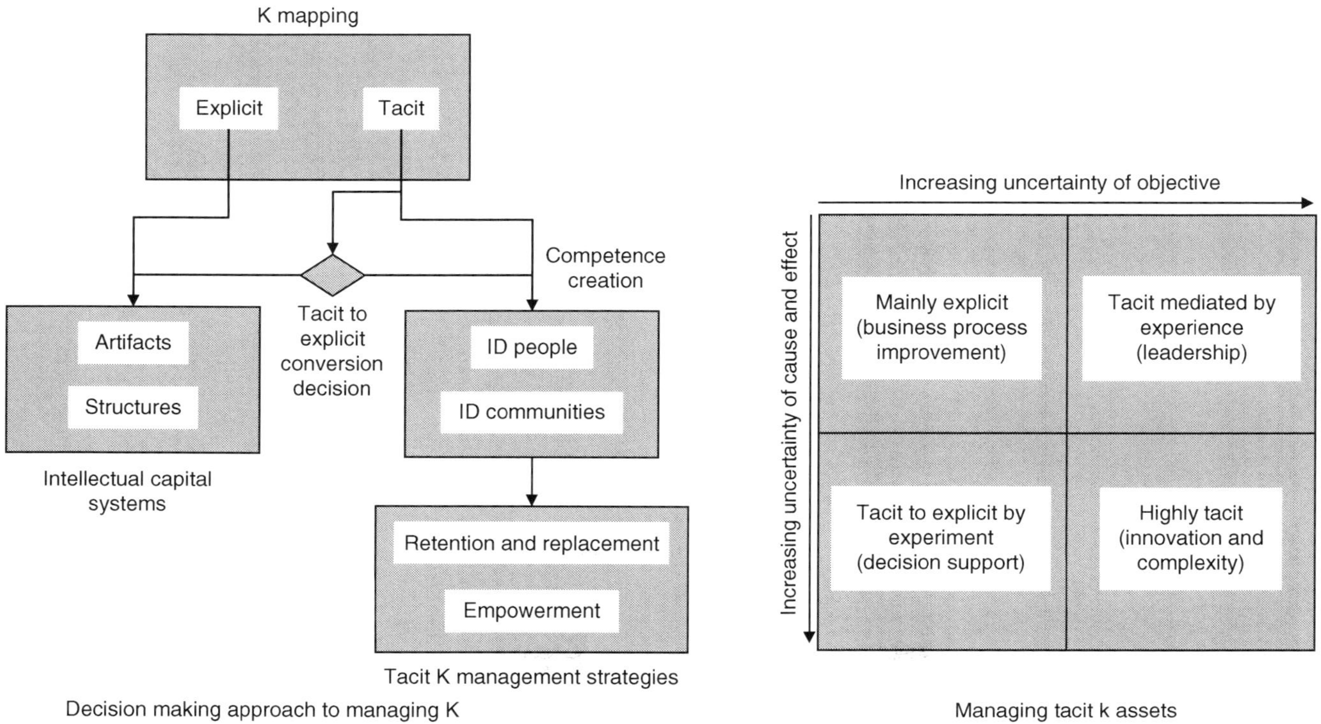

Figure 5.13 KM ecology—Adapted from Snowden (1998b).

specific emphasis on some form of knowledge capability. This included, for example, intangible asset and know-how capital models.

- A variety of *KM maturity* models, as discussed earlier.
- A variety of K representations such as Spender and Leonard discussed earlier as well as several variations to the SECI model.

This leads us to an important observation: there is extensive overlap between IC, K maturity, and KM models depending on the orientation of the model. Each of the models addresses some aspect of KM—categories, contexts, architecture, decisions, IC, and so on. We cannot truly understand the nature of organization K and related cognitive factors by relying on any single model.

5.3.3.1 SECI Model of Knowledge Creation

We have primarily focused on the SECI model of Nonaka and Takeuchi because the model is (1) considered a classic and is largely responsible for the development of the field and (2) actually addresses the way knowledge is created in an organization through the interaction of tacit and explicit K. Furthermore, (3) K creation has implications for design, strategy, leadership, and management in organizations. Table 5.3 and Figures 5.14 and 5.15 illustrate key aspects of the core Nonaka model as follows.

Table 5.3 Characteristics of SECI.

K-conversion mode	Metaphor	Commentary
Socialization	Learning by sharing	Tacit–tacit conversion. A process of sharing experiences in a common field. Triggered by fields of interaction— e.g., communities, teams, open office design.
Externalization	Learning through conceptualization	Tacit–explicit conversion. A process of articulating tacit K into explicit concepts. Triggered by dialogue and collective reflection—e.g., metaphors, testing hypotheses, talking, documenting.
Combination	Learning through networking (linking K)	Explicit–explicit conversion. A process of systemizing explicit concepts into K systems. Triggered through networking and linking K. Accomplished through analyzing, categorizing, and reconfiguring K threads.
Internalization	Learning by doing	Explicit–tacit conversion. A process of absorbing explicit K into tacit K. Triggered by doing—e.g., simulation tools; on-site application of K system.

Adapted from Nonaka and Takeuchi (1995).

- Figure 5.14 is the core SECI model. Note that the K spiral originates in socialization and progresses, in sequence, through externalization, combination, and internalization, where the sequence repeats itself. The spiral continues until sufficient knowledge has been created and embedded into (or as) a product.

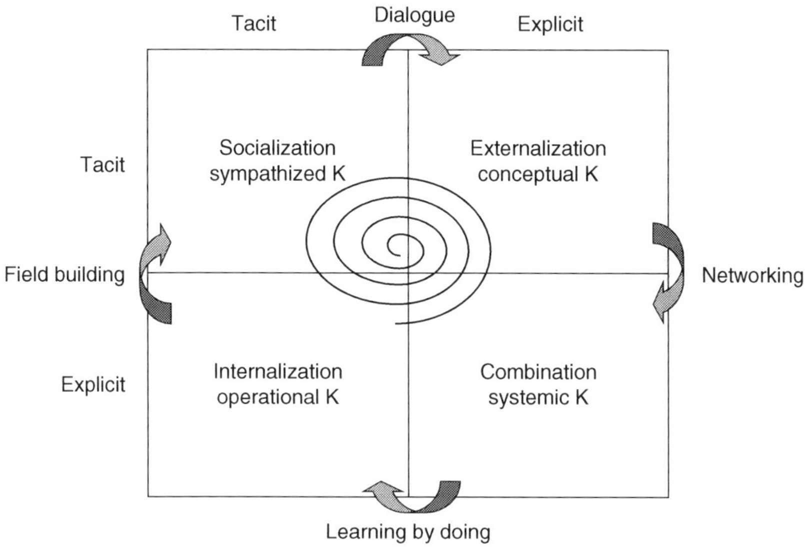

Figure 5.14 SECI model of KM—Adapted from Nonaka and Takeuchi (1995).

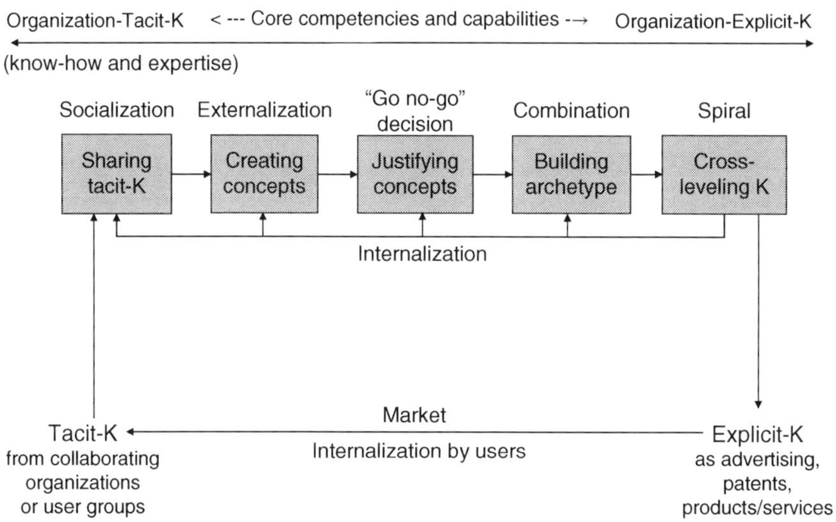

Figure 5.15 Application of SECI to workflow for new product development—Adapted from Nonaka and Takeuchi (1995).

- Table 5.3 lists the characteristics of each K-conversion mode in SECI. Note that we have included the triggers for each mode; for example, socialization is typically triggered through human interaction and conversations, either through discourse structures such as teams and communities or through open office design or other specialized structures that facilitate communications.
- Figure 5.15 is a process map or workflow for new product development. It is assumed that the process creates knowledge that is embedded in a new product, or is the product itself. Observe that "Justifying Concepts" is not part of the SECI model (in Figure 5.14) but corresponds to the "go no-go" decision-making process associated with new product development. This serves as a kind of "gate" that must be passed through each successive spiral. We have used this model with our students to both analyze and reengineer K-creating processes—that is, collections of reengineered tasks should correspond to this high-level map shown. We thus use Figure 5.15 as a *BPR reference model for designing K-creating processes.*

5.3.3.2 Criticism of the SECI Model and Nonaka

As important as the SECI model may be in KM, it is not without criticism. One such criticism is that it is simple and concerns itself with "only" tact–explicit interaction; in contrast, most K models are typically architectural or categorical. A second criticism is the strong focus on organizational K, perhaps at the expense of individual K. A third criticism is that SECI is more of a theory of decision making than one of knowledge creation (Crane 2013).

However, perhaps the most serious criticism is that Nonaka and his colleagues have misinterpreted the works of Polanyi. Polanyi believed that all K is personal and contains a residue of tacit knowing. Is it only tacit K that contains cognitive elements, or can explicit knowing also contain the same elements? The follow-up question, then, is whether full explication of tacit knowing into explicit K is possible? If tacit K is acquired (in part) through social practice, elements reside in that practice and resists conversion—recall the three dimensions of tacit K shown in Figure 4.8.

5.3.3.2.1 Revisiting the Knitting Metaphor

One suggested alternative to K-conversion is the interplay between focal and subsidiary awareness. Recall Polanyi's model of dynamic cognitive experience shown in Figure 1.13. In that model *awareness* is a dimension that roughly corresponds to inputs and has two aspects. The first is focal awareness that is toward the explicit pole and the second is subsidiary awareness toward the tacit pole. We have suggested a knitting metaphor to characterize the assembly of explicit-oriented focal threads into a tacit-oriented subsidiary fabric. But knitting works in both directions: we can *unknot* the fabric into constituent threads. Naturally, the same can be said of the output dimension of *activity*, which contains a conceptual aspect (explicit) and a bodily (tacit) aspect. We thus argue that this interplay between focal and subsidiary awareness, in addition to the interplay between conceptual and bodily activity, need not be considered an alternative to K-conversion, but complementary to it. *Awareness and activity are dimensions of personal K creation; in contrast, K-conversion (SECI) is a dimension of organizational K creation—each operates at a different level of analysis, but they are interrelated through the dynamics of K creation.*

Restating the Issue. Can we restate the issue as one of degree? Can a tacit thread be fully converted into a corresponding explicit thread? Is it the essence of the tacit thread that is converted, or is it something less? Is this question a roadblock to the practice of organizational design or second divide thinking? Also, Nonaka and Takeuchi argue that only tacit K contains cognitive elements and thus is concerned with the future, whereas explicit K concerns the past. Of course, we would argue that tacit K might refer to the past as well as the future; a more intriguing question, however, is whether explicit K is confined exclusively to the past.

5.3.3.2.2 Nonaka Answers His Critics

In 2009 Nonaka (with von Krogh) answered his critics. Among the more salient points are the following:

- Tacit K and explicit K are on a continuum, thus creating a nuanced interpretation of tacit–explicit interaction. Recall that we have included this continuum (as well as an individual–social continuum) in our conception of the information continuum. Thus, knowledge contains varying degrees of tacit and explicit elements and can momentarily take on different forms. K-conversion is representative of tacit and explicit K interaction and takes place along the continuum. When viewed from an architectural perspective, the continuum is important because it allows for the examination of knowledge that serves specific purposes in an organization.
- Justified true belief does not correspond to traditional organization theory but is representative of a pragmatist epistemology that includes justified belief within both an individual and a social context.
- K-conversion is supported through empirical studies that bridge theory and practice. We can state that explicit knowledge is a *representation* of the underlying tacit K on which it is based. As K is transformed into an explicit form, it is enriched. We have made a similar argument earlier in this book: explicitness is a necessary precondition for theory building, refinement, and the construction of complex systems. However, expert K can never be fully captured in systems or software due to the absence of corresponding tacit elements; yet, expert K is the basis for increasingly explicit K on which robots and automated systems are based.

- Similarly, tacit K is acquired through action, practice, and reflection, that is, the "movement" from explicit to tacit K. Hence, tacit K contains elements of explicit K along with situated elements unique to action and practice.
- Within social theory and organizational science, tacit K has come to serve two purposes: as a foundation of social practice and as a foundation for innovation.
- The relationship between organizational K creation and social practices in organizations is not a trivial issue. Organizational K creation requires a space (i.e., ba) that influences the outcome of the K creation process. People who participate in K creation activities often transcend their respective social practices. Fragility in K creation arises from the different biographies, social practices, and tacit K-bases that organizational members bring with them.

The Authors Revisit SECI. Tact–explicit interaction provides specific design guidelines in designing K-creating processes and K creation in general. Organizational design is as much art (practical) as it is science (theoretical). We draw on our model in Figure 5.3 and give greater weight to the practical over the theoretical. In our case, organizational design—strategy, structure, people, process, systems, and so on—are collectively the mitigating factors we must accommodate in our social continuum. We do not discard or dismiss the theoretical but assert that *design is essentially a nuanced expression of experience and intuition, backed by theoretical knowledge.* We believe Figure 5.3 accommodates the many different perspectives of K and is warranted by the necessity of succeeding in the second wave.

5.4 Application of Concepts: Case Study on the PC

The PC has over 7800 volunteers serving in 70 countries. These volunteers bring a broad array of skills and experiences to their posts. During their assignments they acquire valuable additional knowledge on the job—knowledge that is often lost when a volunteer retires after the customary 2 years of service. While the PC is aware of the value of this knowledge and attempts to create conditions in which it can be shared, an extensive case study that involved one of our graduate students and the authors (Gilbert *et al.* 2010) leads to the conclusion that there currently is no formal strategy to manage the flow of information and knowledge between different members of the organization. This is unfortunate since the ability to effectively capture and share information and knowledge could significantly ease the jobs of volunteers, improve their productivity, and promote morale. This case study develops a framework for KM in a volunteer organization such as the PC where high personnel turnover hampers traditional approaches to KM.

There are specific KM-related issues with respect to the PC such as the following:

- High turnover makes the development of organizational memory problematic.
- A key barrier to K transfer is what we have called *organizational opportunity*. Both short duration (2 years) and geographic isolation prevent common techniques of K capture from being employed to their full extent.
- Structured techniques of K capture surfaces both procedural (know-how) and declarative (know-what) knowledge. Neither the opportunity for analysis nor the experiences of a PC volunteer lend themselves to this type of structured analysis and rule-based systems approach.
- The "ba" or K-spaces cannot be designed but are "given" in the field where all the activities take place. Further, all such spaces are social and include primarily practice and cultural spaces.

Our approach is to exploit two types of rich K capture—stories and observational learning, both of which have a large tacit component—and to couple them with rich technical media and social networks. The PC volunteer—highly motivated and committed—operationalizes the K captured through stories and demonstration in the form of photos, voice, and video through a Web portal and a social network.

The framework focuses on how knowledge can be applied to support decisions in the daily activities of the volunteers in the field. Based on this framework, we articulate a set of computer and non-computer-based mechanisms in terms of procedures, policies, and IT tools to capture, share, and transform data and information into usable knowledge. The case study concludes with a description of a hypothetical KM system based on these prescriptions that could be used in the PC or other similar, largely volunteer, organizations with a rapid turnover of personnel. (It should be noted that the work described is not part of an official PC project.)

Illustrated in Figures 5.16, 5.17, 5.18, and 5.19 is our analysis of the PC operating in the Dominican Republic (DR). Here we briefly describe our analysis:

- Figure 5.16 is a process map or workflow. The top flow is from the perspective of a volunteer and the lower flow is from the DR PC perspective. There are several KM-related ideas shown. The first is what we have called knowledge points (KP), which represent opportunities for K creation or utilization (i.e., potential K tasks). The second is to realize that this is a high-level, thin process map; virtually all of the volunteer's work (2 years' worth) takes place within only two activities: *Reside with Local Host* (KP5) and *Execute Projects* (KP6).
- Figure 5.17 illustrates the proposed KM system. Note that the system components follow from the task, KP, and space (ba) analysis. Also, the KM system is divided into hard and soft components, representing lean (traditional)

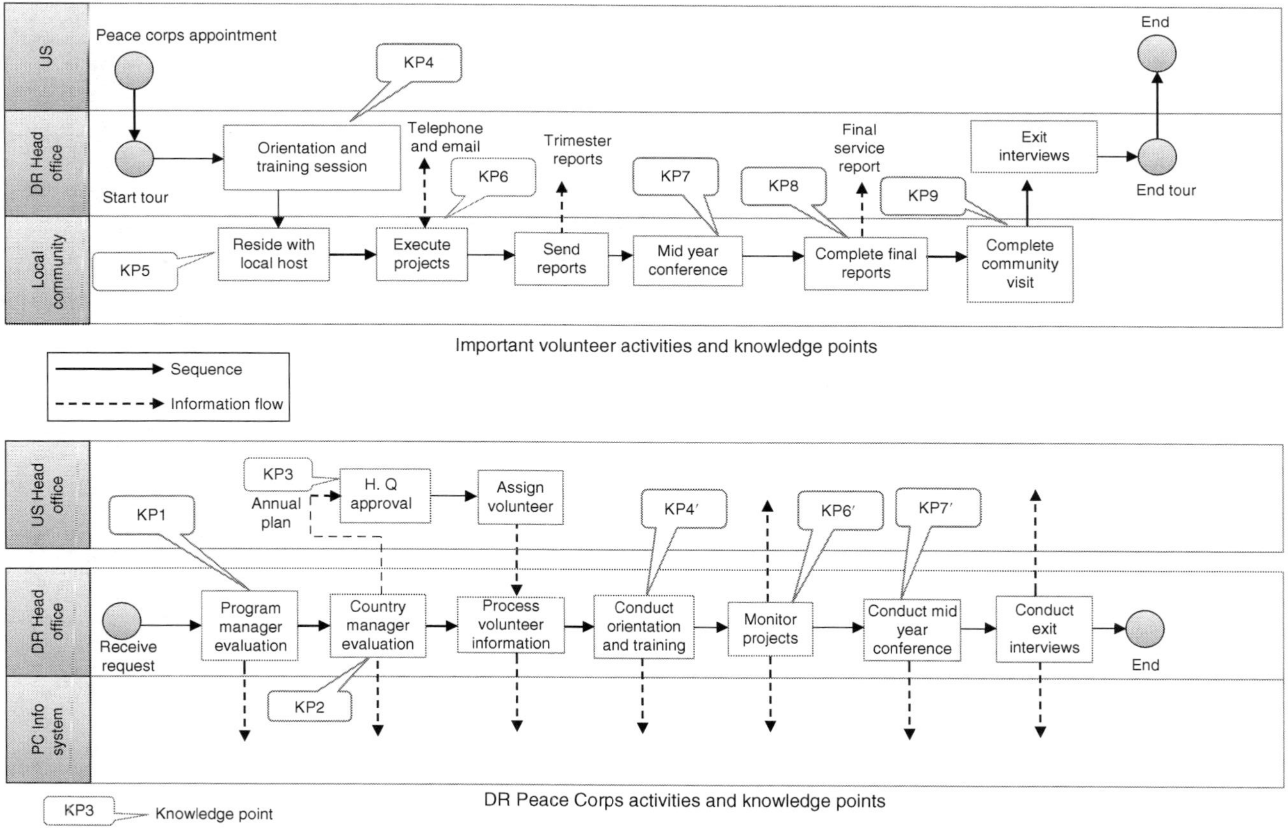

Figure 5.16 Analysis of workflows including knowledge points—Adapted from Gilbert *et al.* (2010).

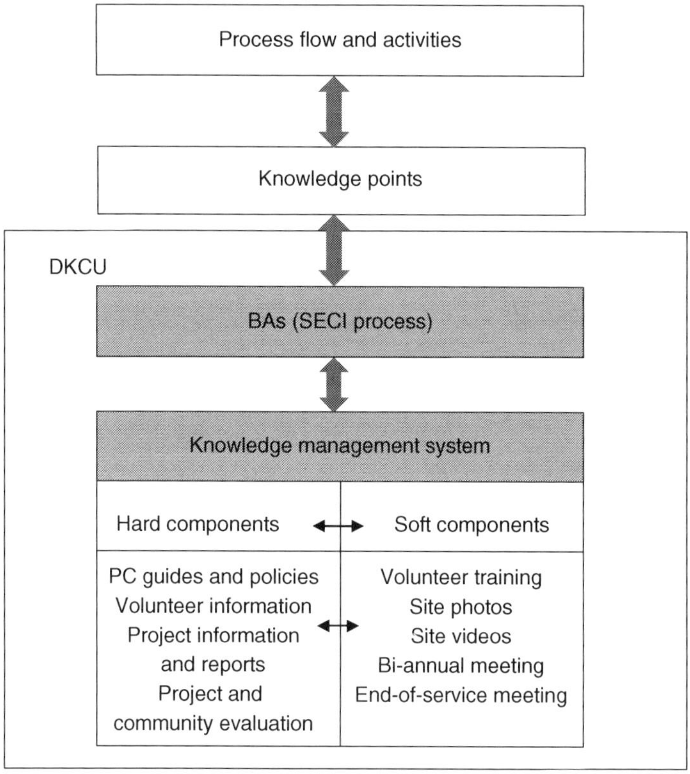

Figure 5.17 Proposed KM systems for the Peace Corps.

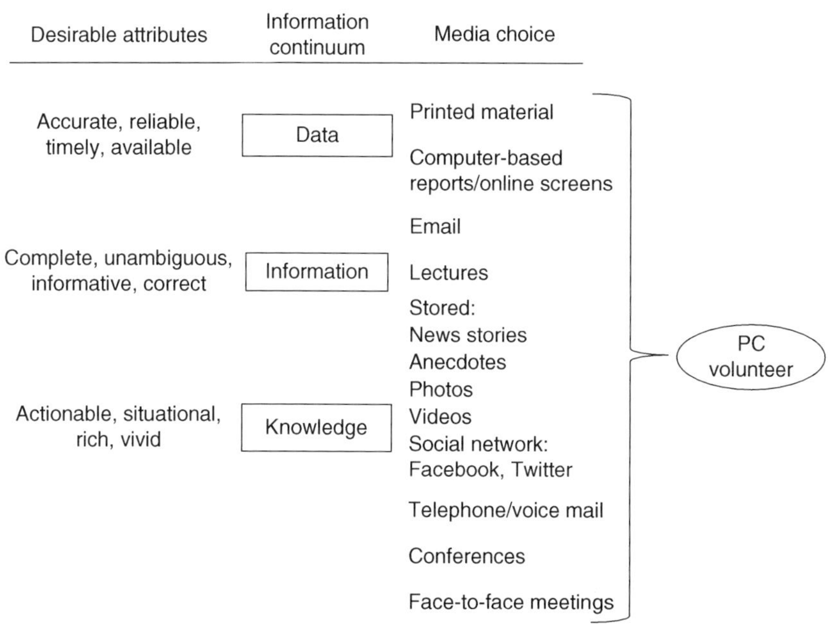

Figure 5.18 Information continuum—communications media linkages in the Peace Corps.

Knowledge point	Analysis
KP1 Site investigation	Externalization: Explicit and tacit; local and technical Interacting Ba: Local community discussions and interviews Hard DKCU: Improved online access to all categories of explicit knowledge
KP2 Consolidation and recommendation	Combination: Explicit and tacit; local, administrative, technical, Political Cyber Ba: Explicit knowledge integration Hard DKCU: Improved online access to all categories of explicit knowledge
KP3 Volunteer assignment	Combination:Explicit administrative, technical, political Cyber Ba: Explicit knowledge integration Explicit: Headquarter system
KP4 Orientation training	Socialization: Elements of internalization explicit admin, technical, political. Tacit local Originating Ba: Elements of exercising Ba: community of volunteers Soft DKCU: Video
KP5 Immersion	Socialization: Tacit local, cultural, political Originating Ba: "Indwelling" with local community Social network: Anecdotes, video, photos

Knowledge point	Analysis
KP6 Project execution	Internalization: Explicit technical, tacit technical, local, cultural, political Exercising Ba: Local community Hard and soft DKCU: Social network: anecdotes, video, photos online access to past and present technical information
KP7 Annual "all volunteer conference"	Socialization: Tacit local, cultural, political, technical elements of externalization explicit admin, technical Originating Ba: Interacting Ba: "All volunteer Community" Soft DKCU: Video elements of hard administration information
KP7* Development area volunteer conference	Socialization: Tacit knowledge in specific expertise area; elements of externalization explicit knowledge in specific expertise area Originating Ba: Exercising Ba: development area volunteers Soft DKCU: Video elements of hard technical information
KP8 Project summary report	Combination: Explicit local cultural, political, technical Cyber Ba: Explicit knowledge integration Hard DKCU: Online access to past reports
KP9 Exit interview	Externalization: Tacit and explicit local, cultural, political, technical Interacting Ba: Dialogue with program manger and country director Soft DKCU: Voice recording, video
KP10* Social network	Socialization: Elements of other SECI components explicit and tacit knowledge in all topic areas Originating Ba: Community of volunteers Soft DKCU: (Private WIKI), e-mail groups

* Added knowledge point

Figure 5.19 Knowledge point analysis in the Peace Corps.

and rich systems media, respectively. Note too that we have used the notion of a dynamic system for knowledge creation and utilization (DKCU) described by Nonaka and Reinmoeller (2000). A DKCU generates and uses creative routines that are embedded in the creation and sharing of individual tacit knowledge. A creative routine, in turn, is a dynamic action pattern for innovation in a particular context, such as a specific space or ba. Further, we have superimposed the idea of hard and soft system components described in previous writings (Morabito *et al.* 1999).

- Figure 5.18 shows the information continuum—specifically, data, information, knowledge components, characteristics, and media (lean and rich) selected for the new KM system.
- Figure 5.19 shows the ba, type of knowledge, and K-conversion analysis for each KP. Note that the key KP of KP5 and KP6 have large measures of tacit K. For example, KP5 is characterized by socialization and indwelling (originating ba) and is best implemented through social media. KP6, in turn, is characterized by internalization, exercising ba, and both hard and soft system components.

5.5 Knowledge Formation

The premise of the SECI model is that knowledge is created when it moves and is transformed. This is known as knowledge conversion. The K spiral emerges as K-conversion is amplified through successive modes of socialization, externalization, combination, and internalization.

As mentioned earlier, one suggested alternative to tacit–explicit K-conversion is the interplay between focal and subsidiary awareness (along the awareness dimension in the Polanyi model). We may make the same claim on the interplay between bodily and conceptual activity (similarly along the activity dimension). Also, we remind the reader that cognitivity is a high-level activity that arises from the interplay of awareness and activity. See Figure 1.13 for an illustration of the Polanyi model and its constituent dimensions. Note too that cognition leads to knowledge and includes perception, exploration, imagination, reasoning, criticism, and testing processes (Bunge).

The idea that knowledge may be created—or more generally *formed*—through interaction along the information continuum has been proposed by the authors (Morabito *et al.* 1999). Figures 1.2, 1.3, 1.4, 1.5, 1.6, 1.7, and 1.8 are illustrations of the information continuum from a variety of perspectives. Observe that the knowledge continuum represented by both figures is a subset of the information continuum and is an illustration of the K continuum described by von Krogh and Nonaka above. Figure 5.20a (high level) and b (details) illustrate K formation along the information continuum and include several ideas discussed previously among which are elements of the Polanyi model, SECI, K-nuggets, and scripts.

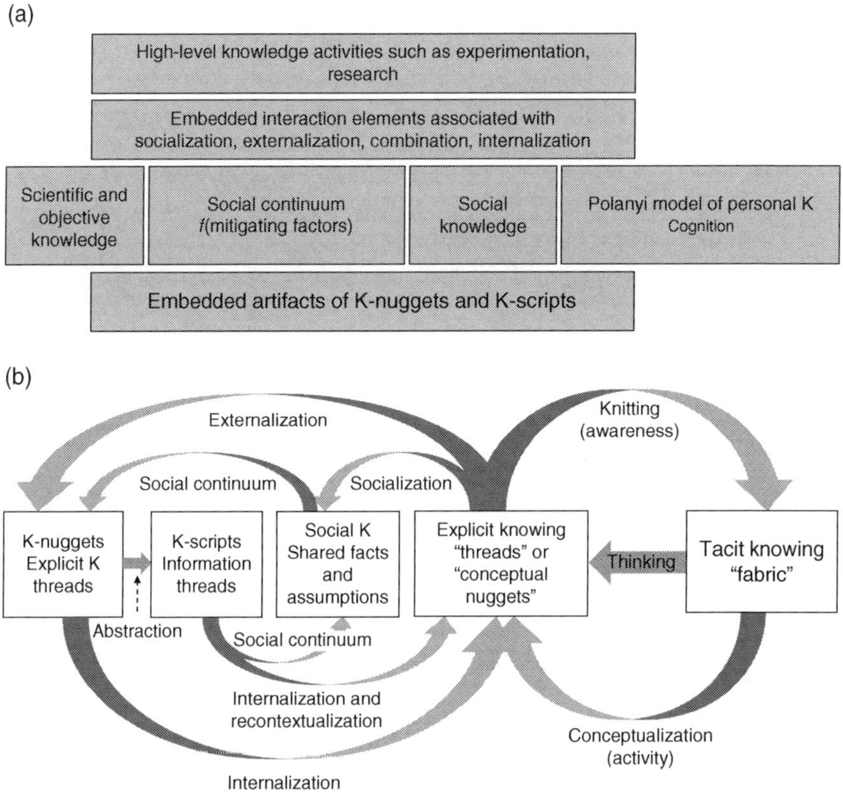

Figure 5.20 (a) Knowledge Continuum and Knowledge Formation. (b) Knowledge Continuum and Knowledge Formation—details.

From these figures we may make the following observations:

- Is K formulated or formed; that is, is new K intended or does it simply arise? This is a similar question that Mintzberg asks in his analysis of strategy making. As with strategy, the answer is either or both. New K may be part of a strategy that includes the development of specific K and expertise, and hence, is intended. On the other hand, K may simply form as part of everyday conversations. In our discussion we are using "K formation" to include both formulated and formed K.
- K formation on the right side of the diagram is internal to the individual and represents cycling through the Polanyi representation. As we move to the left knowledge expands and becomes more social and practical, constrained by the impulses of objective and scientific knowledge from the far left of the diagram.

- K formation is in constant flux, up and down the information continuum.
- There is a triple effect to the Polanyi cycle: knitting, conceptualizing, and thinking. Knitting takes place along the awareness dimension and concerns the construction of a tacit fabric of deep understanding. K is broadened in the fabric but less refined. Conceptualizing takes place along the activity dimension and addresses the creation of specific conceptual threads. With the creation of conceptual threads K becomes more narrowly focused and refined—all necessary to the development of complex theories. There is thus a continual interplay between tacit knowing and explicit knowing embodied in the cycle, which, in turn, is the basis of cognitivity or thinking.
- As Nonaka stated in his 2009 rebuttal earlier, tacit and explicit K are on a continuum, embedded in our representation of the information continuum. K formation (K creation for Nonaka) may take place anywhere along the continuum. Thus, K formation is situated in a particular organizational context.
- Socialization is the complex activity in which social tacit assumptions are shared, but in which cognitive elements—mental models and sensorimotor skills—are not directly transferred. Instead, the cognitive elements are imbibed or absorbed through observation and imitation, joint experimentation, and probably many other unknown processes.
- Social K is K that a group shares and believes to be K. This includes assumptions, beliefs, facts, and the process of *social construction of reality*. Social K is constructed through socialization from individuals and thought leaders, by explicit K threads and information mitigated by social factors in the workings of the social continuum, and by the internal energy and momentum of the social group itself.
- The social–practice continuum connects the reality of scientific and objective truths with social relevance. The location on this continuum is a function of the mitigating factors (see Chapter 4) such as the ennobling goals of the organization.
- Social construction embodies the social continuum and the directionality of K-binding. In particular, the forces of social construction are driven by both experts or others in power (early K-binding) and the individuals of the social group (late K-binding). The interplay between principled K and the demands of the social group are enfolded in the process of social construction.

5.6 Knowledge Exchange and Transfer

K exchange and K transfer are high-level terms that refer to the transmission or flow—sending and receiving—of knowledge. The types of K may be explicit or tacit and their transfer modes include computers, networks, databases, documents, models, mappings, communities and groups, and people and their relationships. Also, Davenport and Prusak (1998) introduce a utility perspective and state that K transfer implies the transmission, absorption, and eventual use of K.

Of particular interest in KM literature is the transmission and codification of tacit K. *Knowledge capture* refers to the acquisition of tacit K and its subsequent codification into an explicit system. In contrast, *knowledge sharing* refers to the transfer of tacit K, though here we may additionally share explicit K. As we can imagine, there is overlap between capture and sharing. For example, is this text a vehicle for capturing or sharing K? From the authors' perspective, we are sharing K with the reader; however, the astute reader may also be capturing our intent and with it a certain level of tacit understanding.

5.6.1 Knowledge Capture

K capture implies the existence of K in a form not readily accessible or usable. The outcome of the K capture process should be an explicit or codified framework that enhances understanding and is useful in the development of a system. Codification may also refer to the resulting system itself. See Dalkir (2011).

Tacit K Capture. Closely associated with K capture is the notion of *K acquisition*, the latter term common in artificial intelligence (AI) and coined by expert systems developers. In this perspective, K capture or acquisition implies the capture of tacit K from an expert, its codification in the form of rules and facts, and the subsequent development of a system. K capture techniques (i.e., implying tacit K capture) include structured interviews (scripted), questionnaires, surveys, observation, and simulation (Parsaye 1988).

Other forms of tacit K capture include interviewing experts (i.e., structured interviews and storytelling), learning by being told, learning by observation, brainstorming, protocol and scenario analysis, blackboard approaches, consensus decision making (see expert consensus in Chapter 2), surveying experts (i.e., Delphi), learning history, and so on.

K-Oriented Use Cases. Learning history is particularly instructive since it typically includes the development of *best practices* and *lessons learned.* Both best practices and lessons learned represent the capture of tacit know-how, which is codified into a description, codified models, graphics, and business rules. Also, in our teaching experience, several organizations are building what they are referring to as "K use cases." In effect, these are K-oriented guidelines of a best practice in use case format. Common K use cases are shown in Figure 5.21.

Tacit K Capture from External Organizations. There are several techniques for acquiring tacit K in external organizations. One technique is to purchase it. The most direct and perhaps most expensive way is through a corporate merger or acquisition. In this case, the parent organization is capturing the entire target organization, including its technology, people, and culture. For example,

• K sharing	• Connecting organizational subunits and people
• Internal and external outsourcing	• K Capture and refinement
• Empowering employees	• Identifying and linking experts
• Retaining experts	• Enabling problem solving
• Enabling collaboration	• Enabling conversations and dialogue
	• Building trust

Figure 5.21 Organizational best practices and K-oriented use cases—student input from organizations.

IBM paid 14 times the Lotus' book value during its acquisition of Lotus Development in 1995 (Davenport and Prusak 1998), which is the markup representing the appraised IC value of Lotus. That is, the organization-level tacit K (e.g., people, culture) was considered more valuable than the tangible product itself (see IC in Chapter 1). Additionally, Spender's model of individual–social K (Figure 5.8) suggests that communal K represents the essence of a group's distinctiveness and embodies the organization's culture. A question that naturally arises is whether it is possible to indeed capture another organization's culture through a merger, which, after all, is an image of the target organization's history. Would not, in time, the acquiring organization's culture impose itself on the target organization?

A related but much simpler way to purchase K is to raid competitor organizations and attract their key K workers. This is a very common practice on Wall Street, where 20–30% turnover rates are not uncommon. Unlike a merger, this technique attracts specific people and captures specific K. Yet another form of external K capture is to use and learn through external benchmarking of a competitor—this is neither purchasing nor capturing K in whole but an indicator to be emulated.

Tacit K Capture Internal to the Organization. Internal organization-level tacit K capture may be through single- or double-loop learning (Argyris and Schon 1978). The former is sometimes called experiencing; that is, experience with established routines should codify a set of rules and procedures. We were exposed to this approach while teaching at a large package delivery organization. Improving an operational process from data and K gathered *within* that process is an example of single-loop learning, where the outcome might lead to reduced cycle time and reduced defects. If, on the other hand, we capture data and K *external* to the process, such as competitor and customer surveys, the outcome might include new assumptions about customer needs, significant change in product features, or entirely new models to replace existing ones. That is to say, double-loop learning involves embedding new K into a process.

5.6.2 Knowledge Codification—Structuring Explicit K

There are several popular schemas that have been used to represent K in an explicit codified form. Among these are taxonomies, models, concept maps, and decision trees and tables. We may also add K maps (i.e., a people Yellow Pages) that do not contain K but point to people who possess it. Recall from Chapter 4 that both *taxonomies (hierarchies of elements)* and *ontologies (taxonomies with embedded rules and other attributes)* are common K representations in the development of cognitive systems.

Parsaye *et al.* (1989) have described the *structuring of K* in the development of expert systems and intelligent databases. We may use this as a guide in developing codification strategies. The structuring of K includes the following elementary components:

- Naming K objects—includes the initial identification of K objects
- Describing—a functional description, including attributes and actual values in the case of formal notation
- Organizing—includes categorization and possession (i.e., inheritance)
- Relating—associations between K objects
- Constraining—rules for relating, including predicates for each attribute

Observe that these elementary K components represent the artifacts associated with the development of systems of any kind, scientific taxonomies, or decision-making processes.

Generalized K Capture and Codification. Common approaches to K codification assume the existence of a K-nugget (which may be very complex) and its eventual representation in the form of a concept hierarchy with dependencies (i.e., K taxonomy), concept map, decision tree, or table. We have extended these common artifacts associated with K codification to additionally include the K structuring associated with intelligent systems described by Parsaye (1988) earlier. This addition introduces elements of ontologies—rules, relationships, and constraints. This is not very different than, say, capturing system requirements and building an information model.

Figure 5.22 illustrates the interconnected activities of K capture, acquisition, codification, and system building. We have generalized to provide for the capture of explicit or tacit K and the subsequent development of a system, whether or not an expert system. Note that the sources include human experts, documents, and databases. A certain level of domain expertise and analysis is necessary for K capture, as is subsequent abstraction, summarization, and model building. An example of explicit K capture (as opposed to tacit K capture) concerns the understanding of new laws, such as the Affordable Healthcare Act (i.e., a US healthcare law), which, given its complexity and importance for both organizations and their employees, requires explicit K capture by domain experts and its codification into comprehensible forms.

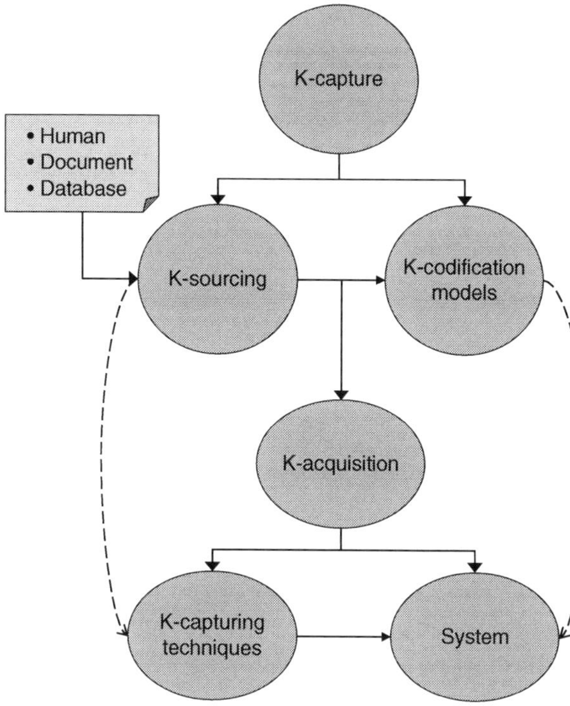

Figure 5.22 Knowledge capture, acquisition, and system codification.

In many ways, this is not very different than traditional analysis where a systems analyst interviews experts and analyzes existing documentation and system artifacts, and develops requirements—a codified data model is one such explicit outcome.

As another example, Figure 5.23 is a codified model that we developed as part of our analysis of adverse events in hospitals. Note that an adverse advent may be simultaneously categorized along two hierarchies (i.e., multiple inheritances) and may arise from any combination of human, system, or organizational errors.

Explicit K Transfer. The transfer of explicit K implies the existence of codified K—perhaps in the form of a system—sourced from tacit K or explicit documentation. Most of the systems discussed in Chapter 4—for example, K repositories, K mapping—are vehicles for explicit K transfer, as are documents, processes, and procedures. Recall too that such systems will contain K threads in the form of nuggets and scripts, making reuse possible—this is one advantage (i.e., efficiency) of explicit rather than tacit K transfer. Finally, note that explicit K transfer implies its capture (from some source), codification, storage, and dissemination or distribution.

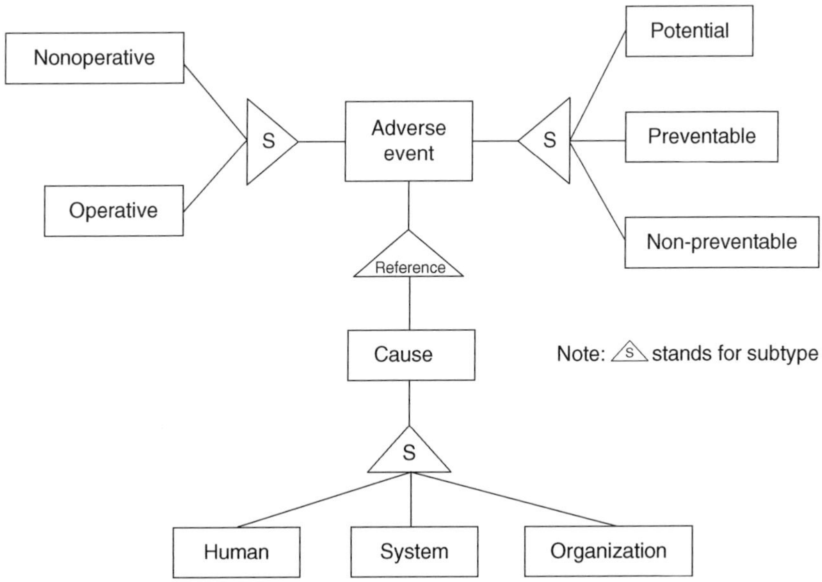

Figure 5.23 Codified information model of an adverse event—Based on Wong *et al.* (2007) and Bohmer *et al.* (2009).

5.6.3 Knowledge Sharing

5.6.3.1 Communities of Practice

Several years ago one of the authors had a conversation with social learning researcher Etienne Wenger concerning CoP, from which we learned the following: The essence of a CoP is the *domain* or a shared area of interest, a *community* or group of members who build relationships based on shared interest in that domain, and a *practice* that exchanges or builds K. The latter includes problem solving, sharing insights and experiences, and building tools and databases. Though a *CoP shares primarily existing K*, the nature of social interaction means that new K may also be formed.

A CoP provides the social heat (recall the coffee metaphor) necessary for K formation and application (i.e., the practice). CoP may exist in organizations, governments, or schools—but is most prominent in organizations. An example of a CoP is illustrated in Figure 5.24. In this illustration, Daimler Chrysler is shown as a product organization with five major divisions, including large cars, small cars, trucks, minivans, and jeeps. Recall from Chapter 3 that in a product organization, functional areas (e.g., R&D, marketing) are distributed among the product divisions, diluting corresponding K. CoP provide an opportunity to generate social interaction and reconstitute expertise across product divisions. For example, Figure 5.24 shows five CoP for key K domains in automobile manufacturing. Observe that, contrary to popular belief, not all CoP are self-organizing but may, in fact, be designed.

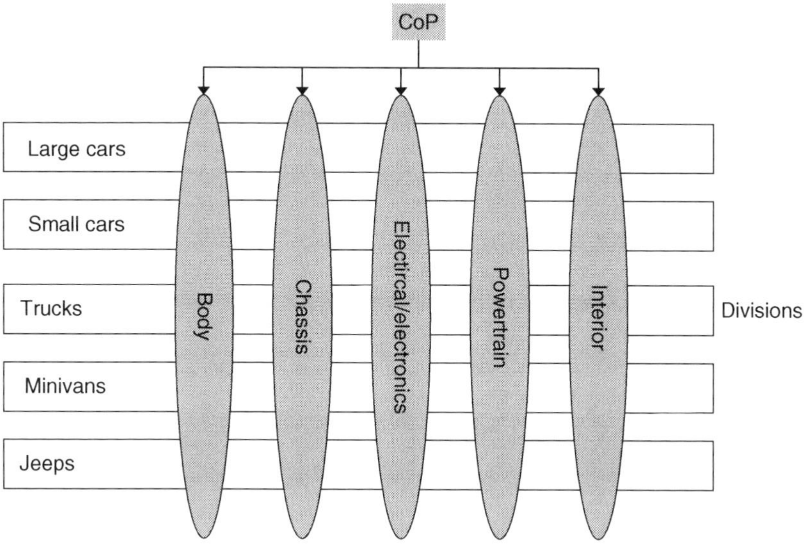

Figure 5.24 Five CoP in a product-organized automobile manufacturer—Adapted from Wenger and Snyder (2000).

There are always organizational constraints in any community. Davenport (2005), for example, points out that extending the Chrysler K communities ("tech clubs") to European divisions of Daimler did not meet with success—issues of geography and culture prevented effective K sharing. Also, while the CoP promoted K sharing in general, *detailed K sharing* among specialists was less than satisfactory and was seen as an obstacle to quality. Thus, we must consider a CoP or similar K communities as one part of an *organizational landscape of communities*—CoP, teams, task forces, lateral processes, self-organizing groups, and so on—each with its own strengths and weaknesses. Certain structures (e.g., teams) embed K sharing in the work activities of the group; in contrast, a CoP exists primarily to build relationships, share K, and construct tools (e.g., database) that facilitate sharing.

5.6.3.2 Organizational Modes of K Sharing

In addition to a CoP, organizations may engage in other activities that promote K sharing. Most of these modes promote the absorption of K and include the following:

- *Shadowing* where a new employee is partnered with an experienced person and thus absorbs know-how through participation in activities with the more experienced person.
- *Job rotation* where workers acquire departmental or functional K and know-how by working in an unfamiliar area. This is a very common training approach where new employees are assigned to a department for a period of

time (usually 6 months) and then rotated to another department, and so on—typically a total of four departments covering 2 years, though other variations are common too.
- *Offsites* or *retreats* where workers engage in socialization and organizational strategizing. This has the dual advantage of encouraging employees to surface their ideas on the future possibilities of the organization and of building relationships in a relaxed environment.
- *Rich components* in a DKCU or other system. See the PC case study discussed earlier. Rich technical elements (e.g., video) provide an opportunity to elicit deeper understanding that lean modes.
- *Open forums* such as company outings with no specific K or business agenda. Company outings create an opportunity to build comradely and communal bonds.
- *Open office design* that promotes human interaction (to be discussed in Chapter 6 on the social aspects of KM—that is, social networking is essential to sharing and building social K beyond the boundaries of the organization).
- *Office technology* that promotes collaboration. This includes social networking (to be discussed in Chapter 6).
- *Culture* where workers imbibe the tacit assumptions of the social group (to be discussed in Chapter 6).
- *Exercise of practice.* Recall from Chapter 4 that a K practice is not an individual endeavor but exists in a larger system where K is shared and tested (Figure 4.9b).

5.6.3.3 Transferring Deep Knowledge

Recall from Figure 2.6 that K is transferred along a continuum between passive reception and active learning. Several of the activities include rules of thumb, practice, observation, problem solving, and experimentation. This leads to an important observation: *K sharing in general often includes both explicit and tacit K transfer.*

5.6.3.3.1 *Tacit K Transfer: Nonaka Approach*

Recall from Chapter 2 that modes for specifically transferring tacit K include *(1) direct observation, (2) narration, (3) imitation, (4) experimentation,* and *(5) joint execution* (von Krogh *et al.* 2000). Observe that several of these K transferring modes are embedded in CoP and the other K sharing activities mentioned earlier.

5.7 Knowledge Base

The heart of a K organization is its K-base. Recall from Chapter 2 that in the hypertext organization the K-base is not a single entity but a collection of systems. This is true in any K organization, independent of its structure.

We may extend the K-base to represent every piece of existing and potential K in the organization and include, for example, its K workers who generate, discover, and use K. From this holistic perspective, the K-base includes much more than systems' content, but also includes strategic visions, cultural assumptions, information linkages (as in departments or communities), K-maps, lessons learned, best practices, and so on.

If the organization is to realize and leverage the benefit of KM, a global or line-of-business perspective is necessary, if only to be aware of the K it already possesses. Only through such a high-level view can K integration and exploitation take place. This requires that *the K-base be managed as a single entity.*

Figure 5.25 is a representation of a K-base and its governance dimensions. Physical integration is not likely; however, organizational integration, effected through a governance structure comprised of key K roles, may be used to produce a single managed entity.

From Figure 5.25 we make the following observations:

- The K-base contains explicit and tacit K threads, hard and soft system components, K visions and strategies, K management systems and K maps, business process and system models, departmental and lateral process linkages, community and practice landscapes, cultural elements, and its K workers. In fact, the K-base, at least conceptually, should contain every possible thread of K, including its structure, rules and facts; situational context (e.g., business process) and K-spaces; people and relationships; and other

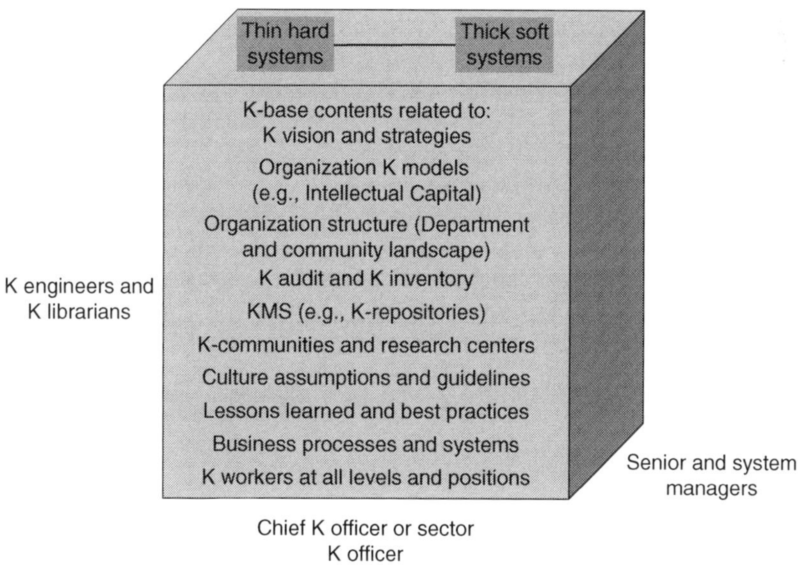

Figure 5.25 Contents and governing representation of an organizational K-base.

artifacts. Also, the K-base should be organization-, sector-, or division-wide; sufficient structural breadth is necessary to promote integration and reuse.

- The governance structure should include (1) a K management officer (KMO) who has overall responsibility for the K-base, (2) the K librarians who perform K-related activities such as extraction and building lessons learned libraries, and (3) senior and key system managers who provide strategic and operational input into the activities of the KMO and librarians. Note that by K librarian we include other specialized K roles such as K engineers and K integrators.
- The K-base is managed by a chief or sector KMO. The KMO is responsible for coordination among senior and key system managers and downstream integration with the organization's K librarians. Among the chief responsibilities specific to the KMO role is the creation and maintenance of the organization's IC model, participation in the organization or business area audit and inventory activities, and development of the organization's K vision and strategy. Naturally, all these activities receive input from and are in coordination with the senior and system managers. Also, the KMO guides the reengineering of K processes and reuse strategies for the organization, inspires and supports key K workers and K managers, and manages the K librarians. Finally, the KMO is responsible for establishing and maintaining external relationships with vendors, research centers, and universities. We consider the K-base and its structure to be an important part of an organization's strategy and will revisit this topic in Chapter 7.
- The K librarians perform specific organizational K activities, such as assisting the KMO in the following: analysis and development of the IC model of the organization, preparing the K strategy, identifying key K threads through an audit and inventory, identifying K gaps, and so on. K librarians also serve as internal consultants for the entire organization and support specific K activities, such as K capture and codification, K refinement and integration, local or global audit and inventory activities, business reengineering of K processes, creating and managing CoP or other K communities, developing lessons learned and best practice libraries, among others. The K librarian thus provides expertise directly or in partnership with organization members and subsequently refines and integrates K.
- *Organizational memory* is a term that is sometimes used in conjunction with OL. We consider organizational memory to be a subset of the K-base—specifically, those areas that represent the output of K-related activities such as project readouts, community (CoP) readouts and output, K mapping, and K system development generally. Hence, organizational memory includes codified rules and models (including those derived from expert systems development), best practices, lessons learned, and employee yellow pages.
- The senior and key managers are the infrastructure of the organization, who provide strategic direction and internal coordination. The development of

the IC model, for example, requires input from the high-level managers of the organization. Also, it is the relationships among the KMO, K librarians, and the management infrastructure that permits navigation to the contents of the K-base, which, as mentioned earlier, is not a single physical entity.

- We suggest that the K-base, the key K workers (e.g., K engineers), and the senior and system managers collectively represent an organizational subunit or department. In Chapter 7, we propose and describe the structure and workings of what we are calling the Chief Knowledge Office.

5.8 Organizational Design Representations

5.8.1 Organizational-Level K in Context

New organizational knowledge is not haphazardly created but arises as an outgrowth of a business or K analysis and strategy. Figure 5.26 illustrates the development of organizational K using common KM terminology from which we make the following observations:

- K content should be driven by a K strategy, such as, for example, Sharp's decision to cultivate expertise in flat panel screen displays.

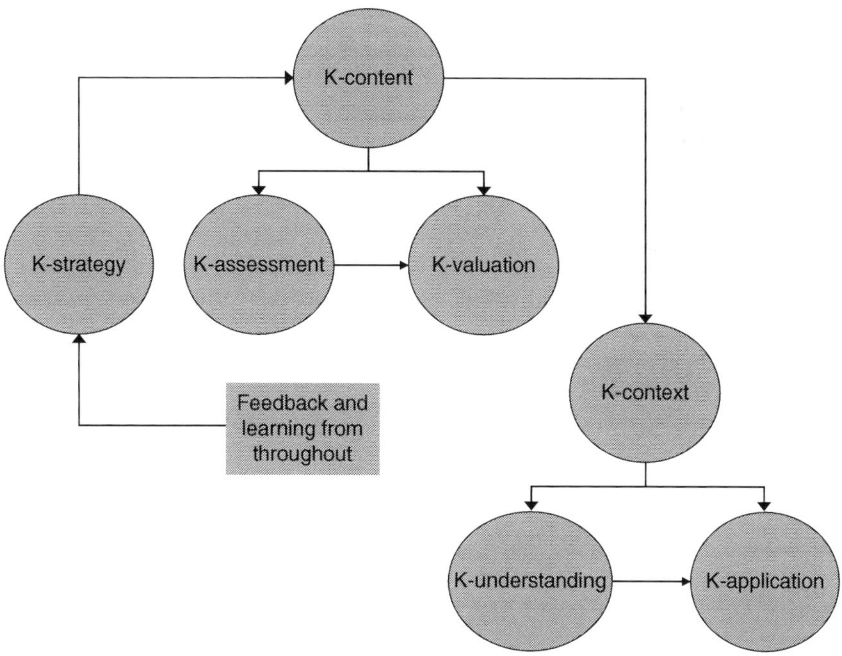

Figure 5.26 Organization-level knowledge planning context.

- New K threads—K content—should be studied and assessed from both a core competence and an economic perspective. This should correspond with the organization's IC model with input from the organization's KMO and K librarians. This analysis forms part of the organization's K-base and is under the governance of the chief or sector K officer.
- K content must be contextualized, that is, situated with respect to specific products and business processes. This is part of the collaboration between the K officer, strategists, and product managers. The output should be a specific K application in context, such as a reengineered business process.

5.8.2 Archetype Design Pathways

Figure 5.27 illustrates three archetype design paths. These pathways correspond to the structural, process or activity, and cognitive architectures illustrated in Figure 1.2. Simply stated:

- Structural approach. Structure drives design
- Process approach. Process drives design
- K approach. K drives design

The key characteristic in traditional or structural organizational design is that, after strategy, structure drives design. (In fact, in classical organizational design, even strategy sometimes follows structure.)

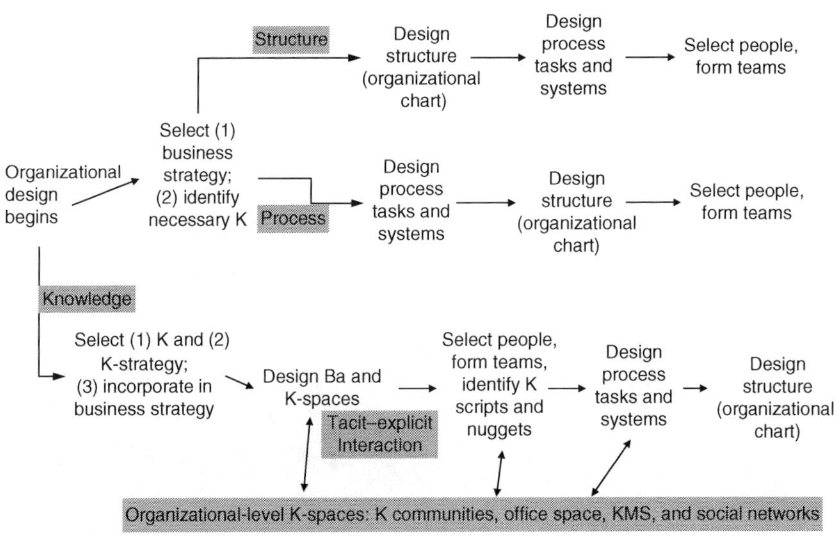

Figure 5.27 Archetype organizational design process.

In process- or activity-driven design, the key characteristic is that design of structure and people follows process design. If we consider an additional constraint—the addition of K—then K may be identified early, along with strategy, or pushed downstream into process and systems design.

In K-driven design, organizational design follows knowledge. This implies that we have an idea of the K to be cultivated. Input will be manifold and include, for example, K strategy components such as the (1) identification of synthesis opportunities, (2) specific competencies to be developed, and (3) competitor and customer analyses. Also, input will come from the (4) organization's audit and inventory activities, which should reveal K gaps, and (5) analysis of new science and technology opportunities. In each of these cases, K may be known early (flat panel screen display components) or roughly understood (new product features). That is to say, specific K threads may not be fully known until, say, product development progresses.

The next phase in K-driven design is that we identify or design (1) the ba or K-spaces necessary for knowledging and (2) opportunities for tacit–explicit interactions. In the PC case study, the physical ba were largely given; in response, our intervention K strategy was to develop a rich technology and social networking ba.

We make the following observations from Figure 5.27: many of the elements in K design are organization-wide. That is, K communities (e.g., CoP) and knowledge management systems (KMS) are global in nature, though a given project may identify a specific requirement. Also note that as we move from a static to a dynamic architecture—structure, process, cognitive—*organizational structure is pushed downstream* in the design process. Finally, it is important to recognize that these three pathways are archetypes and actual design may be a hybrid of each.

5.8.3 A Metaperspective on K Design in the Organization

Figure 5.28 illustrates a meta-framework for organizational K design. We draw on several design ideas that we have discussed as well as those applied in the PC case study.

The key design elements are presented here.

- *A thick context or situation.* This is typically a business process, situation, or opportunity. Recall from Figure 2.19 that the organization may be represented with an interleaving of thin and thick layers. The thick layer is the controlling layer where context and understanding are established. (Also shown in Figure 2.19 are the lower levels: the thin process layer consists of both thick K tasks and thin non-K tasks (what we call administrative tasks). In the PC case study, the thin process layer was operationalized with a process map or workflow diagram and included the identification of lower-level thick K tasks (in the case study, we call these tasks K points).)

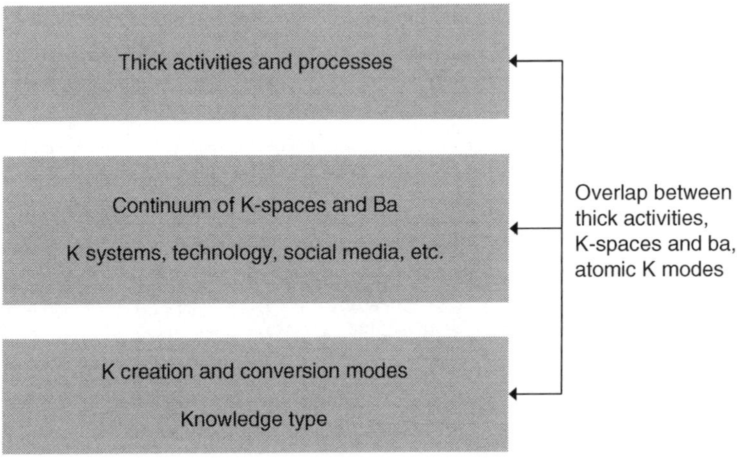

Figure 5.28 Metaperspective on K design.

- *Ba and greater ba.* Recall from Chapter 2 that ba is that which produces a stream of meaning in time and space. It also supplies the energy necessary for "self-organization" and K-creating. Ba may exist at many ontological levels, each of which may be connected to form a greater ba (Nonaka *et al.* 2000). Individuals form the ba of teams, which collectively form the ba *of* the organization; the market, in turn, becomes the ba *for* the organization. K creation is amplified through organic interactions among different levels of ba. Nonaka *et al.* (2000) emphasize that the presence of tacit K means the various ba are not independent and their relationships are neither modular nor easily assembled—they are integral. Unlike modular interactions (e.g., precise contracting), ba interaction is neither predetermined nor clear, but evolves and changes along with the ba themselves. We have previously called this relational contracting. Ba are supported by and their interactions made coherent through a knowledge vision and a trustful sharing of K. Finally, continuous exchanges among ba are necessary to create and strengthen their interrelationships—this is foundational to the relational contracting class of interactions we discussed in Chapter 2.
- *The K-space* is an extension of the ba concept along a continuum of K-oriented streams (see Figure 2.17), that is, a *range or continuum that lends systemization and structure to K design. Whereas ba is conceptual and a high-level organizing idea (descriptive), K-spaces are more detailed and facilitate design (prescriptive).* Recall, for example, in Chapter 4 we illustrated a class exercise on designing the desktop K-space for a specific K worker—this required an integration of several spaces, such a job role and technology.

The K-space continuum extends from the very small to the very large. Whereas Nonaka and his colleagues discuss the importance of ba in terms of self-organization; in contrast, we believe that the *K-spaces must be deliberate (at some level of detail) while allowing for serendipitous collisions and self-organization.* This is another difference between ba and K-spaces. The purpose of creating a range of spaces is to (1) provide a set of concrete design options, which may, for example, lend detailed structure to the hierarchy of ba mentioned earlier, and (2) enhance our understanding of the complexity of designing knowledge systems. Also, we do not restrict the K-space idea to K creation but to knowledging generally.

- *K-spaces interact* with each other and reinforce the application or creation of K. K-spaces include everything from a psychological orientation (e.g., psychological contracting) to office space design to social and cultural spaces. Note too that discourse structures (e.g., teams and task forces, lateral processes, K communities, and CoP) are embedded in the range of organizational and social K-spaces. Technology spaces include KMS as well as social media. As the PC case study illustrates, design often involves combining spaces from several points on the continuum shown in Figure 2.17.
- *Type of K and interaction modes.* We consider K-spaces and ba to consist of several sublevels. The lowest level is the atomic or primitive ba associated with tacit–explicit interactions. The primitive ba correspond with the SECI K-conversion modes: originating, interacting, cyber, and exercising ba, respectively. Higher-level K-spaces embody one or more low-level ba; for example, in the PC case study, the K-spaces associated with the KP5 immersion activity were primarily physical, social, and cultural (all given) and embedded the originating ba associated with indwelling, which in turn, positively reinforced socialization in a group context (i.e., "care" in K creation). Finally, we do not restrict interaction modes to the SECI model, but include the interaction associated with personal K (as shown in the Polanyi model): focal and subsidiary awareness and conceptual and bodily activity interactions, respectively.
- *Design Constraints.* We have been using the terms ba or K-space somewhat freely, as if we can simply design as we see fit. In fact, these spaces (as in the PC) are often a given and resist change. In such cases, a good strategy is to create and superimpose additional K-spaces to compensate for any restrictions in existing spaces; again, in the PC we superimposed rich media and social network K-spaces over the existing cultural and political spaces.

5.8.4 Organizational Structure and Knowledging

Organizational structure may be an effective enabler, or alternatively an inhibitor, to knowledge generation and flow. Recall from Chapter 3 that according to information processing theory, structure is one vehicle that facilitates (or inhibits) the flow of information. The same holds for knowledge. The vertical and horizontal linkages illustrated in Figure 3.5 carry knowledge as well as information—this is particularly true of the horizontal linkages such as the task force and team. The linkages are not a complete list—lateral relations, in the form of task forces or teams, carry and exchange rich information and knowledge. Other vehicles for K generation include lateral processes, K communities and CoP, and several less visible and less formal spaces (though significant) such as social and cultural spaces.

The organizational forms described by Galbraith (see Chapter 3) represent, as it were, a progression or evolution as each succeeding form accommodates the organization to specific requirements; for example, the product structure has a short product development lifecycle and is thus popular with industrial organizations. Hybrid designs are common too—for example, product and function—to leverage the benefits of each respective form.

Where knowledge is concerned, functional design is best. We find this interesting since the functional form is the earliest form of organization. Specifically, function promotes expertise through (1) concentrated study, (2) domain-specific work, and (3) peer interaction. German organizations, though huge with a large number of products, are typically designed as functional structures. Recall too that Google also designs itself along functional structures (Schmidt and Rosenberg 2014). The common denominator in this design philosophy is the central position that knowledge holds in their respective companies. A functional structure promotes K generation, where K discovery, peer discussion, and application are part of the everyday activities within the functional department. Also, the organization need not be exclusively functional; recall from Chapter 3 that a large international bank is organized along three forms: geographic, product, and function.

5.8.5 Thin and Thick Organizational Design

Recall from Chapter 2 our discussion of thin and thick aspects of organizations, societies, and cultures. Table 2.1 is a comparison of thin and thick organizational characteristics. Figure 2.19 is a representation of thin and thick organizational layers. And, Figure 2.20 illustrates two generic applications in a film and research organization, respectively. Observe that the distinguishing feature in both the film and research organization is the predominance of thick activities.

Figure 5.29a illustrates thick dominance and is representative of K and craft organizations. Examples of thick-dominance organizations are hospitals and universities that have created layer upon layer of practice over the decades and centuries. Recall too from Chapter 3 that hospitals and universities are examples of professional bureaucracies. In these two cases, the organization built one K practice upon another. The film and research organizations built their thickness directly—as a function of the immediate expertise necessary for their activities. Another example is a consulting organization, which is essentially a

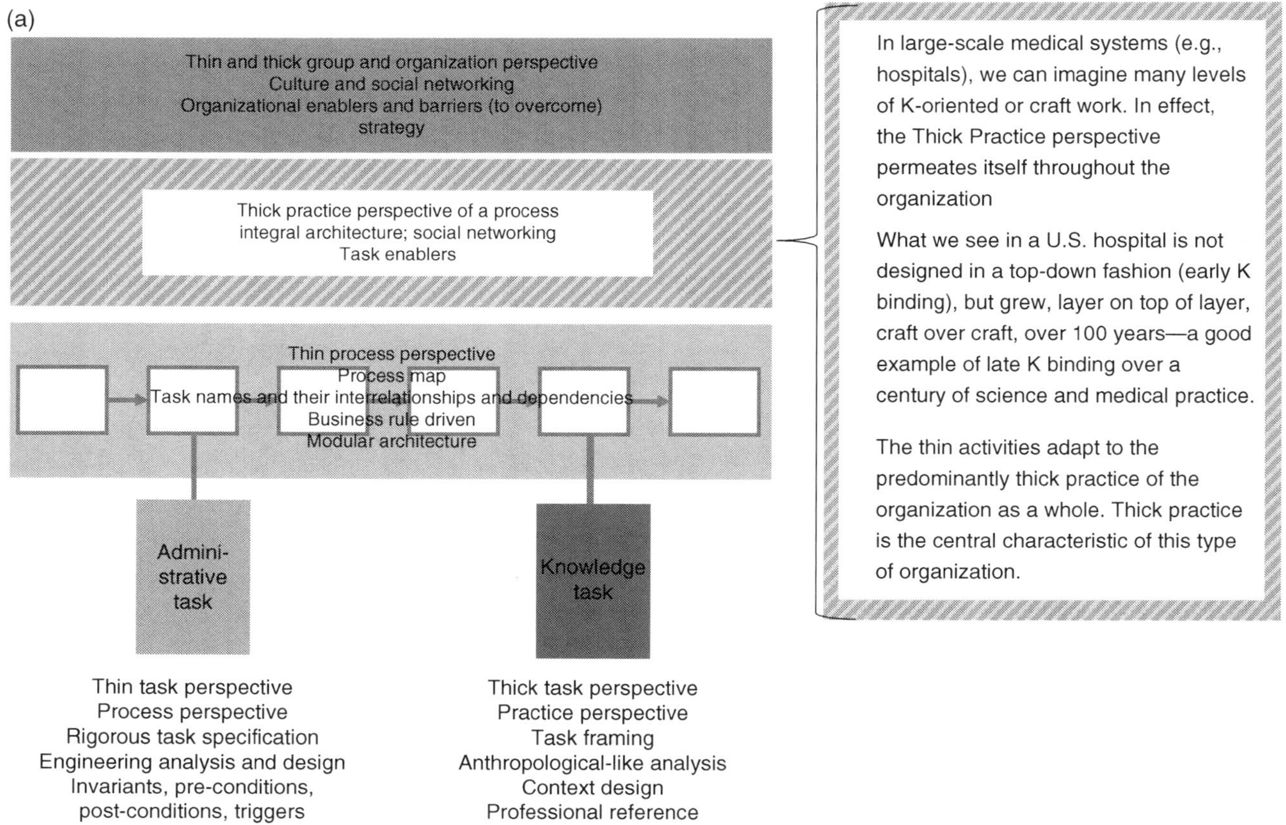

Figure 5.29 (a) Archetype thick dominance organization: practice-oriented craft K systems. (b) Archetype thin dominance organization: process-oriented automating–informating systems. (c) Distribution of knowing in a large K organization.

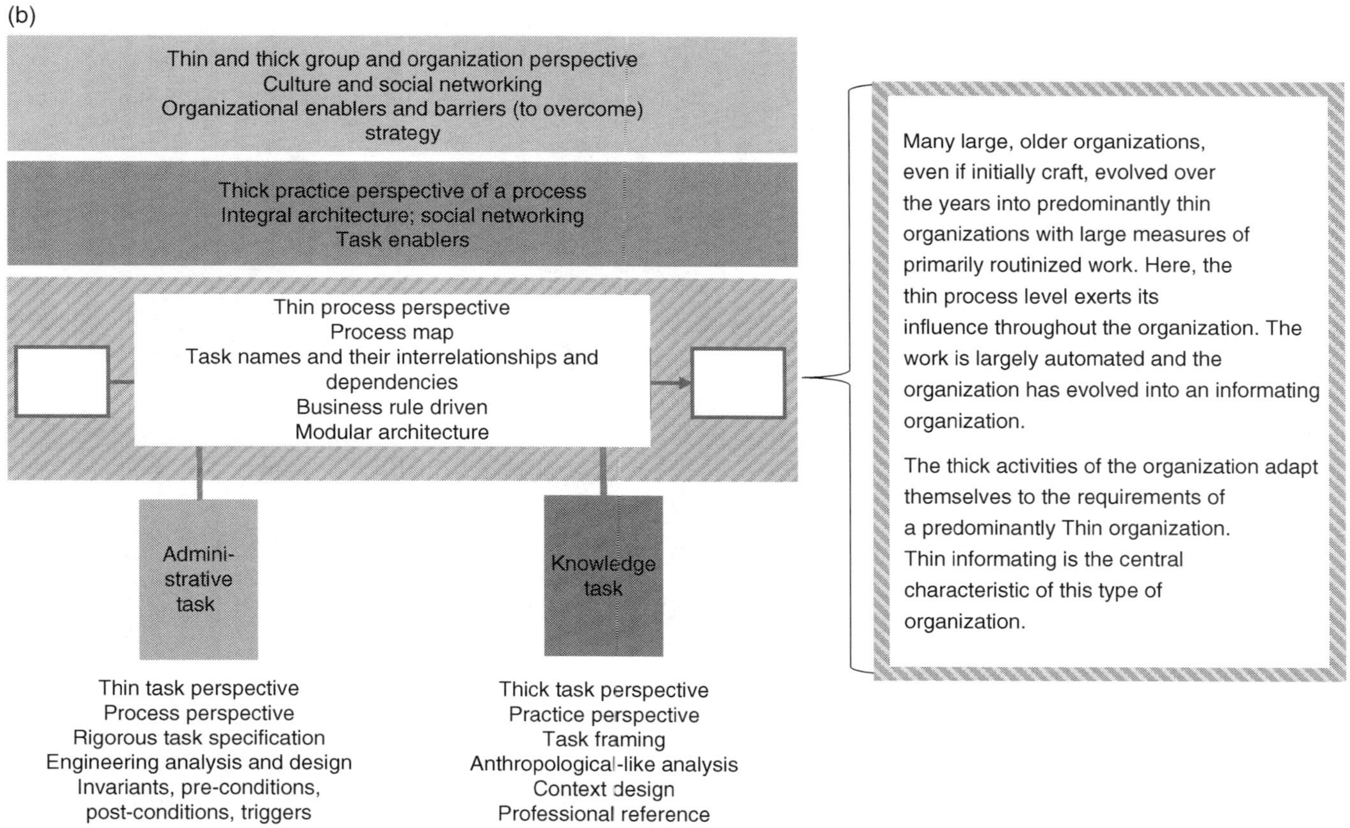

Figure 5.29 (Continued)

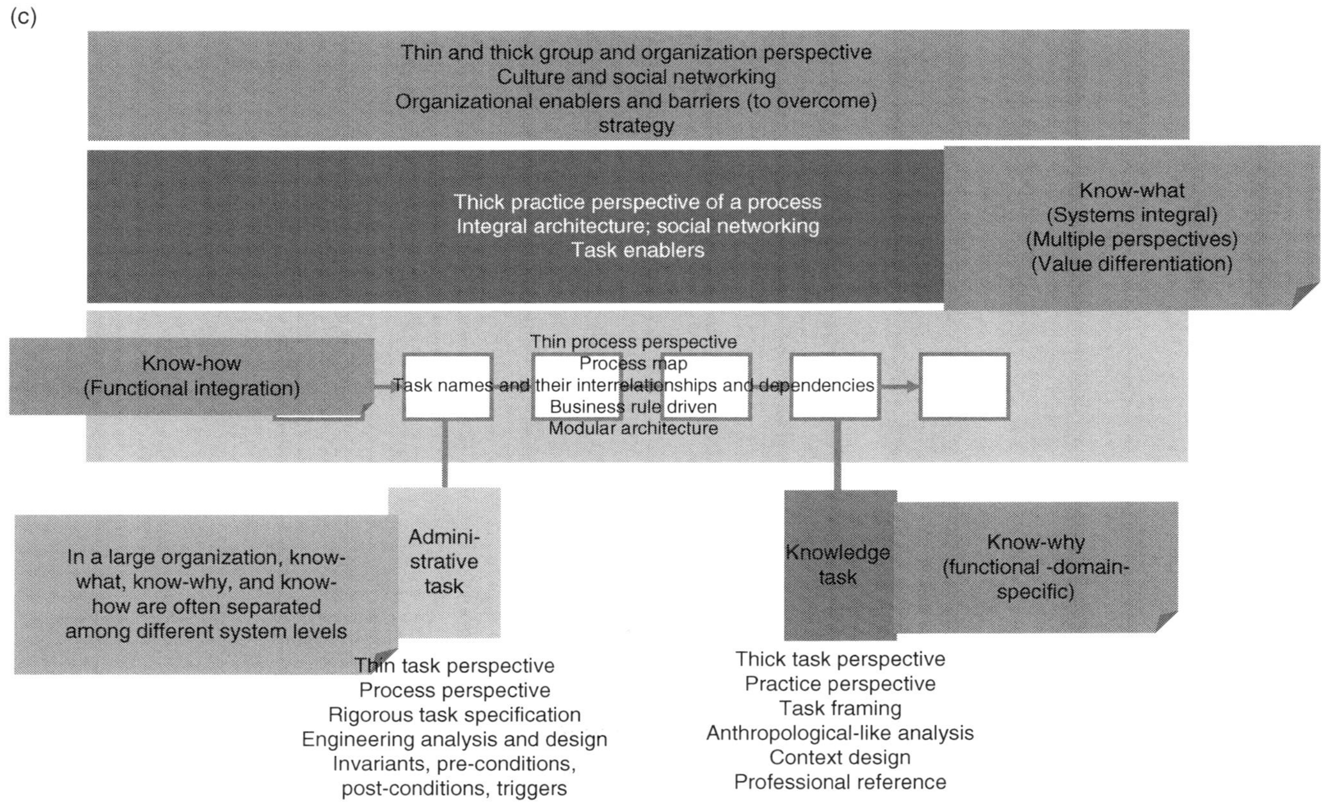

Figure 5.29 (Continued)

layered thick organization. The central characteristic of all such organizations is the predominance of thick K practice activities and its influence over the entire organization, including thin activities.

In contrast to thick organizations, Figure 5.29b illustrates thin dominance and is representative of industrial and service (non-K) organizations. Many large organizations fall into this category. Even if originally a craft, such organizations have routinized and automated large chunks of their work while superimposing bureaucratic policies, standards, and procedures. Whereas craft and K work may exist, it is the THIN aspects than permeate the entire organization. As an example, many government institutions (e.g., veteran hospitals) *do* contain thick work—that of their medical staff. However, the measures and logic of the THIN organization contaminate its thick aspect; hence, we see the artifacts of the industrial organization burdening thick medical practitioners, such as performance reviews and corresponding bonuses. The physicians are forced to adopt the culture of the thin organization; the results include documented lists of patients "on hold," "cooked books," and far worse.

Figure 5.29c illustrates three forms of *knowing*—know-why, know-how, and know-what—superimposed over the generic thin–thick layered organization. Each of these forms of knowing are often found in the head of a single K worker, but in a large organization, there is differentiation in thin and thick layers. For example, know-why corresponds to functional work and domain-specific K. Know-why is at the lowest level and resides in the head of the individual K worker (e.g., a film actor). Know-how is embodied in a thin layer of process and represents how the functional pieces fit together (e.g., the process of making a film). Know-what is located in the thick practice perspective (e.g., a film director) and represents the integration of know-why and know-how with multiple, high-level perspectives (e.g., producers, writers, audience). It is in the thick practice layer (know-what) that everything comes together (e.g., middle–up–down) and a unique system (e.g., a film) is crafted, driven by the creativity of the thick practitioner. (For another example, see Kusunoki (2004) in his case study of Sony's product differentiation.)

5.8.5.1 Balancing Thin and Thick Layers

Organizations have a mix of thin and thick layers with prominence sometimes determined by industry. It should be not be a surprise that hospitals and universities are predominantly thick or that an automated manufacturing facility is primarily thin.

Thin systems promote efficiency, whereas thick systems embody and promote expertise; every organization needs a good measure of each. A hospital or university with no thin activities may produce high-quality services but at a high cost. On the other hand, the dysfunction associated with a veteran's hospital may be traced to a thin layer that permeates the thick. In general, K organizations are confronted with similar balancing problems—enhancing expertise but controlling costs. In, say, a consulting firm, the market guides the balance. On the other hand, a film organization comes closest to an entirely thick adhocracy (Chapter 3) where thick dimensions are gathered for a specific project—an arrangement that many (Drucker 1988) believe will become the predominant structure in the K era.

5.8.6 DNA Design Metaphor

An organization-level metaphor that is similar to the thin–thick layered model is Toyota's DNA representation of itself (Liker 2004) as shown in Figure 5.30. In this representation, Toyota identifies both tacit and explicit elements that collectively form a thick and thin strand, respectively (thin and thick being our terms). The thin strand contains visible elements and artifacts and is known as the Toyota Production System (TPS). In contrast, the hidden elements often associated with culture represent the Toyota Way. In the East, the "Way" has a spiritual connotation; hence, we refer to the Way as the thick strand. As with a DNA molecule, each strand must be connected through any number of cross-linkages. In Toyota, the cross-linkages represent the stuff of design and management—they are the initiatives, actions, policies, systems, and structures that simultaneously connect and strengthen thick and thin elements.

Toyota is rightly given credit for developing the TPS elements that have come to be known as Lean Production. TPS elements are central to the Lean

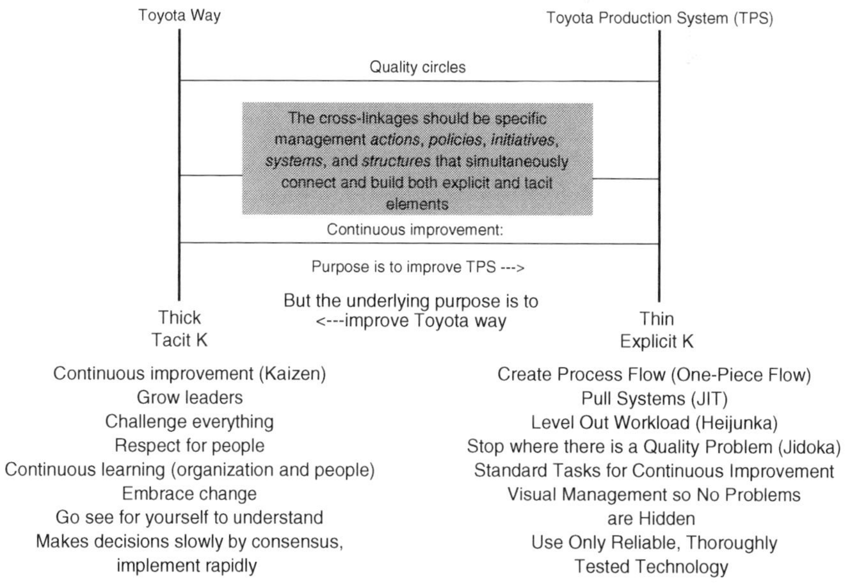

Figure 5.30 DNA representation for Toyota—Adapted from Liker (2004).

concept and include, for example, Just-in-Time and Jidoka (stopping where there is a quality problem). Others, such as Heijunka (leveled production), are complex to implement but are foundational to the "TPS House." Additionally, Toyota points out that an isolated TPS strand is not sufficient to capture the way Toyota works—the Way strand and the cross-linkages must connect to, and support, the TPS strand. That is, the interconnected elements that constitute the entire molecule are necessary to understand Lean and Toyota in general.

Observe that Toyota's DNA molecule is an organization-level elevation of Polanyi's model of individual K creation. Note too that this model is not very different than the thin–thick layered model reduced to a single thin and thick strand, respectively. However, one advantage is that it illuminates which specific systems, structures, and other cross-linkage elements may be used to connect and reinforce specific thin and thick elements, thus bringing coherence to specific designs. For example, what better way to promote the thick elements of respect for people, embracing change, and growing leaders than to have employees participate in continuous improvement activities such as quality circles? From this perspective, thin and thick organizational elements cannot be separated any more than explicit and tacit knowing can be separated in the Polanyi model. The cross-linkages—design and management—bring everything together.

5.8.6.1 A New Silver Bullet

The idea of a "silver bullet"—a single paradigm—to address organizational design has existed for many decades. The one best way to organize is one such example, popular with management students. More recently, the silver bullet has come to take on an IT flavor. In particular, there arose the "killer-app" idea such as American Airline's SABRE reservations system. Today, we have a suite of killer-apps, centered on data analytics.

In fact, there is nothing "killer" about these trends. Toyota has proposed a silver bullet—problem solving—that it uses and, we believe, comes closest to simultaneously supporting the critical K activities of the organization, building trust, and engaging K workers. Problem solving is foundational to K work, and ranges from analytics to exploration and experimentation. Developing a new standard operating procedure or going to Mars are examples of problem solving. At the organizational level, problem solving simultaneously connects and builds both explicit and tacit K. *Problem solving is the central cross-linkage in designing K work and in engaging and building K workers.*

5.8.7 Organizational Metaschema

Illustrated in Figure 5.31 are four interrelated structural configurations of the K organization. Included are discourse structures (Chapter 2 and later),

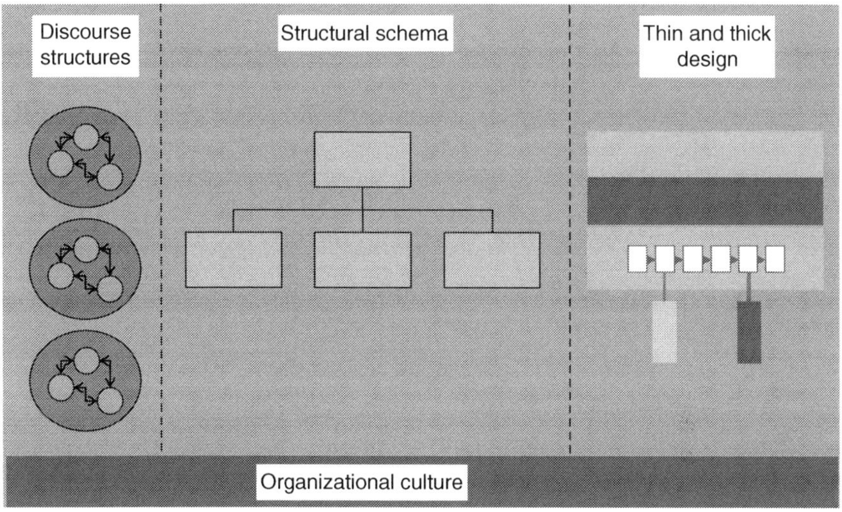

Figure 5.31 Organizational metaschema.

structural design (organization chart), and thin–thick layered design (discussed earlier and throughout the book), all on a bed of shared assumptions and practices that constitute an organization's culture (to be discussed in Chapter 6).

Recall from Chapter 2 that by *discourse structure* we mean symmetrical configurations that promote discussion, debate, dialogue, and thickness in general. This ranges from serendipitous collisions, self-organizations, facilitated dialogue sessions, CoP, social networking, and collaboration. Discourse structures are embedded in K-spaces, the organizational chart, and throughout the thin and thick layers of design (see Figure 2.22).

Figure 5.32 illustrates the core structural design of a K organization. It consists of a functional departmental structure, its business processes and primary coordinating structures (lateral processes, teams), and those K-spaces that add thickness and are closest to the workings of a K organization. As we mentioned earlier, the primary structure need not be fully functional, but a hybrid oscillating around function. In the next chapter, we will discuss office space design and the factors that promote social interaction and collaboration.

5.9 Architecture of the Learning Organization

A colleague of ours observes that OL has always appeared somewhat slippery and difficult to grasp. Yet, continuous improvement programs, such as Six Sigma and Lean, depend on learning to be successful. Additionally, the

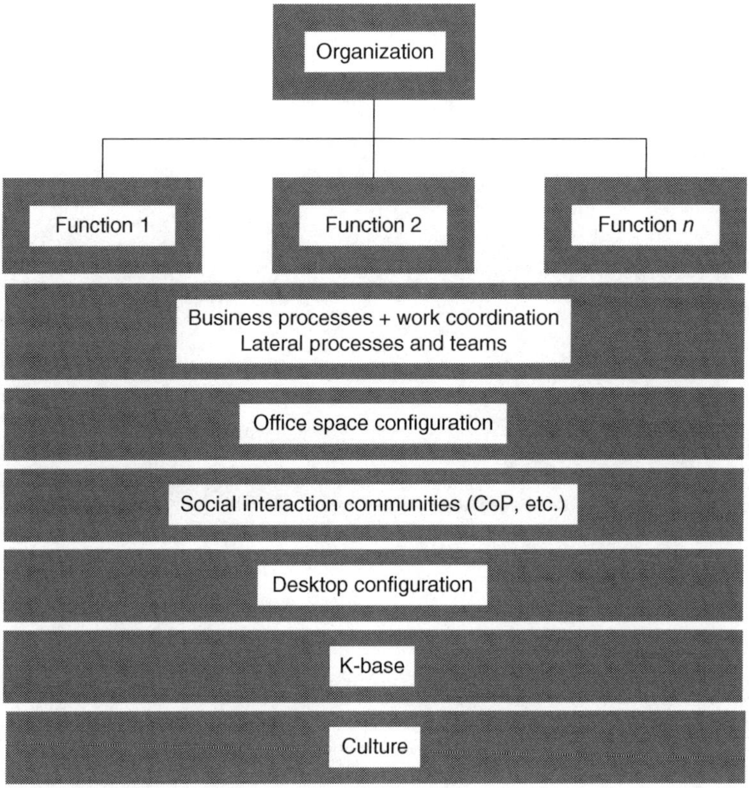

Figure 5.32 Structural schema with selected K-spaces.

literature contains several models of OL that are characterized in terms of K and KM. Let us look more closely at OL and its relationship to KM.

5.9.1 Toward a Definition of the Learning Organization

Garvin (1993) reinforces the idyllic-but-difficult-to-act image associated with OL by quoting Peter Senge's description of a learning organization as a place "where people continually expand their capacity to create the results they truly desire, where new and expansive patterns of thinking are nurtured, where collective aspiration is set free, and where people are continually learning how to learn together." This sounds perfect but is indeed slippery. (Garvin (1993) makes a similar comment on Nonaka's description of a K-creating company. In this regard, however, we disagree. The strength of Nonaka's model is that the theory and artifacts (e.g., ba) are sufficiently rich to guide management action and design—for example, Toyota DNA cross-linkages).

Whereas there exists significant disagreement as to what constitutes a learning organization, Garvin (1993) observes that most scholars agree that OL is a *process that unfolds over time, with links to knowledge acquisition and improved performance.* On the other hand, scholars differ on other important matters, such as the need for change in patterns of thinking and behavior, the role of information processing, organizational routines, and organizational memory, among others. Finally, Garvin (1993) defines a learning organization as one that is "*skilled at creating, acquiring, and transferring knowledge, and at modifying its behavior to reflect new knowledge and insights.*" Learning requires new K, but it must be accompanied by changes in the way work gets done if improvement is to be realized; that is, learning involves the *translation of new K into new ways of acting.* Garvin (1993) further observes that many organizations—universities and consulting firms in particular—generate new K but fail to translate that K into new behaviors and processes.

Evolution of OL

Argyris and Schon (1978) refer to single-loop learning as that in which individuals or organizations (e.g., their processes) make adjustments to their behavior based on information (typically internal operational data), but within a fixed framework of assumptions, goals, and norms. In contrast, double-loop learning results in behavior change based on new K (including external K) and in which the original framework is also open to change. The former improves what is already being done and the latter tests the assumptions underlying what is being done.

Senge (1990–1991) has extended the concepts of single- and double-loop learning into *adaptive* and *generative* learning, respectively. Adaptive learning is about "coping" and generative learning is about "creating." Generative learning requires new ways of looking at things—within the organization and in the world itself. The Japanese experience with TQM illustrates the evolution of adaptive (passive) to generative (active) learning: Initially, TQM was about "fitness to standard," or meeting design specification and thus making a product reliable. This was followed by "fitness to need," which implied understanding and meeting customer needs. Finally, the last phase in learning is "latent need," or anticipating customer needs, even if the customer does not think or ask for them. This evolutionary sequence relies on the KM activities of continuous experimentation and feedback.

5.9.2 Two-Part Learning Cycle

5.9.2.1 Nevis, DiBella, and Gould

The research of Nevis *et al.* (1995) suggest a cyclic character to learning. They have identified a three-stage cyclic model: K acquisition, K sharing, and K utilization. In this perspective, learning is neither a linear nor a planned process but, in fact, may be informal and even unintended. Furthermore, K may be generated in every stage and is not confined to organizing an "acquisition effort." Learning environments are often fluid and chaotic with ill-defined and subtle embodiments.

Nevis *et al.* (1995) have also proposed a two-part learning model. The first part concerns what and where learning takes place and is based on an organization's culture, experience, and competencies. The second concerns that which promotes learning and is based on best practices and common organizational processes. The former is descriptive and concerns an organization's *learning orientations* or style, and includes, for example, its K source (internal versus external) and dissemination mode (formal versus informal). The latter is normative and contains an organization's *facilitating factors*, such as concern for measurement and climate of openness. Each learning orientation and facilitating factor is situated within a specific K acquisition, K sharing, or K utilization stage.

5.9.2.2 Senge, Kleiner, Roberts, Ross, and Smith: Fifth Discipline Field Book

Similarly, Senge *et al.* (1994) have stated that the essence of learning is a cycle of skills and capabilities, awareness and sensibilities, and attitudes and beliefs. Deep learning involves development of (1) new skills and capabilities, which in turn, alter (2) awareness as people start to perceive and experience the world differently, which again, (3) changes beliefs and assumptions about what can be done—and the cycle continues. For example, new skills and capabilities include aspiration, reflection and conversation, and conceptualization—each of which reflects a change in individual, group, or organizational patterns of behavior.

Senge *et al.* (1994) go on to identify a "domain of action" architecture as a triangle containing (1) guiding ideas (i.e., primacy of the whole, community nature of self, generative power of language), (2) theory, methods, and tools, and (3) innovations in infrastructure (e.g., quality circles is one such innovation).

Figure 5.33 illustrates the interaction of the deep learning cycle and the action architecture of learning. The deep learning cycle is tacit and discipline-based—learning is subtle, abstract, and intangible (thick). In contrast, the architecture triangle is structural, explicit, and the focus of management activity (thin). The central causality of change is in the cycle and is largely cultural; management action is in the triangle and is tangible.

5.9.3 Building Blocks of OL

Figure 5.34 shows the key building blocks associated with OL. Observe the complementary nature of the elements and their close affinity to KM. Here we briefly describe each set of OL elements.

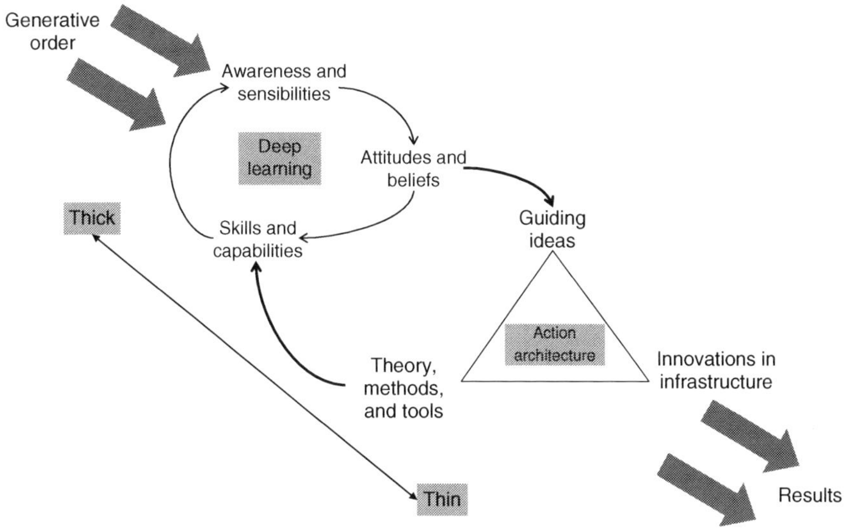

Figure 5.33 Organizational learning—Adapted from Senge *et al.* (1994).

1. Systems thinking
2. Personal mastery
3. Mental models
4. Building shared vision
5. Team learning

Five disciplines of org learning
—Senge *et al.* (1994).

1. Systematic problem solving
2. Experimentation
3. Learning from past experiences
4. Learning from others
5. Transferring knowledge

Building blocks of a learning organization
—Garvin (1993)

Learning orientations

1. Knowledge source: internal–external
2. Product–process focus: what?–how?
3. Documentation mode: personal–public
4. Dissemination mode: formal–informal
5. Learning focus: incremental–transformative
6. Value-chain focus: design–deliver
7. Skill development focus: individual–group

Facilitating factors

1. Scanning imperative
2. Performance gap
3. Concern for measurement
4. Experimental mindset
5. Climate of openness
6. Continuous education
7. Operational variety
8. Multiple advocates
9. Involved leadership
10. Systems perspective

Org learning systems—Nevis *et al.* (1995)

Figure 5.34 Key components of organizational learning.

Senge (1990–1992) has stated that organizations must simultaneously overcome learning disabilities and strengthen their capacity to learn. Accordingly, Senge (1990) has proposed five OL "disciplines." The most important discipline is *systems thinking* that integrates the other "core" disciplines into a unified whole. That is, OL requires a shift from isolated analyses to integrated solutions; from a belief that problems are not fully external to a realization that the organization participates in making its own problems. The solution is for the organization to discover and create its own reality.

The development of a learning organization requires that its managers embrace *systems thinking*, encourage *personal mastery*, surface and challenge prevailing *mental models*, build a *shared vision*, and facilitate *team learning*. Observe the significant overlap with KM. In fact, Nonaka has stated that Senge's model is not very different than his own, and, additionally, Nonaka claims that the K-creating organization builds in double-loop learning. Nonaka's major criticism of Senge is that the latter rarely discussed K creation and its relationship to OL.

Observe that Senge's five disciplines contain large measures of thick properties, as does (to a lesser degree) Nevis' learning orientations. Collectively, the elements shown in Figure 5.34 comprise many best practices associated with KM. Several of these practices, in turn, have a significant body of knowledge and implementation guidelines associated with them. Again, note both the significant and reinforcing crossover between these practices and their relationship to KM.

5.9.4 OL and KM: Together

OL has been considered somewhat idyllic in principle, with a huge assortment of implementation practices—the combination often seems like a hodgepodge of management slogans. We believe OL is easier to grasp if it is considered a manifestation of KM. As with knowledge itself, we may break down the learning shown in Figure 5.33 into thin and thick aspects—the challenge to management is to bring balance to this interaction of thin and thick learning. Also, as with the creative process of knowledging (Figure 2.21), thin structures follow thick assumptions, skills, and beliefs. Whereas attention is necessarily on the thin structural side of learning, managers must attend to and be mindful of the generative elements in the thick, deep learning dimension. The best practices listed in Figure 5.34 are primarily thin but several are sufficiently broad and cultural to include thick aspects (e.g., climate of openness, experimentation); all are interrelated as thin and thick reinforce each other (returning to Figure 5.33).

In practice, it is difficult for us to separate OL and KM—Toyota's DNA model and reliance on problem solving illustrates the intertwined character of OL and KM. While Figure 5.33 illustrates thin–thick interaction, Toyota's DNA model makes the interaction more manifest and tangible. Thin learning (and design) is explicit, structural, visible, easily altered, and not very substantial. Thick learning is tacit, deep, substantial, generative, and enduring. Recall too from Figure 1.14 that deep learning is meaningful and requires time to cultivate as mental models combine into patterns of patterns; in contrast, rote learning is transitory with less development.

5.10 Closing Remarks

Knowledge and the Organization

Knowledge is fundamentally a foreign idea in the world of organizations. Though K must exist in some form (otherwise an organization would have no products or services to market and sell), organizations are not K-oriented. Organizations are primarily information entities; in contrast, human beings are K entities. The challenge in designing a K organizational design (or in developing a K-based theory of the organization) is to overcome this essential tension and bring these two realities together.

Organizations are information entities Human beings are knowledge entities

The key source of innovation and sustainable performance in an organization is tacit K (Nonaka and Takeuchi 1995), Again, design requires bringing people (tacit K) and the organization (information) together—this is the implication of layered thickness in organizations. Thickness embodies expertise and relationships—all those people, policies, and activities that promote "care" in K creation. Recall from Chapter 2 that care requires the organization to attend to people and their relationships (Von Krogh 1998). Without care, individual K creation is reduced to "capturing" and social K to "transacting." In contrast, in an organization with high care, individual K is characterized by "bestowing" and social K creation by "indwelling." See, for example, KP5 in Figure 5.19 in the PC case.

The 21st century organization is faced with many design challenges, the most important being the creation, discovery, and application of K. The first divide of KM—traditional management and corporate technology—reinforces the organization's informating capabilities and facilitates knowledging more by accident than by design. In this book, we have advocated the second divide and suggested constituent waves from a wide variety of disciplines. For example, the concepts of thin and thick are ethnographic in origin; furthermore, the design idea of interleaving thin and thick layers is to bring balance and harmony to the efficiency and informating concerns of the organization with those of K and the human being. Further, our rejection of performance reviews (to be discussed) and other such incentives is that they superimpose a thin industrial construct on a thick human one. In such a case (e.g., veteran hospitals), individual K workers will strive to practice their respective professions, but they will also adopt those behaviors as the external situation demands. Hence, there is only superficial balance between organizations and humans, as thick is sacrificed for thin. Justification in a thin environment makes knowledging a fragile process.

We have concentrated on philosophy and a variety of models to bridge the gap, not merely between the first and second KM divide, but between managers and K, and between the organization and people. Managers do things—in the world of fact-based decisions, it is easy to lose sight of the big picture. Senge reminds us that systems thinking promotes a larger perspective; that is, in an organization the results of our actions are often not visible, making learning difficult. The value of creating models and second wave ideas is to surround the manager with richness and largeness—to lower risk by creating a framework by which thick, complex decisions can be made.

Many have argued that the greatest human invention is writing—writing makes explicit knowing both manifest and permanent (explicit knowledge). Complex scientific models can only arise from explicit K; on the other hand, social K is tacit-oriented. Recall that the cognitive and skill dimensions of individual tacit K are not easily shared, but its practice and social assumptions are almost always shared. Thus, each individual shapes, and is shaped by, the social context in which his or her K develops.

Wisdom too, as a special dimension of K, never fully develops, but is constantly forming and reforming. *Wisdom is a journey that strives to simultaneously balance the demands of scientific truths with social realities.* Wisdom manifests itself in the way scientific truth or theoretical K is applied in the social world. Similarly, managers are not machines but should bring something more than numbers to their decisions—that is, management practice. Practice (as a manifestation of wisdom) is an application of theory to concrete specifics; it can never be deduced from dogma or theory. While practice is sourced in theory, it lives in the social world and is thus more complex and nuanced than theory.

Continuums of Information and Knowledge

There has been a huge amount of published work on K and KM. Our discussion of K and K practice brings balance between *K as such* and the *organization in practice*. K is distinguished by any number of continuums; generically, we call this the information continuum. Examples or instantiations include the interactions within the explicit–tacit, individual–social, individual–organizational, and internal–external continuums. Figure 5.3 is an integrated model of K and presents a larger continuum between theoretical and social K. Creating models of K is a function of the language we choose to communicate our thoughts; hence, even our K models are closer to social belief than to objective K. Similarly,

Figure 5.20a and b illustrate several interrelated K continuums. Explicit K always contains a residue of tacit K; that is, as we illustrate in Polanyi's model, explicit knowing cannot exist without tacit knowing, suggesting the existence of cognitive elements in explicit knowing or knowledge. Explicit K structures things, but its content is attached to tacit K (what we have called the tacit hook). Finally, Figure 5.28 illustrates a metadesign schema that embodies a human–organization continuum: a range from K and primitive ba to a business process. Like steam in a teakettle, K rises from the human enclosed in primitive ba through successive levels of K-spaces to a business context. The lesson of the information continuum is that K simultaneously embraces many continuums (not just poles within a given continuum)—individual and social, theoretical and practical, and so on.

The internal perspective of knowing is localized, experiential, and particularized to the activity at hand, and concerns, in Nonaka's words, the "here and now." We can further state that tacit K has no global map—it is below consciousness and contains a stream of unconscious cognitive elements. In contrast, the external perspective is global and generalized and concerns "there and then."

KM and Design

KM is not a toolkit with tools such as CoP and big data analytics. It is essentially a system of complex human interactions that spiral to embrace other parts of the larger system, including the physical, technological, and social. The literature on KM sometimes reinforces the notion that KM tools will lead to more efficient operations and greater creativity. The centrality of people is often lost. When looked at more broadly, KM is about applying the principles of K, rooted in philosophy and culture, to the practice of people in their organization.

The PC case study illustrates an important dilemma in design. We may speak freely about systems, ba, and K-spaces, but in the real world we are limited as to what is possible. Existing contexts sometimes function more as rigid bodies than levers, and hence resist change. In such a case, we may move the larger system toward knowledging by adding and superimposing K-spaces over existing ones. This is one value of social networking, communities, and rich media—they are not yet fixed and constrained by the culture of the organization.

Social Knowing and Organizations

The Nonaka approach to KM is focused predominantly on the organization. Knowledge creation takes place in well-defined organizations, and hence is somewhat organized and controlled through individual, group, and organization ontological levels. K creation is directed and the K more homogeneous than not. On the other hand, in an interconnected world the organization is increasingly permeable to outside influences. As K seeps into the organization (weakly or strongly—to be discussed), K is likely to be more heterogeneous where knowing is more social than individual in character. This is more than simple knowledge transfer across organizational boundaries but a type of social absorption and organizational integration of external and internal knowledge. Knowledging in this new world is a complex mix of information exchange with tacit knowing and shared assumptions (we will revisit interorganizational K creation in Chapter 6). Social K is increasingly engulfing organizations.

Toward a K-Based Theory of the Firm

The essence of the K organization is *knowing* as opposed to knowledge itself. The activities normally associated with KM, such as K capture, codification, and distribution, though essential, are background activities. Recall from Figure 1.13 that the Polanyi model of personal K is characterized by knowing (tacit and explicit)—the K organization elevates the Polanyi representation and attempts to extract and spiral that human knowing into products (explicit K) or services (tacit know-how). See Spender (1996) for further exposition on a K-based theory of the firm.

5.11 Class Exercises

1 Discuss a dialectic situation or strategy in your organization, and explain how you would create a K synthesis. Use Tables 5.1 and 5.2 as a guide.

2 Describe the lifecycle of a K thread in your organization or school. Use Figure 5.11 as a guide.

3 Continue Exercise #7 in Chapter 4 and map the major tasks of the selected job to the Wiig cycle. Use Figure 5.10 as a guide.

4 Discuss whether your organization is predominantly thin, thick, or a mix. Demonstrate your answer by constructing a layered thin and thick representation for your organization, identifying which areas or departments are thin or thick. Include the location of significant *knowing*: know-why, know-how, and know-what. Use Figure 5.29a–c as a guide.

5 Integrated design project. Using the PC case study as a guide, perform an organizational analysis of an important business process and its K-spaces. Develop a complete analysis, including a process map or workflow, K tasks

(K points) identification and analysis, ba and other K-spaces, organizational and environmental contexts, and a proposal for an integrated KM system using both hard and soft components. Your solution should look similar to the representations in Figures 5.16, 5.17, 5.18, and 5.19.

6 Identify selected best practices in your organization or school, and discuss how they are operationalized. Which additional best practices would you include?

7 Assume you have been appointed the Chief K Officer of your organization. Using Figure 5.25 as a guide, design the organization's K-base, identifying its contents and governance structure. How would you operationalize your design?

8 Construct a DNA molecule for your organization or school. Include key thin and thick elements and the cross-linkages that hold the strands together. Describe the changes you would make to both the thin and thick elements and the cross-linkages.

9 Discuss your organization's structure and its corresponding K-spaces, using Figure 5.32 as a guide. Identify the organizational form (function, product, hybrid, etc.—see Chapter 3), the office space layout, the number and type of CoP, your desktop configuration, and so on.

10 Describe a learning experience in your organization or school. Making reference to Figure 5.33, discuss the interrelationships between deep learning and the action architecture elements (e.g., attitudes and beliefs, guiding ideas) and how they affected and changed each other.

References

Akutsu, A., and Nonaka, I. (2004), "Branding Capabilities: A Look at Sony's Capabilities in Brand Knowledge Creation." In *Hitotsubashi on Knowledge Management*. Eds. Takeuchi, H., and Nonaka, I., John Wiley & Sons (Asia) Pte Ltd.

Ankeny, J. (2015), "20/20 Visions: How the Next Five Years Will Revolutionize Business." *Entrepreneur*. January 2015.

Argyris, C., and Schon, D. (1978), *Organizational Learning: A Theory-in-Action Perspective*. Addison-Wesley.

Autor, D.H. (2014), "Polanyi's Paradox and the Shape of Employment Growth." Kansas City Federal Reserve Bank Symposium. Jackson Hole, WY. September 3, 2014.

Bernstein, J.H. (2014), "Disciplinarity and Transdisciplinarity in the Study of Knowledge." *Informing Science: The International Journal of an Emerging Transdiscipline*, 17: 241–273.

Bohmer, R., Bloom, J.D., Mort, E.A., Demehin, A.A., and Meyer, G.S. (2009), "Restructuring Within an Academic Health Center to Support Quality and Safety: The Development of the Center for Quality and Safety at the Massachusetts General Hospital." *Academic Medicine*, 84(12): 1663–1671.

Boisot, M. (1998), *Knowledge Assets*. Oxford University Press.

Boisot, M. (2002), "The Creation and Sharing of Knowledge." In *The Strategic Management of Intellectual Capital and Organizational Management*. Eds. Choo, C., and Bontis, N., Oxford University Press.

Cortada, J.W. (2011), *Information and the Modern Corporation*. The MIT Press.

Crane, L. (2013), "A New Taxonomy of Knowledge Management Theory: The Turn to Knowledge as Constituted in Social Action." *Journal of Knowledge Management Practice*, 14(1), March 2013.

Curtis, B. (2004), Overview of the Process Maturity Model (PMM). Teraquest 2004.

Dalkir, K. (2011), *Knowledge Management in Theory and Practice*. Second Edition. The MIT Press.

Davenport, T.H. (2005), *Thinking for a Living*. Harvard Business School Press.

Davenport, T.H., and Prusak, L. (1998), *Working Knowledge*. Harvard Business School Press.

Despres, C., and Chauvel, D. (2000), "A Thematic Analysis of the Thinking in Knowledge Management". In *Knowledge Horizons: The Present and the Promise of Knowledge Management*. Eds. Despres, C., and Chauvel, D., Butterworth Heinemann.

Drucker, P.F. (1988), "The Coming of the New Organization." *Harvard Business Review*. January–February 1988.

Earl, M. (1998), "*What on Earth Is a CKO?*" Earl and Scott, IBM Survey. London Business School. In *Knowledge Horizons: The Present and the Promise of Knowledge Management*. Eds. Despres, C., and Chauvel, D. Butterworth Heinemann.

Foroohar, R. (2015), "What Starbucks Knows About America." *Time*. February 16, 2015.

Garvin, D.A. (1993), "Building a Learning Organization." *Harvard Business Review*. July–August 1993.

Gilbert, E., Morabito, J., and Stohr, E.A. (2010), "Knowledge Sharing and Decision Making in the Peace Corps." *Knowledge and Process Management*, 17/3: 128–144.

Khalili, H.A., Maleki, A, Sharifi, A.S., Salmani, M.H., and Farshami, S.R. (2012), "Evaluation of Knowledge Management Maturity." *International Journal of Machine Learning and Computing*, 2(6), December 2012.

Kilov, H., and Sack, I. (2005), "Exploiting Reusable Abstractions in Organizational Inquiry: Why Reinvent Square Wheels?" In *Inquiring Organizations: Moving from Knowledge Management to Wisdom*. Eds. Courtney, J., Price, D.B., and Haynes, J.D., Idea Group Publishing. pp. 337–359.

Kuriakose, K.K., Baldev, R., Murty, S., and Swaminathan, P. (2011), "Knowledge Management Maturity Model: An Engineering Approach." *Journal of Knowledge Management Practice*, 12(2), June 2011.

Kusunoki, K. (2004), "Value Differentiation: Organizing 'Know-What' for Product Concept Innovation." In *Hitotsubashi on Knowledge Management*. Eds. Takeuchi, H., and Nonaka, I., John Wiley & Sons (Asia) Pte Ltd.

Leonard, D. (1995), *Wellsprings of Knowledge: Building and Sustaining the Sources of Innovation*. Harvard Business School Press.

Liker, J.K. (2004), *The Toyota Way*. McGraw-Hill.

Morabito, J., Sack, I., and Bhate, A. (1999), *Organization Modeling: Innovative Architectures for the 21st Century*. Prentice Hall.

Nevis, E.C., DiBella, A.J., and Gould, J.M. (1995), "Understanding Organizations as Learning Systems." *Sloan Management Review*. Winter 1995.

Nonaka, I., and Reinmoeller, P. (2000), "Dynamic Business Systems for Knowledge Creation and Utilization." In *Knowledge Horizons*. Eds. Despres, C., and Chauval, D., Butterworth-Heinemann.

Nonaka, I., and Takeuchi, H. (1995), *The Knowledge-Creating Company*. Oxford University Press.

Nonaka, I., Toyama, R., and Konno, N. (2000), "SECI, Ba and Leadership: A Unified Model of Dynamic Knowledge Creation." *Long Range Planning*, 33: 5–34.

Nonaka, I., and von Krogh, G. (2009), "Tacit Knowledge and Knowledge Conversion: Controversy and Advancement in Organizational Knowledge Creation Theory." *Organization Science*, 20(3): 635–652.

Osono, E. (2004), "The Strategy-Making Process as Dialogue." In *Hitotsubashi on Knowledge Management*. Eds. Takeuchi, H., and Nonaka, I., John Wiley & Sons (Asia) Pte Ltd.

Parsaye, K. (1988), "Acquiring and Verifying Knowledge Automatically". *AI Expert*, 3: 48–63.

Parsaye, K., Chignell, M., Khoshafian, S., and Wong, H. (1989), *Intelligent Databases*. John Wiley & Sons, Inc.

Schmidt, E., and Rosenberg, J. (2014), *Google: How Google Works*. Grand Central Publishing.

Sen, A., Sinha, A.P., and Ramamurthy, K. (2006), "Data Warehouse Process Maturity: An Exploratory Study of Factors Influencing User Perceptions." *IEEE Transactions on Engineering Management*, 53(3): 440–455.

Senge, P.M. (1990), "The Leader's New Work: Building Learning Organizations." *Sloan Management Review*. Fall 1990.

Senge, P.M., Kleiner, A., Roberts, C., Ross, R.B., and Smith, B.J. (1994), *The Fifth Discipline Field Book. A Currency Book*. Doubleday.

Snowden, D. (1998a), "Thresholds of Acceptable Uncertainty." *Knowledge Management*, 1(5): 1–9.

Snowden, D. (1998b), "The Ecology of a Sustainable Knowledge Programme." *Knowledge Management*, 1(6): 15–20.

Spender, J.C. (1993), "Competitive Advantage from Tacit Knowledge? Unpacking the Concept and Its Strategic Implications." Academy of Management Best Papers Proceedings 1993.

Spender, J.C. (1996), "Making Knowledge the Basis of a Dynamic Theory of the Firm." *Strategic Management Journal*, 17(Winter Special Issue): 45–62.

Von Krogh, G. (1998), "Care in Knowledge Creation." *California Management Review*, 40(3): 133–153.

Von Krogh, G., Ichijo, K., and Nonaka, I. (2000), *Enabling Knowledge Creation*. Oxford University Press.

Wenger, E.C., and Snyder, W.M. (2000), "Communities of Practice: The Organizational Frontier." *Harvard Business Review*. January–February.

Wiig, K. (1993), *Knowledge Management Foundations*. Schema Press.

Wong, D.R., Torchiana, D.F., Salm, T.J.V., Agnihotri, A.K., Bohmer, R., and Ali, I.S. (2007), "Impact of Cardiac Intraoperative Precursor Events on Adverse Outcomes." *Surgery*, 141(6): 715–722.

6

Social Aspects of Knowledge Management

Chapter Preview

Introduction

The central theme of Chapter 6 concerns the material aspects of social interaction. By material aspects, we mean the shape and manifestation of social interaction in an organizational context. Naturally, social interaction carries with it a strong strain of socialization. Recall our coffee metaphor: in the same way that heat brings forth the flavor in coffee beans, social interaction is necessary to elicit and cultivate K. In fact, we have been introducing *social heat* throughout the book.

- In Chapter 1, we introduced the *K-triangle*, which consolidates the ancient view of K along social, moral, and political lines of thought. Also, we discussed *relationship capital* in organizations as a constituent of intellectual capital models.
- In Chapter 2, we discussed *thickness* and the cognitive elements of design, such as *relational contracting*.
- In Chapter 3, we discussed the trend toward *functional design* as a technique to promote social interaction among K-oriented peers. Additionally, we discussed *symmetrical* structures as necessary for dialogue and collaboration. Closely associated with dialogue are the local and micro-interactions (e.g., *self-organization*) that characterize a *transformative teleology* (see Table 3.4) and, with it, the development of a K organization (see Table 3.5).
- In Chapter 4, we discussed the Kantian interaction of *theoretical* and *practical K* and the development of *social K*. We further developed tacit K to include a *practice dimension*, that is, mental models are in part manifestations of the social contexts in which they were developed. Practice is the application of K in a social setting, which, in turn, serves to reinforce and build both individual and social K. *A practice always produces a change in the real world.*
- In Chapter 5, we integrated several models (e.g., Plato, Polanyi, Kant) to illustrate *K and the emergence of wisdom* through the interaction of individual

Designing Knowledge Organizations: A Pathway to Innovation Leadership, First Edition.
Joseph Morabito, Ira Sack, and Anilkumar Bhate.

and social K and theoretical and practical K, fortified by the wisdom of the ancients. We also introduced a CoP as one manifestation of social heat in an organizational context.

In this chapter, we continue to explore the social aspects of K through five organization-related contexts. Social heat takes its form and structure through people living and working in a social environment characterized by the interaction within these five areas:

1) Social networking. A social network is a collection of actors (i.e., processing elements, such as people, groups, organizations, computers) interacting toward a purpose. The interaction takes the form of information or K flows along the links between actors. Social networks may be formal or casual and typically exist within and between organizations.
2) Culture and trust. A community is best characterized by its culture—the beliefs, values, and actions that bind the community members. Culture colors our world, shapes our identity, and molds our beliefs. Trust, particularly mutual trust or trust as a social phenomenon, is embedded in a community's culture. Trust is an aspect of relational contracting; that is, trust embodies integrity, benevolence, and altruism with knowledging.
3) Collaboration. We use the term collaboration to mean high-level interaction that builds and exchanges K. Characteristically, collaboration takes place among members from several institutions and is typically enabled by technology. Collaboration is the fundamental structure for research, particularly in the academic, scientific, and medical fields.
4) Office design. Organizations have often viewed physical space as a benign cost center. In fact, the reverse is true: the layout of an office is a major driver for culture development, trust, socialization, and the development of K generally.
5) Promotion of dialogue and conversations. As Nonaka and others have stated, managing conversations is the single most significant skill for managers in the K organization. Further, in a hyper-networked world spanning cultures and organizations, the importance of enabling and managing dialogue in a variety of social contexts is amplified above that found within a given organization.

Food for Thought

1) In what way does your ability to interact with others affect your work?
2) Do you trust the people you work with? Do you trust your company or school?
3) Do you share the same beliefs and values that are commonly found in your organization?
4) How do you modulate your interaction with people? Is there a difference when collaborating with experts or highly creative people?

5) Do you like the way your office is designed and your place in it?
6) How do you normally converse in your organization?

Topic Layout

The figure below illustrates the topic layout of Chapter 6. The left section informs the organization on the concepts associated with social networking and culture. We introduce social network analysis (SNA) and social network structure. The left side of the figure also shows culture, its characteristics, and a representative model. The right side of the figure shows the artifacts associated with culture, organizational and cultural dysfunction as well as a cultural comparison between tacit- and explicit-oriented organizations. The right side also includes personality factors and their effect on social interaction and K exchange. Additionally, we discuss large-scale collaboration and the issues associated with creative problem solvers and artists. The center of the figure represents the organization in practice. We start with a discussion of K-networks and the factors that affect their design. We also discuss two examples: the first is an industry example of culture in practice, whereas the second is a representative framework of interorganizational K creation based on network ties and tacit and explicit K. We conclude with two applications and proposed models: the first is a discussion of office design with several student examples, and the second is a discussion of the factors that affect conversations.

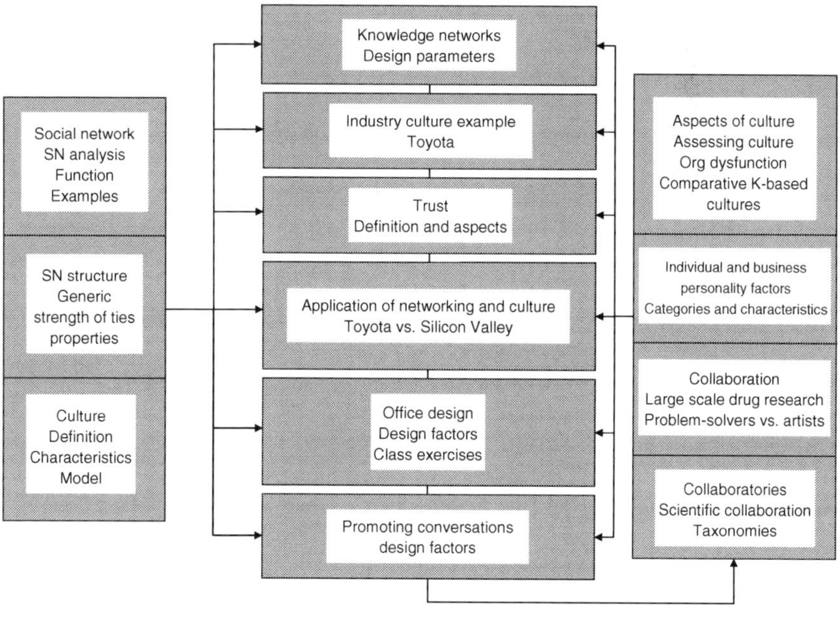

Concept roadmap to Chapter 6.

6.1 Social Networks

Social network analysis (SNA) examines the nature of relationships and flows of information or K between actors—people, groups, organizations, computers, and other processing elements—in a particular layout or topography. Graphically, the actors are shown as *nodes* and their relationships as links or *ties*. Each topographic collection of nodes and ties will be somewhat organized (designed or emergent) around a given theme; also, several networks may be interconnected. Furthermore, the *ties* between actors are of differing strengths, manifestations (e.g., physical proximity, electronic communication), and levels of formality, which collectivity guide the types and amounts of flows.

6.1.1 Examples of Social Networks

The organization chart (org chart) is a common network with which we are all familiar. The org chart is formal and represents a hierarchy of legitimate power relationships, which in turn governs the type and amount of flows. In the case of the org chart, mostly lean information flows from the top down the hierarchy, with light flows going up. Other networks are far less formal; our families, acquaintances, and friends, for example. In fact, we are embedded in any number of such networks, several of which we may not be conscious of, with different levels of importance, intensity, and formality. Note too that a hierarchy (e.g., the org chart) may be reconfigured as a collection of interconnected hub-and-spoke structures. This is a particularly useful perspective in analyzing decision making in organizations.

It is virtually impossible to provide an exhaustive listing of networks. A brief sampling from the Internet suggests the following: political party affiliations, lobbying and government influence links, academic networks, coauthorship networks, business and economic networks (local and global), online and Internet networks, computer networks, opinion leaders, clubs, sports teams, and many others. There are also hidden networks, such as the baseball steroid network or the Bernie Madoff Ponzi scheme (see orgnet.com).

Applications

Applications of SNA are also innumerable and include, for example, tracking the spread of contagions or disease or uncovering terrorist networks. In organizations, SNA may be used to map and discover communication and collaboration holes in decision-making or product development processes, among others. Consider, for example, the possible gaps or missed opportunities in the new product development process, which typically involve pieces of most, if not all, of an organization's networks.

6.1.1.1 SNA Research: Driving Organizational Performance Through Networks

The analysis of an organization's social networks may be used to map and discover (1) effective collaborations, (2) communications gaps, and (3) key network junctions. The K flows within the network landscape reveal patterns that provide senior leaders with insights that may be used to reduce collaborative costs and improve operational efficiency (Cross *et al.* 2010). Their research of IT subunits in large organizations reveals that SNA may produce the following opportunities for organizational improvement:

- *Improve scale through global collaboration.* In a global organization, operations are often locally optimized, sacrificing benefits of scale for the organization as a whole. By analyzing functional and geographic interactions, organizations may leverage diverse expertise and drive the adoption of ideas across geographies.
- *Drive workforce engagement and performance.* Analyzing network behavior of high performers will yield guidelines that may serve as a model for employees with similar job roles.
- *Align collaboration with internal partners and external stakeholders.* Analysis of cross-departmental relationships may reveal the locus of innovation, where support exists as well as where investments should be made.
- *Minimize network inefficiencies and costs.* Analysis of networks may uncover *too much* collaboration at too many levels. The research reveals the importance of minimizing network connectivity points where collaboration fails to produce value.

The results of their research indicate that traditional methods of driving operational excellence are not sufficient. The most effective organizations analyze and leverage employee networks to improve efficiency and promote innovation.

6.1.1.2 K Worker Networks

K workers typically are embedded in specialized networks that cover most of an organization's networks and include the following:

- Departmental affiliations
- Work groups or project teams
- Lateral processes
- Expertise and practice communities
- Research and collaboration networks
- Decision-making networks, depending on position (formal and informal)
- Professional associations
- Career and mentor guidance networks
- Casual conversational networks

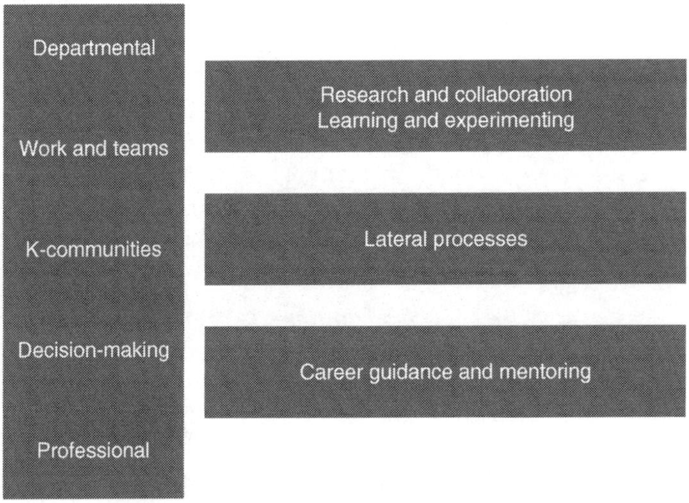

Figure 6.1 Primary and derivative organizational networks.

There may additionally be derivative or offshoot networks that combine pieces of other networks, such as *learning, experimental,* or *innovation* networks. Figure 6.1 illustrates several common primary and secondary networks in which organizational members, particularly K workers, participate. Each is a K-network (discussed here) to some degree—research and collaboration networks, for instance, are K-intensive. Note too that each of these networks may be virtual or face-to-face, which in turn may affect the type and amount of flows within the network. As we know from our study of knowledge management (KM), face-to-face interactions, for example, are more suited to sharing tacit K than are virtual networks. Also, recall from Chapter 5 that K organizations prefer functional design; that is, functional departments may serve as expertise networks.

We may additionally identify *social media networks* as those that depend on Internet technology. Such networks extend the network beyond the boundaries of the organization and include acquaintance, expertise, authorship, and research collaboration networks, among others. This leads us to discuss research and collaboration networks.

6.1.1.3 Network Landscape

Figure 6.2 illustrates a common composition of several overlapping networks in and between organizations and online communities. Observe the many network relationships in which an organizational member participates: a departmental member may belong to any number of project work, K community, and casual groups and vice versa. For example, in Figure 6.2 organizational member "D" participates in one department, two project teams, one CoP, and

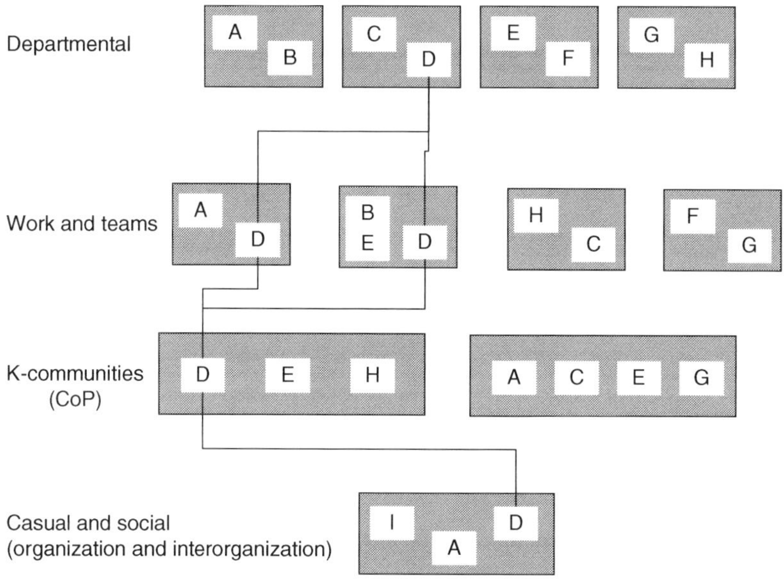

Figure 6.2 Sample organizational network compositions.

one social network. Also, note that K communities and casual groups may consist of members outside of the organization and include partners, customers, or less formal, online communities.

As an example, consider the network landscape for a professor. A professor typically belongs to a department or school, is part of several class structures in a given semester, is a member of several academic committees, and participates in research projects and coauthorship networks, with perhaps the latter two networks extending beyond the boundaries of the university. Observe that the departmental or school affiliations are generally stable, whereas the other networks are relatively dynamic and constantly changing.

Network Interaction During Execution of a Task

Closely associated with the network landscape is the interaction between members, which superimposes additional complexity. Members of work groups may interact with each other, business departments, vendors, customers, and databases. Figure 6.3 illustrates a common situation for a K worker during execution of a task, where participation in a project team often requires interaction with members of other work groups, CoP, and social networks.

6.1.2 Common Network Structures in Small Groups

Figure 6.4 illustrates three small-group networks as typically found in an organization. The first is the *chain*, which is characterized by a rigid structure

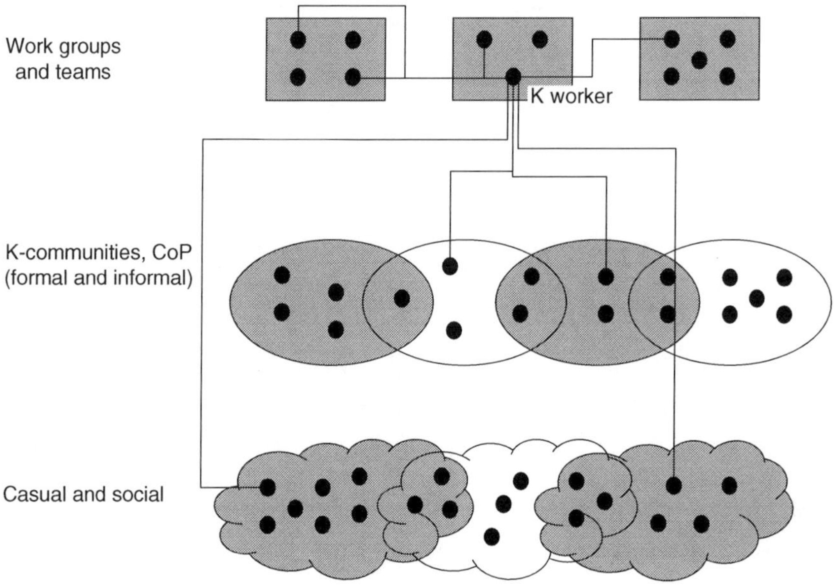

Figure 6.3 K worker interactions and task support.

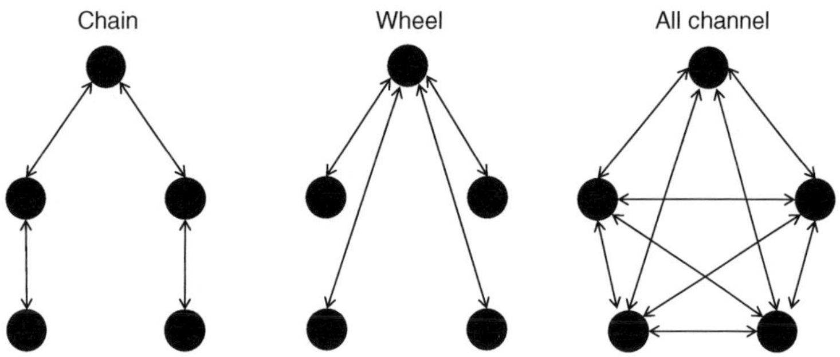

	Chain	Wheel	All channel
Speed	Moderate	Fast	Fast
Accuracy	High	High	Moderate
Leader emerges	Moderate	High	None
Member satisfaction	Moderate	Low	High
K-binding	Early	Early-late	Late
Example	Organizational chart	Team	Self-managed team

Figure 6.4 Common organizational communications networks for small groups.

and operations. The organizational chart is an example of a *chain* network. The second is the *wheel* network, which relies on a central figure that acts as an information conduit. An example is a team with a strong leader. The third is the *all channel* network where there is interaction among all members of the network. An example is a self-managed team.

Recall from Chapter 3 that Mintzberg described five generic organizational forms. We may additionally characterize several of these generic organizational forms by the three common networks shown in Figure 6.4. The workings of a machine bureaucracy, for example, correspond to that in a chain network; the adhocracy is an example of an all channel network with elements of the wheel—as in, for example, a film production company where actors and members of a film crew dynamically interact in all directions while taking their cue from the director.

Network Manifestation

As another example, the authors have worked in laboratory environments where all three of the common network structures appeared (on the surface) to be working, depending on the situation. One way to understand what is happening is to visualize a set of interconnected networks superimposed on each other (as opposed to some hybrid), with the situation guiding the operative network. A second factor is the function of the network, which is connected to the structure. For example, the authors' experience with K communities in a lab environment suggests an all channel structure rather than a chain structure.

We may superimpose these networks (in terms of structure or function) in any number of ways, depending on the context. As another example, in an organization we may visualize a primary hierarchical structure and other secondary structures (e.g., project teams) superimposed, known as a *matrix* structure. Figure 5.32 illustrates an organizational structure in a K context: a primary hierarchical structure (functional) with any number of K-spaces, several of which represent specific forms of social interaction.

Observe that social networks may be characterized in terms of *formality* and *structure*, which, in turn, is interrelated to *function*. All govern the type and amount of information or K flows.

6.1.3 Social Networking Theory

Ties between actors are of varying strengths, though for analysis purposes we often classify ties as either *strong* or *weak* (Granovetter 1973, 1983). Table 6.1 shows several structural and interaction characteristics of weak and strong ties.

Figure 6.5 illustrates the interaction characteristics between clusters in a social network.

Table 6.1 Sample characteristics of weak and strong ties.

Item	Weak ties	Strong ties
Primary embedded K	Explicit	Tacit
Key strength	Carries novel ideas; idea and innovation diffusion	Accessible; motivation to help; security
Social integration	Bridge building	Isolation
Enablers	Technology; common goals or interests	Face-to-face; common culture; common goals
Pattern formation	Promotes new patterns	Reinforces existing patterns

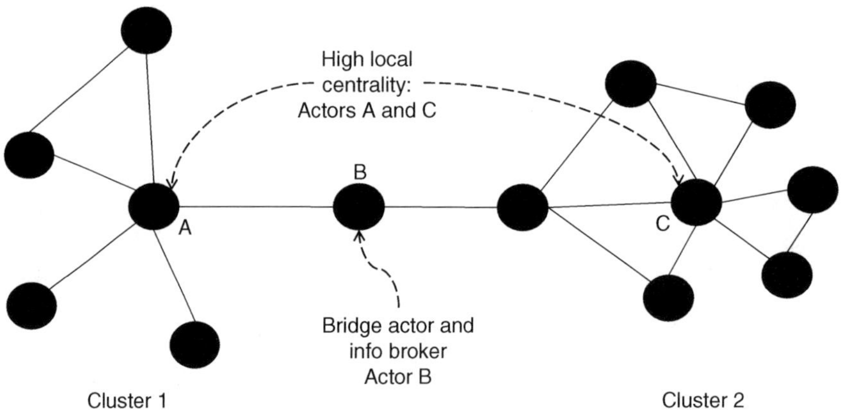

Figure 6.5 Social interaction between clusters.

From Table 6.1 and Figure 6.5 we make the following observations:

- Acquaintances (i.e., actors with weak ties) are less likely to be socially involved with one another than are close friends (i.e., actors with strong ties).
- Weak ties typically carry primarily explicit, sometimes *novel* ideas. It is weak ties that promote the diffusion of new ideas and innovations. Social networks that are large and diverse tend to be rich in weak ties.
- Strong ties are primarily tacit and characterize small or isolated groups. A weakness of strong ties is that they produce a kind of steady state driven by experience with little novelty. The strength of strong ties is their very presence—members are accessible and usually motivated to help; hence, there is more security among members with strong ties than those with weak ties.
- Individual social groups carry specific overlapping characteristics, for example, size, type and strength of ties, level of trust, and diversity. Also, within a social group certain members are more centrally located than others

(measured by the number of ties), giving them wider access and influence (actors A and C in Figure 6.5).

- Certain members often act as bridges between otherwise separate groups (actor B in Figure 6.5). These bridges act as information relays and introduce otherwise novel ideas to each participating cluster or group.
- For example, consider the self-imposed isolation of China during the Neo-Confucian period. During the prior Tang dynasty (618–907 A.D.), China was generally considered the most highly developed nation in the world—in terms of commerce, science, technology, and philosophy. Then, under the Sung dynasty (960–1279 AD), isolation in the form of Neo-Confucianism arose. Isolation created a cultural wall that severed and prevented the development of weak, foreign ties carrying novel ideas (in all directions). In its place, the strong ties of a common culture with common goals were reinforced. This existed for several centuries and only recently was broken with the rise of the Chinese Republic in 1912.
- Similarly, Japan was experiencing a self-imposed isolation during its medieval period until the mid-19th century. Japan too was characterized by a *steady state* that characterizes tacit-oriented societies with few weak ties. However, in 1853, Commodore Perry entered Japanese waters and forever changed Japan and the world. The arrival of a foreigner—a weak tie—brought new ideas and technology and disoriented the establishment. New ideas often appear threatening to those in power within a group with strong tacit ties. However, Japan quickly adopted selected foreign ideas and technologies—for example, modeling its navy after the British Royal navy and its army after the German army.

(Aside: Whereas Japan is still characterized by a tacit-oriented culture, it selects appropriate weak, foreign ideas as necessary, particularly in the area of technology. See section 6.3.)

6.2 Knowledge Network Design in Organizations

Social networking in a K organization manifests itself along several major overlapping themes:

1) *K reuse* through the capture, codification, representation, and distribution of K, in a network of K engineers and K workers.
2) *Coordination of work* through lateral processes, teams, and other *work structures.*
3) *Discourse structures* that inform both decision making and K activities in general.
4) *K communities* within and between organizations. This includes research and *collaboration* communities.

5) The construction of *social capital* (and possibility social K). The development of social K may extend the social network beyond the boundaries of the organization. Recall from Chapter 1 that social capital is an important component in most models of intellectual capital.

6.2.1 Knowledge Networks

We may define *knowledge networks* as "collections of individuals and teams who come together across organizational, spatial and disciplinary boundaries to invent and share a body of knowledge. The focus of such networks is usually on developing, distributing and applying knowledge" (Pugh and Prusak 2013). Observe that there may be extensive functional and structural overlap with the networks mentioned earlier; for example, a K community, in the course of developing, sharing, and applying K, supports learning and problem solving. That is to say, high-level applications are supported by the primitive operations associated with cultivating K.

Most researchers agree that members of a K-network share common goals and interests as well as social and operational norms. Further, we may state that K-networks may be specifically designed. Recall from Chapter 5 that CoP are not necessarily self-organizing but may be designed for a specific purpose; for example, Figure 5.24 illustrates five CoP in an automobile manufacturer.

Intended or Deliberate K-Networks

Pugh and Prusak (2013) have described a three-part approach to designing K communities: goals, a leverage framework for effective K-networks, and design dimensions as follows:

Goals. From their research, four common goals (i.e., "designing for outcomes") have been identified to promote K community energy and performance: (1) coordination, (2) learning/innovation, (3) translational/local adaptation, and (4) support of individual members. For example, organizations may *coordinate* global practices around specific K domains; *learn* through accumulation and distribution of K or learn systematically about network itself or processes; a K-network may identify and *adapt* knowledge to specific *local* situations; and a K-network may develop, accumulate, and adapt knowledge to *support* its members and colleagues' work. The research showed that the latter was the most common goal—support of individual members—within corporate K-networks.

Leverage framework. The idea behind a leverage framework is to identify a chain of factors that may be leveraged to promote a given goal (Figure 6.6). We read left to right to understand the chain of influences: *design* (specific conditions explicitly put in place), *dynamics* (feedback loops (systems and structures) to sustain behaviors), *behaviors* (those behaviors, such as

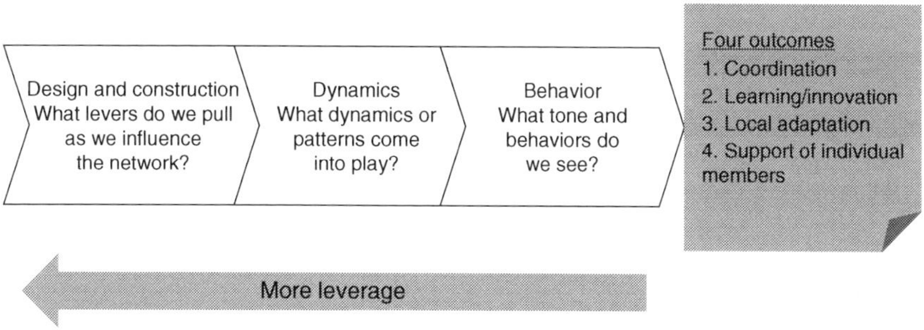

Figure 6.6 Framework for K-network effectiveness—Adapted from Pugh and Prusak (2013).

demonstrating trust, which are conducive to the desired outcome), and the specific *outcome* of the network (one of the four common goals). Reading right to left in Figure 6.6, we peel back the layers that lead to the desired outcome. Note that this is somewhat similar to the K-based approach to design (see Figure 5.27) where we identify the desired K and K-spaces and then proceed in sequence to design the other elements (e.g., people, teams, communities, departments).

Network dimensions. Network leaders attempt to create a consistency between network structures (e.g., charter and technologies) and strategies (e.g., purpose and network composition). The researchers were able to identify eight dimensions that cover a range of factors, as follows: (1) leaders shared theory of change, (2) objectives and outcomes, (3) role of expertise and experimental learning, (4) inclusion and participation, (5) operating model, (6) convening structures and infrastructures, (7) facilitation and social norm development, and (8) measurement, feedback, and incentives. Observe the progression from strategic (dimensions 1–4) to structural (dimensions 5–7) to tactical (dimension 8) considerations.

Emergent K-Networks

Individuals are engulfed in any number of loose networks that arise for any number of purposes. The need for information often drives the development of informal networks; reciprocity is the operational glue that holds the network together. Recall from the earlier section that the most important goal in the design of a K-network is to support individuals. That is to say, the needs of colleagues—individuals—drive networks where in turn reciprocity is common reason to share K (Davenport).

It is the combination of intended (designed) and emergent networks (e.g., self-organizing, serendipitous collisions) that create a *network landscape* that promotes knowledging in an organization.

6.3 Culture in the Knowledge Organization

Culture is about the beliefs, values, and practices of people that characterize a given community; in fact, we may say that culture represents the essence of a community. Edgar Schein describes culture as "a set of basic assumptions about how the world is and ought to be that a group of people share and that determines their perceptions, thoughts, feelings, and, to some degree, their overt behavior."

Cultures do not exist in isolation. Every culture is connected to some other culture; we may, for example, start with the microculture of our families and extend into our professions or occupations, our larger work organizations and industries (e.g., university, company), and eventually the entire global culture. We may superimpose additional major cultural communities—for example, religious, ethnic, and national cultures. The point is that each of us is embedded in an intertwined *cultural landscape* that largely shapes our identity and molds our belief systems and behaviors.

When we enter an organization, we are expected to imbibe the culture of that organization, though we carry both smaller and larger cultures within us. It is thus a characteristic that culture—as with our model of tacit K (Figure 4.8)—carries cognitive and behavioral dimensions that cannot be disconnected from both a narrower and a wider social context. For example, the structure and workings of Japanese organizations—their assumptions, values, and behaviors—reflect the larger, tacit-oriented Japanese culture in which they arose. We may thus visualize a set of concentric circles, each of which represents a community's culture that is both context and constituent to other cultures. Culture is additionally characterized by what Gareth Morgan calls a "whole in the part"—that is, as we imbibe a group's culture, our very presence in turn affects that culture in some way.

Aspects of Culture

Is culture a set of symbols that carry deep meaning? Is culture a historical record that reflects the values and behaviors of an organization's founders and key leaders? Would this record, for example, include the art and music of a society? Is culture a social network characterized by its strong ties? In fact, each of these aspects (and many more) represents an important characteristic of culture.

6.3.1 Manifestations of Culture

Figure 6.7 is Edgar Schein's model of culture. Observe that culture manifests itself at three levels: "the level of deep tacit assumptions that are the essence of culture, the level of espoused values that often reflect what a group wishes ideally to be and the way it wants to present itself publicly, and the day-to-day behavior that represents a complex compromise among the espoused values, the deeper assumptions, and the immediate requirements of the situation."

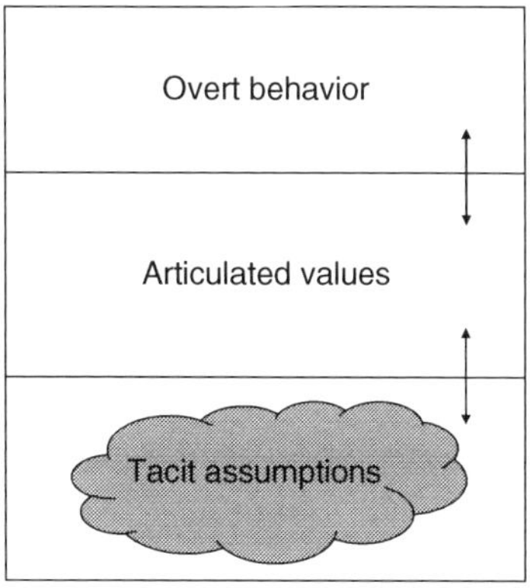

Figure 6.7 Model of organizational culture—Adapted from Schein (1999).

- *Shared assumptions and beliefs*. This represents the core of culture and comprises tacit and taken-for-granted beliefs, perceptions, thoughts, and feelings. Assumptions and beliefs embody reality for an organization's members and are thus "true" and not open to discussion. According to Schein the dimensions associated with assumptions include the nature of (1) an organization's relationship with its environment, (2) human activity, (3) reality and truth, (4) time, (5) human nature, (6) internal diversity (versus homogeneity), and (7) the nature of a human. As illustrated in Figure 5.3, assumptions (i.e., tacit knowledge) are largely socially constructed—a balance between justified belief and social K—and thus evolve over time.

  As an example, let us consider the nature of man, particularly in a work environment. McGregor (1960) has observed two sets of assumptions held by managers with respect to employees. The first is termed *Theory X* and holds that people are lazy and must be motivated and controlled. In contrast, *Theory Y* assumes people are basically self-motivated and therefore need to be challenged and channeled (and not controlled). The former assumes that people are in conflict with their organization, whereas the latter assumes that organizations may be designed so that the needs of people may be congruent with those of the organization. McGregor (1960) also observes that people respond adaptively to the assumptions about them; hence, in an organization run by Theory X managers who view employees as untrustworthy, the employees will respond and behave in terms of those expectations. Today, we generally recognize that human nature is complex, dynamic, and more variable than the X–Y characterizations. Consider, for example, how our motives change as we mature or how we adapt to changes in our social and organizational context. Human variability thus makes it essential that organizations develop a consensus on what their own assumptions are or should be—management strategies, control, and incentive systems are built of those assumptions (Schein 1992).
- *Espoused values*. An organization's values, norms, and actions arise, ultimately, from its assumptions and beliefs. Whereas shared assumptions

are hidden, values (principles and goals) and norms (unwritten rules) are secondarily visible in terms of either corresponding explicit behaviors or that which is articulated. For example, organizations typically conduct periodic "town hall" meetings where there is an open airing of values and norms, almost always from the senior managers. This may be seen as an attempt to portray the organization in a certain light or to springboard the organization in a new direction.

For example, recall from Chapter 4 that John Mackey developed a moral basis for the culture of Whole Foods, what we have previously referred to as ennobling goals. Those ennobling goals—including ending poverty, for instance—imbue Whole Foods' values, norms, and policies. The culture of Whole Foods is thus based on the vision of the founder and its subsequent history.

- *Artifacts and overt behavior.* Schein states that culture originates with shared tacit assumptions and expresses itself outward through an organization's values and norms, which in turn influence and guide those behaviors that support and reinforce the organization's assumptions and values. Note in Figure 6.7 the double arrows between levels—each level has an effect on some other level. Thus culture may be thought of as a complex entity with identifying artifacts that distinguish an organization.

  Hatch (1997) has categorized and identified several visible artifacts that constitute culture, as follows:
  - *Physical manifestation.* Art/design/logo, building/décor, dress, material objects, and physical layout.
  - *Behavioral manifestations.* Ceremonies/rituals, communication patterns, traditions/customs, and rewards/punishments.
  - *Verbal manifestations.* Anecdotes/jokes, jargon/names/nicknames, explanations, stories/myths/history, heroes/villains, and metaphors.

  Schein (1999) has created additional categories that may be used to identify the artifacts within an organization, as follows:
  - Levels of formality in authority relationships; meetings (how often?, how run?, timing?); how are decisions made?; communications: how do you learn stuff?; disagreements and conflicts: how handled?
  - Working hours; jargon, uniforms, and identity symbols; dress codes; rites and rituals; balance between work and family; social events

Note that Schein's list of artifacts, while overlapping with Hatch (1997) in several areas, is closer to the operations and day-to-day activities of an organization (Schein's first bullet). Observe too that several artifacts (Hatch or Schein) encompass those activities that stimulate social networking, motivation and relational contracting, and knowledging. In particular, the organization's layout of its offices and buildings and its rewards and punishments (i.e., performance reviews) is under the control of the organization's leaders and has a direct effect on K exchange and creation. Each will be subsequently discussed.

6.3.2 Assessing Organizational Culture

Schein (1999) states that culture cannot be assessed through a survey. Since culture encompasses all aspects of what an organization has learned over its history, a survey would be unlikely to contain meaningful questions. Further, asking about shared processes and phenomenon is inefficient and possibly invalid. Culture is a group phenomenon and hence its assumptions are best ascertained in a group setting by asking broad questions about different areas of organizational functioning and seeing where there is broad consensus among members of the group.

The approach that Schein has proposed is to drill down through the basic culture model shown in Figure 6.7. That is, in the same way that culture emerges from the bottom layer through the top, culture analysis should proceed from top to bottom.

- *Define a "business problem"* to a group. By business problem, we mean a concrete operation that needs to be fixed or some new strategic intent.
- *Review the concept of culture*—this discussion focusing on Figure 6.7.
- *Identify artifacts.* Start with everything that is identified—the list may be quite extensive.
- *Identify the organization's values.* That is, identify the *espoused* values of the organization, covering its strategies, philosophies, and goals. Specifically, values cover common language and concepts, group boundaries and identity, the nature of authority and relationships, and the allocation of rewards and status, among others.
- *Compare values and artifacts.* Link espoused values and artifacts and assess whether they correspond to each other or not. For example, if an espoused value is customer focus, identity the organization's systems of rewards or accountability (artifacts) and whether they support customer focus. If there is a mismatch, you have identified an area where a deep tacit assumption is operating and driving the system and creating an internal conflict. You now search for that assumption. There are many other such examples; for instance, the inherent conflict between an espoused value of open communications and the overt punishment of whistle-blowers or employees who bring bad news.

  In general, deep tacit assumptions are revealed by (1) identifying inconsistencies between artifacts (i.e., overt behavior, policies, rules, and practices) and espoused values (i.e., as formulated in vision statements, policies, and other management communications). It is sometimes the case that the espoused values are stated in a large organizational setting such as a town hall meeting. The next step is (2) identify what is driving the overt behaviors and artifacts, and, lastly, (3) list the deep assumptions as you uncover them.

- *Repeat the process with other groups.* Your analysis will be more meaningful by repeating the process several times.
- *Assess the shared assumptions.* Assess the pattern of shared basic assumptions in terms of how they aid or hinder the accomplishment of the goals from the first step. Then focus on identifying those assumptions that can help the organization—this should be the focus of management attention. If, in fact, specific assumptions are constraints, then you make a plan to change those elements, if possible, by exploiting the positive elements of your culture. "Try to see your culture as a positive force to be used rather than a constraint to be overcome" (Schein 1999).

Our experience has been that this exercise and other strategic exercises that involve the sharing of ideas are effectively addressed in an offsite.

6.3.3 Organizational Dysfunction and Learning

Organizational dysfunction may arise if the three levels shown in Figure 6.7 do not correspond to each other. For example, Schein (1996) points out that many organizations espouse teamwork (the middle level of articulated values), but the incentive and control systems (the top level of explicit behavior) of the organization reflects the shared tacit assumption (the bottom level) that "only individuals can be accountable and that the best results come from a system of individual competition and rewards." Thus, there is a mismatch between what the organization claims and what it actually believes and acts on.

6.3.3.1 Organizational Subcultures: Occupational Subcultures

Organizations are composed of so many subcultures that it is amazing that anything gets done. The organization operates because the different beliefs and values of each subculture, for the most part, have little to do with the work of the organization. In our school, for example, there exists any number of religious and national cultures (as seen in the many holidays and festivals), none of which interferes with the learning process. In fact, the opposite is true—cultural diversity is an opportunity to learn and enrich our lives!

However, the existence of *occupational* subcultures in a work environment is a different matter. Our occupations represent a lifetime of education, training, work experience, and relationships—all of which are part of the fabric of an organization. A problem arises because, by definition, each occupational subculture has a unique set of assumptions that may interfere with organizational communications and learning.

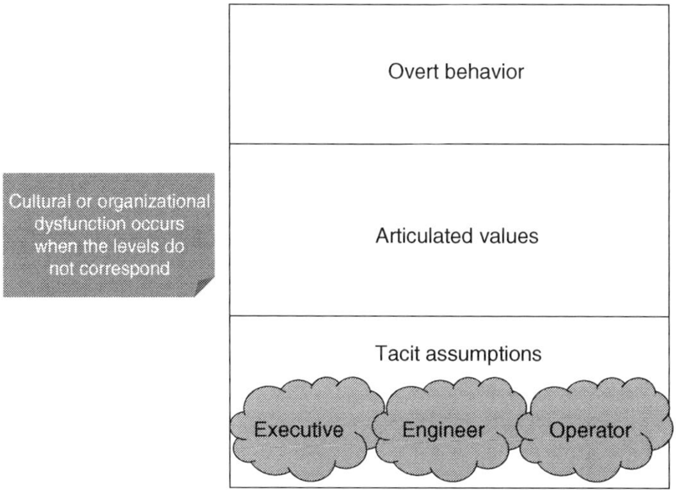

Figure 6.8 Generic occupational subcultures—Adapted from Schein (1999).

Schein (1996) has identified three occupational subcultures that are generic to organizations: the *operators, engineers,* and the *executives.* Each has its own set of contradictory assumptions that prevent learning across the organization (see Figure 6.8 and Table 6.2).

Observe that the operators are closest to the work and value social interaction and K exchange. In contrast, the engineers design the work. They have a technical perspective and view K as embedded in the formality and technology of work; hence, engineers value automation. The executives are furthest removed from the work and are concerned with elements that surround and sometimes run contrary to K work. In sequence, the operators, engineers, and executives practice late to early K-binding, respectively.

Though the assumptions listed in Table 6.2 may seem harsh, they encapsulate the culture of the industrial society; in fact, industrial society owes its success to the engineer subculture and the notion of early K-binding. In contrast, *the requirements of the K organization are forcing a shift in design assumptions and outcomes.* Today, the design of K organizations promotes human interaction and K exchange; that is, a move away from the assumptions of the engineer and executive subcultures (without undoing their obvious advantages) to one which lends greater emphasis to the assumptions of the operator subculture. The K era requires that we attend to the K worker, that is, to those people that are closest to the work.

Industry Example: Occupational Subcultures

Before the divesture of AT&T (i.e., the breakup of the telephone monopoly in the United States), AT&T was a single entity with many subsidiary companies.

Table 6.2 Sample occupational assumptions.

Operators	Engineers	Executives
Organizational success depends on people's skill, K, learning ability, and commitment	Engineers are stimulated by puzzles and problems and are pragmatic perfectionists who prefer "people-free" solutions	Focus on financial survival and growth to ensure returns to shareholders
The required K and skill are local and based on the organization's core technology	The ideal world is one of elegant machines and processes working in perfect precision and harmony without human intervention	People are a necessary evil, not an intrinsic value
No matter how carefully engineered the process or rules specified, operators must have the capacity to learn and to deal with surprises		Organization and management are intrinsically hierarchical; the hierarchy is a measure of status and success and the primary means of control
Operators must be able to work as a collaborative team in which openness, trust, communications, and commitment are values		Executives cannot get reliable data from subordinates so they must trust their own judgment

Adapted from Schein (1999).

Each had a specific function with a unique set of assumptions. Research was carried out by Bell Laboratories, which was composed of scientists and engineers; that is, the engineer occupational subculture. Western Electric was the manufacturing arm and had primarily an engineer subculture. We may distinguish Bell Labs and Western Electric in that the former was primarily research oriented (i.e., a subtype of the engineer subculture), whereas the latter was essentially an engineer subculture. The operating companies handled the day-to-day operations of a telephone company and had the operator subculture. Finally, AT&T corporate managed the relationships between the other subsidiary companies as well as with the government and was thus characterized by an executive subculture. This arrangement worked amazingly well, primarily because the different occupational subcultures were physically separated (except for controlled interactions), and thus each was able to flourish.

With the breakup, the operating companies took control of the research arm and the results were less than desirable. The collision of engineer and operator subcultures, each with unique assumptions, prevented effective interaction and learning. Furthermore, the operator subculture had primarily calculative or coercive psychological contracts, whereas the research arm had mostly cooperative contracts. Neither could flourish in the same way as previously, and the research arm was eventually sold.

6.3.4 Comparative Cultures

Table 6.3 is a comparison of tacit- and explicit-oriented cultures and includes both assumptions and visible artifacts. This was compiled primarily from Nonaka but also from others, such as Spender. There are several key observations, such as the different K creation modes that are emphasized, the different emphasis (experience versus analysis) and with that, different risks. Overadaptation—a risk associated with tacit-oriented cultures—is explained with an example here.

Historical Example: Overadaptation

During Japan's modernization in the Meiji Period following its opening to the world, the Japanese navy was influenced by the writings of Alfred Mahan, a US naval theoretician. Mahan advocated a single decisive battle to ensure victory. This strategy was immensely successful against the Russian navy in the early 20th century. With a single victorious battle against the Russian navy, Japan became the first Asian power to defeat a Western empire. This same strategy was successfully used again in the duel attacks on Pearl Harbor and Southeast

Table 6.3 Comparing tacit- and explicit-oriented cultures.

Tacit oriented	Explicit oriented
• Strong on socialization and internalization	• Strong on externalization and combination
• Emphasis on experience	• Emphasis on analysis
• Risk: overadaptation to past successes	• Risk: analysis paralysis
• Ambiguous organizational intention	• Clear organizational intention
• Group autonomy	• Individual autonomy
• Creative chaos through overlapping tasks	• Creative chaos through individual differences
• Redundant information	• Less redundant information
• Middle managers operate as knowledge engineers	• Middle managers operate as information relays
• Middle–up–down processes	• Top-down management processes
• Organization as a knowledge creating entity	• Organization as an information processing mechanism

Adapted from Nonaka, Spender, etc.

Asia. Japan had anticipated that the United States, as Russia decades earlier, would sue for peace; instead, they awakened a "sleeping giant." The Japanese navy spent the rest of World War II searching for another decisive battle. Two such encounters took place in the Solomon Islands and at Midway. With Japan's defeat, the war was essentially over. Overadaptation was a silent strategic partner in the US victory.

6.4 Industry Example of Culture: Toyota

Table 6.4 and Figures 6.9, 6.10, 6.11, 6.12, and 6.13 illustrate interconnected dimensions of Toyota's culture (Liker and Hoseus 2008). We briefly describe each here:

- *Toyota cultural manifestations*. Figure 6.9 shows an instantiation of Schein's model of organizational culture. Observe the broad social perspective with its emphasis on people and community. Note too that several artifacts in the top layer are Toyota's representations of itself and appear in the figures later; that is, these artifacts are working manifestations of assumptions, policies, and designs.
- *Toyota DNA*. Figure 6.10 is a representation of Toyota's DNA and is a variation of Figure 5.30. We see that the two strands of the Toyota Way and the Toyota Production System (TPS) (each a cultural artifact in Figure 6.9) represent the people value stream and the product value stream, respectively. The product value steam is typical of lean organizations and is often characterized with a value stream map or what Toyota calls one-piece flow.

Table 6.4 Blended culture.

			Blended and balanced
Western culture: what people want!	**Eastern culture: what people want!**	**Rewards and recognition in Toyota Japan**	**Rewards and recognition in Toyota USA**
Individually recognized for their accomplishments	Give the group credit for success and take personal blame for failure: individual recognition is embarrassing	Pay is the same for a given job classification and grows with seniority; any individual recognition is deflected to "we did it as a team"	Pay is based on job classification, seniority, company performance, plant performance for production workers and adds individual bonuses for management

Adapted from Liker and Hoseus (2008).

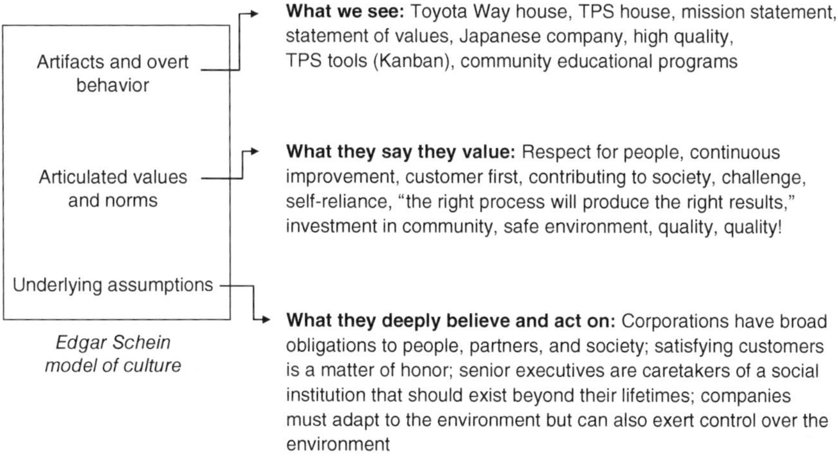

Figure 6.9 Toyota cultural manifestations—Adapted from Liker and Hoseus (2008).

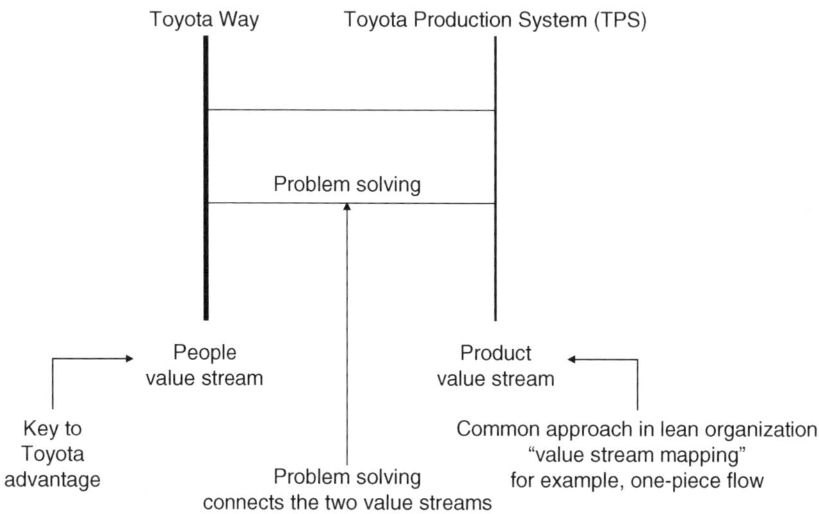

Figure 6.10 Toyota DNA—Adapted from Liker and Hoseus (2008).

The people value stream is a formalized approach by Toyota to cultivate its people. The two value streams are connected with a *problem solving* cross-linkage. That is to say, problem solving is the strategy that may be used to simultaneously cultivate people and product value streams. As we discussed in Chapter 5, we believe problem solving directly supports the critical K activities of the organization, builds trust, and develops and engages K workers.

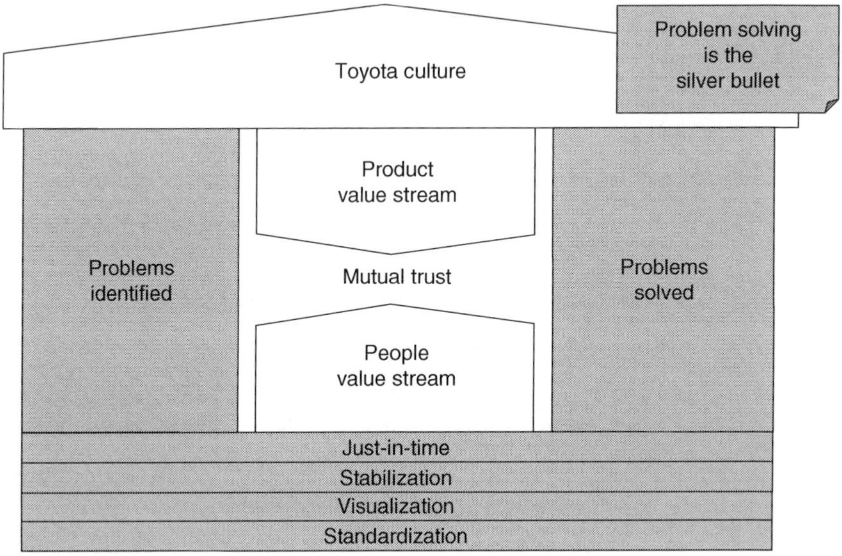

Figure 6.11 Toyota Way House: culture with embedded DNA—Adapted from Liker and Hoseus (2008).

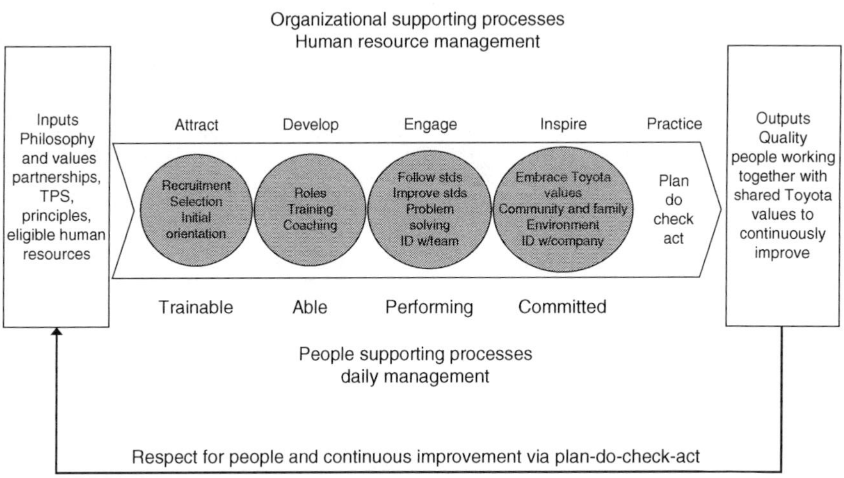

Figure 6.12 People value stream—Adapted from Liker and Hoseus (2008).

- *Toyota culture with embedded DNA*. Figure 6.11 is known as the Toyota Way House and is a representation of Toyota's culture. The four foundation principles (i.e., just-in-time, stabilization, visualization, and standardization) are part of another house—the TPS House. Observe that resting on the TPS principles is problem solving. Further, the basis for problem solving—connecting the

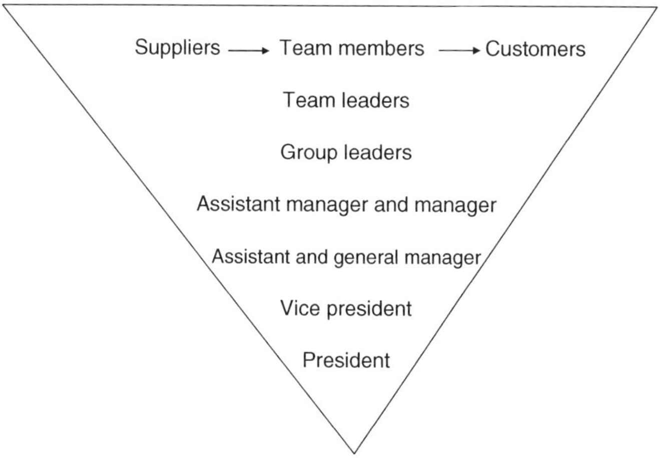

Figure 6.13 Servant leadership—Adapted from Liker and Hoseus (2008).

product and people value streams—is *mutual trust* (an important ingredient for K activities). It is instructive to note that Toyota employs two "house" artifacts: the TPS House emphasizes the lean production process (DNA thin strand) and the Way House emphasizes culture and people (DNA thick strand).

- *People value stream.* Figure 6.12 is a representation of Toyota's people value stream. This is a realized artifact—an actual process and policy that Toyota uses with its employees. The people value stream is a central component in Toyota's cultural profile and is foundational to the Toyota Way (thickness). Its emphasis is on the education and training of people and includes, for example, attracting the right people, and developing, engaging, and inspiring them. This is validated (with feedback) through their practice, which in Toyota is "plan–do–check–act." Finally, the people value stream is supported with two sets of processes: organizational-level HR processes and low-level people supporting processes. We will revisit hiring and educating K workers in the next chapter, but for now it is important to understand that *a people value stream of some sort is the most critical component of a K culture. Manifest investment in people is perhaps the single best way to both elevate and engage K workers.*
- *Servant leadership.* Figure 6.13 is an illustration of a leadership artifact. Observe that the traditional hierarchy is upside-down. This means that responsibility for motivating, training, and engaging people (i.e., realizing the people value stream) is in the hands of superiors and direct managers. The philosophy behind servant leadership includes "the team member is the expert, focus on the problem not the person, mistakes are okay as long as people learn from them, take care of the people building the cars, and you work for your team members." There are two key ideas to servant leadership.

The first is that the team members are in the value stream. The second is that the further removed from the value stream, the less that person directly adds value. In the latter case, value is added by supporting those above you in the upside-down pyramid. The ideas illustrated with the servant artifact embody what we have called late K-binding and thickness in general.

- *Blended culture.* An organization operating in a global environment must adapt itself to customers and employees of differing cultures, each with its own set of tacit assumptions. How do international organizations manage cultural integration? Table 6.4 is a sample of the cultural assumptions and its policy implications in Toyota USA. Observe that Toyota is attempting to blend and balance its practices based on differing cultural assumptions. This is a Toyota perspective—each global organization must make its own judgment on the assumptions of its customers and employees and what to do about them. For example, our research indicates that several Japanese organizations are moving away from the seniority-based reward system common in Toyota to one based on performance (as in the United States).

6.5 Trust

Trust is central to the structure and workings of an organization. The concept of trust underlies all relationships at every level of analysis, including the social interactions and cultural assumptions that we have been discussing.

How would we define trust? Is there a difference between trust and, say, cooperation, confidence, or predictability? Also, what causes trust and what are its expected outcomes? Should my union trust our company? Is my performance review honest, or are there other, even ignoble factors at work? Can I trust the members of my team—in all aspects of work, commitment, and protection? Can I trust my manger to look out for me in difficult times? And so on.

6.5.1 Trust Between Individuals

The definition of trust proposed by Mayer *et al.* (1995) is "the willingness of a party to be vulnerable to the actions of another party based on the expectation that the other will perform a particular action important to the trustor, irrespective of the ability to monitor or control that other party." Mayer's research identifies the components or factors of trust as *ability* (competence or expertise), *benevolence* (having the interest of the trustor in mind), and *integrity* (ethical behavior that the trustor finds acceptable). Thus the question "do you trust them?" must be qualified and replaced by "do you trust them to do what?" That is, our trust in someone depends not only on our assessment of his or her integrity and benevolence but also on the ability to accomplish a given task.

In general, trust in one area is not transferrable to dissimilar tasks or situations. From this model we may explain any number of common interactions: whom do we seek for advice in a particular situation, for emotional support, for political support, for allies in a particular initiative or project, or simply for information?

Note too that this research concentrates on trust between two individuals in *one direction*: from a specific trustor to a specific trustee. Thus, the authors' research addresses neither mutual trust between two individuals nor trust as an aspect of a wider social system.

6.5.2 Trust and Relationships: Social Architecture

Whom would you trust: a man who thinks honesty is the best policy or an honest man? The question is not inconsequential. The former is based on *policy*, meaning it can change (sometimes very quickly) for any number of reasons. The latter is a *characteristic* of a person and is more resistant to change (Kay 1996). Stated another way, policies are thin, but human characteristics are thick. Similarly, organizations contain thin and thick characteristics in unequal amounts—for example, in your organization, is customer relationship management a policy or an organizational characteristic?

In contrast to the claim of economists that people and institutions maximize things, Kay (1995) asserts that in the real world neither people nor institutions maximize anything. They try to satisfy both *objectives* and *outcomes*, and when in conflict, they attempt a balance and do not try to maximize one or the other. For example, as professors, do we excite and stimulate students or do we impart information? We try to do both—and good professors know how to strike a good balance.

Understanding and balancing both policies and characteristics, or objectives and outcomes, are behind relational contracting. Recall from Chapter 2 that we distinguished between transactional and relational contracting (see Table 2.4). Note, for example, that altruistic behavior is significant with relational contracting, whereas none is expected with transactional contracting. Relational contracts also tend to have long durations, whereas transactional contracts are short and apply to the transaction at hand.

The network of relational contracts, embedded in the culture of an organization, may be thought of as the *social architecture* (our term; Kay (1995) simply refers to architecture) of an organization. Social architecture requires that we distinguish between the attributes of the organization and the attributes of individuals. Organizations in certain industries (e.g., investment banking) sometimes depend on star performers. In explicit-oriented societies, such as the United States, most interactions are governed through transactional contracts including the legal constraints that accompany such agreements. On the other hand, tacit-oriented societies rely strongly on their social architecture. We may extend this thinking: transactional contracting characterizes the industrial era, whereas relational contracting and social architecture characterize the K era.

The success of most K organizations rests on its social architecture—the dependency of an organization on its people and their networks as a whole, because the output of the organization is the product of those people as a group. Kay (1995) further describes internal architecture as the structure of relational contracts between the firm and its employees and among the employees; external architecture is found in relational contracts between firms. Both internal and external architectures promote the sharing of K and, as we will see later, are of different forms, where each form is dependent on the character of network ties and culture within the context that characterizes social architecture.

6.5.3 Trust and Culture

Whole Foods builds trust into their concept of a "conscious culture." The qualities of a conscious culture are TACTILE: trust, accountability, caring, transparency, integrity, loyalty, and egalitarianism. Building trust is slow and difficult. Whole Foods builds high trust by (1) being truly purpose-motivated, (2) fostering conscious leadership, (3) recognizing that trust is reciprocal, and (4) transparency. Observe that these attributes enable K sharing and have been discussed previously. For example, in terms of reciprocity, trust is, in effect, a relational contract between management and employees and among the employees. Leaders must trust (and entrust) employees by granting them the necessary autonomy for their work. We have seen this before. Recall the pencil metaphor described in Chapter 4: a given assembly line worker in Toyota isn't simply given the responsibility for continuous improvement—the organization has its (reciprocal) role too: to constantly train, to accord authority and responsibility, to trust, and to grant autonomy to make decisions.

It is clear that trust in the social context of an organization is intertwined with and is a dimension of culture. Recall that Figure 6.11 is a cultural artifact for Toyota, known as the Toyota Way House, and is a representation of Toyota's culture. This is an important artifact to Toyota (one of only two "houses"). Observe that *mutual trust* holds a central position, acting as the only linkage between the product and people value streams and between identifying and solving problems. As with Whole Foods, Toyota's approach to building mutual trust is multidimensional. Toyota emphasizes communications as part of its people value stream. Toyota also provides benefits in keeping with a tacit-oriented Japanese culture; that is, benefits that encourage stability, long-term security, and trust. The assumption is that long-term mutual prosperity will lead to mutual trust.

6.6 Illustrative Example: Interorganizational K Exchange and Creation—Effects of Ties and Culture

Ahmadjian (2004) applied the core Nonaka concepts of SECI (socialization–externalization–combination–internalization) and ba to a study of interorganizational K creation in two very different environments: Toyota and Silicon Valley (SV). Toyota's competitive advantage comes in part from its ability to create K-networks with its suppliers; in contrast, SV's innovation originates from multiple networks of individuals, organizations, and universities. In this analysis, we will uncover the interconnected effects of ties, culture, ba, and K creation across organizational boundaries.

The ba associated with Toyota reflects a tacit-oriented Japanese culture. The ba consists of relatively few organizations (primarily suppliers) emanating from the central Toyota hub. The ties are intentionally facilitated and are few, strong, long-lasting, and stable. Central to the energy associated with the Toyota ba is the exclusiveness of being a Toyota supplier and partner and the shared sense of history and identity. In fact, the Toyota ba and structure of its organizational arrangements characterize the Japanese economy generally. The K created is similarly tacit and focused on the processes between suppliers and Toyota. Also, shared tacit understanding allows Toyota to formulate only general specifications for parts; it is left to the individual suppliers to define specific specifications, which are then shared across all suppliers. This is effected through a variety of techniques, such as an exchange of employees ("shukko" which is similar to job rotation) where tacit K is both shared and created and is subsequently codified into explicit routines and procedures.

In contrast, the SV ba is driven by a shared culture, values, language, education, and physical space. The ba is constituted of many weak and dynamic ties—the ties develop as a consequence of the environment. Specifically, the SV environment includes a common region, university affiliations and education experiences, and companies. Whereas Toyota shares people (tacit K) across organizational boundaries, SV shares product ideas and features. Innovation in SV derives from its many networks that *recombine explicit threads of K*. Whereas Toyota focuses on process, SV focuses on product. Also, in SV, knowledge is a property of the region and industry and not a company.

We can see the impact of culture on these K-creating arrangements. Japan is a tacit-oriented culture whose ba reflects that culture—Toyota's success in K creation is concentrated in developing interorganizational tacit K. In contrast, SV's ba drives explicit exchange and K recombination. There is also an interesting paradox: the Toyota model is tacit-oriented but characterized by early K-binding (i.e., intended, strong ties); in contrast, SV is explicit-oriented but largely characterized by late K-binding (i.e., dynamic, weak ties).

Finally, the research suggests that Japan may be losing its competitive innovation capabilities. The Toyota model contains a small number of interactions and is in contrast to the free-flowing exchange of (explicit) K that characterizes the SV model. The key challenge for Japanese companies is to collaborate outside the boundaries of closely linked business groups. That is to say, the risk of a tacit-oriented network and culture is its reliance on strong ties, which trends toward a steady state with few innovations. In contrast, a culture characterized by many weak ties is characterized by K recombination and innovation.

From this case study, we may make an additional observation: *The ba and associated K-spaces, formed in part by history and culture, drive the strength or weakness of ties, which in turn largely dictate the character of K flows and the type of K created.*

6.7 Collaboration

Here, we briefly discuss the role of culture, trust, structure, and systems in collaboration between organizations. The foundation for this discussion is largely taken from the literature on drug discovery—a large-scale scientific endeavor with enormous impact on our lives. Over the past 100 years, drug discovery has shifted in character from the serendipitous to the structured, the latter being a highly structured process with a 99% failure rate that is largely automated, regulated, expensive, and lengthy. In response, organizations are shifting to an approach that *externalizes research*—collaborations between life science partners throughout the R&D cycle (Waller *et al.* 2011).

6.7.1 Collaborative Innovation

"Innovation has shifted its venue from the individual to the group: almost all innovation today is done collaboratively, in teams, networks, or alliances" (Lynch 2011). Further, to solve what we have previously called messy problems, multidisciplinary teams are essential. Several bodies of K and perspectives are necessary; it is therefore an imperative, particularly in drug development or other forms of scientific research, for companies, universities, and laboratories to work in a seamless, synchronistic, and synergistic manner.

The driving force behind multidisciplinary teams is "differentials." That is to say, differences in perspectives and expertise lead to new questions that challenge the status quo. Creativity requires divergent thinking, that is, generating new ideas, whereas innovation demands convergent thinking, that is, combining those ideas into the optimum or best result. Collaboration among multidisciplinary teams with divergent ideas triggers creative friction, which in turn leads to a stream of ideas; innovation, on the other hand, requires the convergence, integration, and alignment of those ideas. The goal here is simple—to produce a breakthrough in thinking and results. Collaboration as a *central organizing principle* for all research and development will lead to greater success than individual efforts.

Example: The joint insight by Watson and Crick regarding the double-helix structure of DNA is an example of the type of collaboration we have been discussing. Watson and Crick each came from a different perspective and both were open to new ideas. They never conducted any experiments; instead, they looked at the data of other researchers and interpreted that data from a fresh perspective. By integrating the work of others, Watson and Crick saw patterns in data that enabled their insight into the structure of DNA.

6.7.2 Culture and Collaboration

The characteristics of culture that support collaboration, particularly large-scale collaborative innovation are as follows:

- *Selecting the right people.* This includes selecting people with the appropriate level of competence, high character (i.e., trustworthy), willingness to collaborate (not necessarily easy to locate in the scientific community), creative, courageous, and who bring (collectivity) the necessary cognitive diversity (i.e., differentials).
- *Building a system of synergistic trust.* Lynch (2011) describes a "ladder of trust" that includes, going up the ladder from the first level (transaction) to the highest level of trust (creationship). Observe that the ladder of trust is actually a progression along the contract continuum, as shown in Figures 2.16 and 2.24:
  - Transaction—mutual self-interest
  - Relationship—listening, respect, integrity, and fairness
  - Guardianship—honor, predictability win–win, and security
  - Companionship—teamwork, camaraderie, and excellence
  - Fellowship—compassion and belonging
  - Friendship—loyalty and personal commitment
  - Partnership—shared vision and risk–reward
  - Creationship—innovation and synergy

Associated with the development of trust are the following aspects of organizational design.

- *Eliminate fear.* There is a paradox in scientific analysis and human behavior: on the one hand legitimate inquiry (asking probing questions) and on the other skeptical, almost inquisition-like questions. One solution is to have the work group think of a new way to frame a problem or to find a new problem that no one had noticed before. Inquiry versus inquisition through open or closed questions reveals several interesting observations. See Figure 6.14.
  - Quadrant 1. Traditional scientific inquiry. Data and analytic driven.

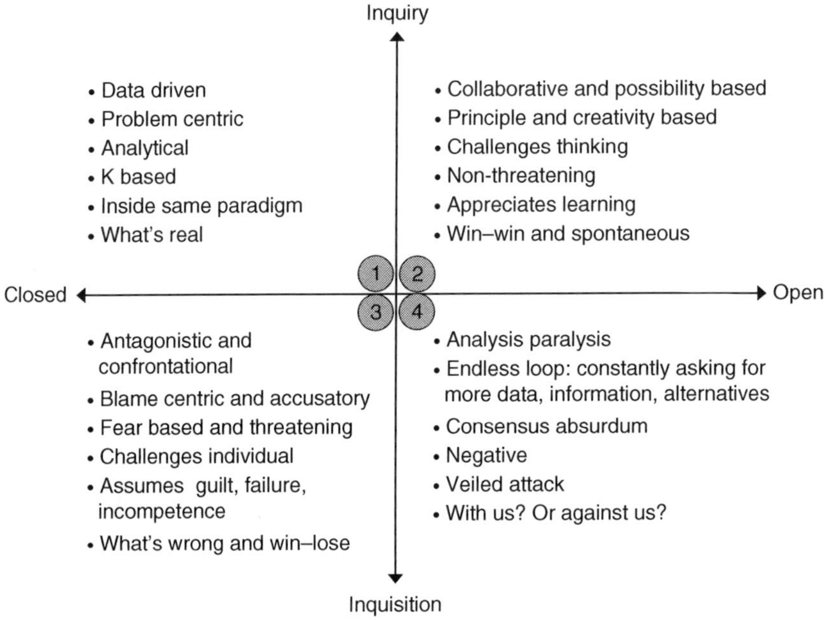

Figure 6.14 Inquiry versus inquisition: open and closed questions—Adapted from Lynch (2011).

  - Quadrant 2. Open inquiry. Exploratory. Joint creativity.
  - Quadrant 3. Closed-ended questions requiring yes or no answers.
  - Quadrant 4. Questions try to discover who is at fault. Forensic work.
- *Empower innovation champions.* The effective champion bridges ideas from different domains, exploits the network structure of ideas, and builds communities.

6.7.3 Collaboration Structures and Databases

The shift to a collaborative model implies a shift from a closed to an open model of R&D among several organizations. The closed model is premised on the discovery and development of new drugs and patenting those drugs to secure a monopoly on the profits of the newly developed drugs (Mayrand-Chung *et al.* 2011). In the past, partnerships were primarily concerned with sharing reagents, such as cell lines or antibodies.

In contrast, a consortium is a multistakeholder effort developed in order to undertake a large-scale initiative that no single organization could carry out alone. A key question is whether the collaborative initiative will be organized as a competitive or precompetitive initiative. In the case of a precompetitive initiative, the outcomes are intended to benefit the scientific community as opposed to the participants in the effort. Such consortia often include government agencies, such as the US National Institutes of Health (NIH).

For example, the Alzheimer's Disease Neuroimaging Initiative is the NIH's largest public–private partnership focused on brain research. Structurally, the initiative includes a steering committee, an industry scientific advisory board, and a variety of research-related "cores," such as genetics, biomarker, and clinical cores. Other similar collaborative initiatives, for instance, the Osteoarthritis Initiative, are emerging.

Database requirements. One systems requirement is for the capability to gather and merge data from many different groups. This requires the capture and integration of metadata from different groups and the creation of a database with the ability to join tables from different source schemas or databases.

Observe that these requirements correspond to classical data warehousing. This is one tool, then, that we may use to support collaboration.

6.7.3.1 Collaborative Database Infrastructure: Case Study

A collaborative team was established across the R&D group of the pharmaceutical companies of Johnson & Johnson. The purpose of the team was to establish a translational research infrastructure. The team was extended with open innovation partnerships with Rutgers University and the Cancer Institute of New Jersey (Szalma 2011).

The central paradigm was the development of a data warehouse—the translational medicine data warehouse (tranSMART)—for all the traditional reasons that organizations build data warehouses: the development of a consistent database with clean and transformed data from multiple sources. The data warehouse was established to answer translational use case questions, such as "what is the correlation between animal models and human data?" or "what is the best biomarker strategy for a given compound?" The data sources were internal (e.g., clinical data), master (e.g., gene and protein names), and public source (e.g., nonclinical gene expression data). The tranSMART data warehouse integrates data from multiple data silos across the drug discovery and development process. The data warehouse also contained internally developed tools for search and analysis (i.e., Dataset Explorer).

6.7.4 Collaboration Tools and Systems

Recall from Chapter 4 that we discussed the design of systems and tools with respect to a job role. For example, Figure 4.14a–g illustrates a portion of the analysis that a student team performed for the Knowledge Management Officer (KMO) at a large international bank. Figure 4.14g shows the key components (e.g., data warehouse, collaboration tools, K repository, etc.) of the KMO workstation. Recall too that both the workstation (1) components and the (2) contents of each component follow from the preceding K analysis for specific job functions. The collaboration component contained specific and sometimes generic tools such as instant messaging, online chat, and e-mail, among others.

We envision a variation to this design approach to support collaboration. That is, it is necessary to identify a specific problem or research area (as opposed to jobs) and individual and organizational responsibilities and constraints, categorize K and their sources, and the types of analyses, expected outcomes, and decisions required. Each environment must thus be specifically configured and designed to the purpose, knowledge, and function at hand.

In the case of a collaboration environment, the identification of K is complex—it requires a body of K that supports a major function (e.g., cancer research) and all the necessary interactive elements. As an example, the cancer biomedical informatics grid (caBIG) community is organized around three classes of workspaces. The first are domain workspaces organized by scientific discipline such as clinical trials. The second are "cross-cutting" workspaces that provide services such as common data elements. The third are strategic workspaces that facilitate planning, documentation, and training (Komatsoulis 2011).

Among the technology support are vocabulary services, a cancer data standard repository, and a software development kit for the development of application programming interfaces (API) and other information systems. The services include data and analytical, metadata, client application, higher (e.g., workflow), and security services. This type of analysis is not very different from that we described in Chapter 4, except that, in place of a specific job function (e.g., KMO), the function here (cancer research) is much broader and requires a research architecture to segment the K areas; hence, the idea of workspaces.

6.7.5 Collaboration and Collaboratories: Taxonomies

Among the major obstacles faced by collaboration groups are coordination challenges, trust building, and the communication of tacit K (Bos *et al.* 2008). While scientific communities are more egalitarian and international than most communities, they have specific problems that include the following:

- *Scientific K is difficult to aggregate.* Unlike information, K is difficult to transfer. A scientific idea is often cutting-edge, requires specialized expertise to understand, is difficult to represent, is constantly changing, and may be tacit—all dimensions that make scientific K difficult to communicate and work with in a collaborative environment.
- *Scientists work independently.* As with K workers generally, scientists like working independently in both their day-to-day activities and the larger direction of their work. Scientists resist controls that are commonplace in corporations. This culture of independence benefits science but makes it

difficult to aggregate the output of scientists' labors. Scientific collaborations require a particularly strong effort to maintain open communications channels, adopt common tool sets, and remain focused on common goals.

- *Difficulty of cross-institutional work.* Interorganizational work is more than crossing physical boundaries. It requires crossing cultural boundaries that includes overcoming policies and practices, issues with intellectual property, and so on.

The National Science Foundation funded the study of large-scale, Internet-based research collaborations across several disciplines (e.g., physics, medicine, and environmental sciences). The outcome of their study included theories about a new organizational form—*collaboratories*—and guidelines on how to construct successful collaboratories. Here we summarize the key finding—a seven-category taxonomy of collaboratories (Bos *et al.* 2008). Observe that this classification works well with most other forms of collaboration and serves as a useful management guide. Note too that the taxonomies correspond with the information continuum.

<u>Tool-oriented</u>

1) *Shared instrument.* This type of collaboration's primary function is to increase access to scientific instruments such as telescopes. Direct instrument access is often supplemented with communication tools—videoconferencing, chat, and so on.

   Among the technology issues are synchronous (real-time) communications, remote access, and managing large-output datasets. The key organizational issue is the allocation of access and oversubscription.
2) *Community infrastructure project.* The community infrastructure project seeks to develop an infrastructure—common resources—to support work in a particular domain. Community infrastructure projects are often interdisciplinary.

   The key technology issue concerns the development of standards for data and data collection. The key organizational issue is the negotiation of goals among the diverse disciplinary partners.

<u>Data-oriented</u>

3) *Community data system.* A community data system is a shared information resource that is created, updated, and maintained by a geographically distributed community. The resources are semipublic and of wide interest. Note that a project team with an online data system of documents would not be considered a community data system. The community data system may be viewed as a *public goods project.* An example is the Protein Data Bank, which is a single worldwide repository for large molecules and nucleic acids.

A key technology issue is the development of data standards that facilitate data storage and search as well as the development of modeling and visualization techniques. The central organizational issue is the motivation of contributors and the development of large-scale decision-making methods.

4) *Open community contribution system.* The open community contribution system concerns the aggregation of *work* from geographically separated individuals working toward a common research problem. This is in contrast to the community data system, which concerns common *data.* An example of an open community contribution system is Open Mind, which is an online system for collecting commonsense judgments from the general public. Open Mind collects and aggregates data, which is made freely available to commercial and noncommercial interests.

   The key technology issue is the development of a system that operates across platforms and is easy to use. The key organizational issue is the maintenance of quality control among a diverse and widely distributed body of contributors.

Knowledge-oriented

5) *Virtual community of practice.* A virtual CoP is a not very different than a standard CoP that we have discussed in previous chapters—a network of individuals who share a research area and communicate about it online. As with a traditional CoP, a virtual CoP shares K in the form of insights, advice, and techniques. Also, the CoP (virtual or otherwise) concentrates on the development of K and does not engage in joint projects. An example is Ocean US, a CoP for researchers studying oceans with a focus on US coastal waters.

   The main technology issue is usability, and hence the virtual CoP utilizes Internet standard technologies such as bulletin boards and Web technologies. The key organizational issue is the effort involved in maintaining enthusiasm and participation, and for a virtual CoP, a shifting set of participants.

6) *Virtual learning community.* The main goal of a virtual learning community is to increase the K of participants and not to conduct original research. The virtual learning community typically involves formal education by a degree-granting institution. An example is the Ecology Circuitry Collaboratory project, which educates and encourages young ecologists to combine empirical research with quantitative modeling.

   The key technology issue is the disparity of technology infrastructures between large universities, especially universities across the world. The key organizational issue is the alignment of educational goals and assessments across multiple universities.

7) *Distributed research center.* The distributed research center attempts to aggregate scientific talent, effort, and resources beyond that of individual

researchers, and thus may be seen as a university research center across distances. Like a CoP, the distributed research center is focused on a specific topic of interest where most communication is person-to-person; however, unlike a CoP the distributed research center undertakes specific projects. Note too that the participant base is more narrow than that found in an open community contribution system.

The key technology issues are similar to all the other collaboratories, including the standardization of data and long-distance tech support. The key organizational issues concern the maintenance of effort among diverse participants, standards, decision making, and administrative support.

6.8 Collaborating with Creatives

A word of caution when working with what Google terms "smart creatives." Certain types of strongly creative people tend to resist input from their colleagues; they are better at giving ideas than taking them (Elsbach *et al.* 2015). Such creative people sometimes think of themselves as "artists," separating themselves from other creative talent.

Elsbach *et al.* (2015) distinguish between creative people and artists, at least in a business setting. According to their research, 15–20% of people who do creative work see themselves as "creators of unique outputs that embody personal, artistic visions." That is, artists are distinguished by having a very personal stake in their work, preferring to work independently, and exhibiting high energy, all of which in turn makes them resistant to useful feedback. In contrast, most other creatives self-identify as "problem solvers" who are open to the ideas of colleagues, embrace working in collaborative groups, and help channel projects to commercial ends.

In their research, Elsbach *et al.* (2015) have identified three elements associated with artists in a business context: a creative signature style, control of how ideas are executed, and noncommercial motivation. Consequently, four tactics are suggested when interacting with artists: offer broad suggestions, temper your enthusiasm, delay decisions giving artists time to react, and show respect and like-mindedness.

We see a characteristic that emerges from this research: artists are less likely to be embedded in networks (in a work context) than creative problem solvers and other team players. Seclusion (to some degree) seems to better describe *artistes* than collaborators. In designing the K organization, therefore, certain artifacts surface; for example, artists seem better suited to private workspaces than open warehouse designs. Also, scientists and academics share K and interaction characteristics of both problem-solvers and artists. This will be discussed more fully later. See Figure 6.15.

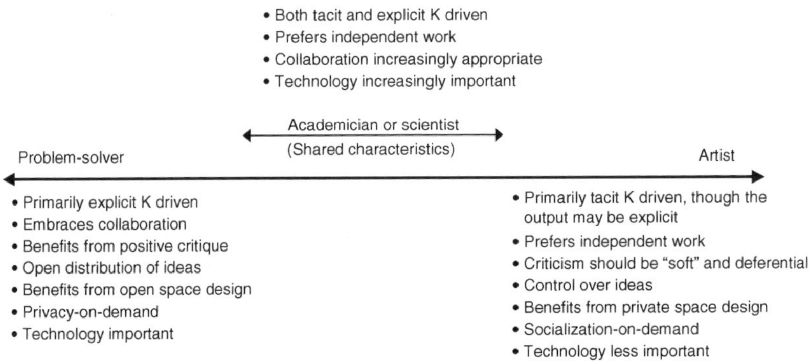

Figure 6.15 Design considerations with creatives.

Application: Artists

Let us consider filmmaking and the respective role of director and actor—each an artist. The director's primary prerogative (or responsibility) is telling the story. This means finding a structure to the script and setting up story events. The actor is given direction in order that the actor's actions and interactions illuminate and create those events. The actor's prerogative is to create truthful behavior in the context of the script and the director's direction. "Actor and director must respect each other's creative territory" (Weston 1996).

There is an important lesson here, even in a business context. Creativity can be structured through the prerogative or rights accorded the artist in his or her job role; that is, artistic prerogative should be structured and designed. The project "story"—in organizational terms, the strategy—and its events are crafted by the "director," whereas the individual artist/K worker is to give life to the story through his or her design of the product and its features. Individual actors interact, but only within the prerogative of each in their respective roles.

6.9 Office Design

In many of the organizations in which we teach, employees had been required to spend a significant amount of time working from home. There is an implicit trade-off operating here: the real estate cost of office space versus the benefits of social interaction among employees. The benefit associated with the former is realized almost immediately, whereas the benefits associated with the latter are manifold and often deferred. Among the benefits is the transfer and creation of K, particularly tacit K, which is optimally transferred (or absorbed) face-to-face. Another benefit is the development of social ties—weak and strong—and the solidarity that frequently follows. Finally, the effect on developing a K culture and a learning organization—for example, developing common assumptions—is enhanced with social interaction. We can only imagine that the senior managers in such companies believe that technology can sufficiently replace face-to-face interaction.

Good news! There is growing recognition that periodic interaction is not very different from social isolation and is *exactly opposite* as that required for a K organization. Today, many organizations are spending their efforts bringing employees back together through new office configurations to promote interaction and hence the development and transfer of tacit K, social ties, and K culture.

6.9.1 The Growing Shift to Open Office Design

In the following discussion, we make reference to three articles that appeared in the *Harvard Business Review* in October: (1) Congdon *et al.* (2014); (2) Bernstein (2014); and (3) Waber *et al.* (2014). These articles confirm our observation that there is a shift to open office design. In fact, Congdon *et al.* (2014) state that "more than 70% of employees work in an open office environment, and the size of their individual workspaces is shrinking." There are also cultural differences in space size: the average space allocated an employee is 320 ft^2 in Germany, 190 ft^2 in the United States, 70 ft^2 in India, and 50 ft^2 in China.

Is there an additional trade-off—privacy—to open office design? Specifically, the privacy issue includes control of the physical space, information, and stimulation. *Physical space* refers to the traditional perspective of privacy; *information control* concerns access, including personal and professional; and *stimulation control* concerns interruptions, distractions, and noise. Sections 6.9.1.1–6.9.1.3 are a brief summary of their findings. (The key limitation in this research concerns the effect of open design on creative thinking.)

6.9.1.1 Focusing Attention

There are several ways of thinking about attention. The first is *controlled attention* where work on a task requires intense focus, such as writing or thinking, while avoiding unrelated thoughts and inhibiting external stimuli. The second is *stimulus-driven attention* where focus is switched when something catches our attention. This is particularly common with routine work. The third is *rejuvenation* where there is a respite from thinking for both our body and brain.

Our experience is that all three modes may be necessary. As we will see, the factors include the personality of the person involved, the type of work being performed, and in the case of K workers, the type and intensity of K being exchanged or created.

6.9.1.2 Strategic Space Planning

The research discusses several space planning models as follows:

- Distributed model—spaces that support stimulation control are blended into areas for both private and group work. In this case, different types of spaces are placed in close physical proximity and the worker may move from one type of space to another.
- Zone model—defines certain locations within the larger workspace as private, quiet spaces, either a specific area or an entire floor as in a "library." In this case, it appears that the basic space is open, with reserved areas for private work.
- Ecosystem—the most successful environments provide a range of spaces, that is, an ecosystem, that allow people to choose where and how they get their jobs done. *The primary advantage is that workers control stimulation* through their choice of where to sit.

We will see each of these design philosophies in the student examples here.

6.9.1.3 Transparency Zones

Another factor in office design is *transparency.* It is anticipated that transparency promotes creativity and productivity; on the other hand, there is no evidence that an open office (one way of operationalizing transparency) actually improves either creativity or productivity (Waber *et al.* 2014). Also, recall from Chapter 2 that creativity, though related, is separate from K creation. A second type of transparency implementation is to use sensor technology to monitor and track worker communications. For example, warehouse employees in Amazon carry hand computers that track their every move. As a general concept, transparency takes many other forms; for example, individual compensation in Whole Foods is transparent to everyone in the company.

A zone is a type of boundary that seeks to balance privacy and transparency. The four popular zones include (1) boundaries around teams where there is full transparency within a team but limited transparency between teams. Also, implied in this type of arrangement is that each team selects the type of open office for itself. Another boundary zone concerns (2) feedback and evaluation, where each is separated from other organizational activities. A third zone is (3) the separation between decision and improvement rights—here too each is separate and distinct. Finally, there is (4) a boundary around time where organizations create a window of privacy and individuals are allocated a percentage of weekly time for individual work, a not uncommon occurrence today.

6.9.2 Student Exercise: Examples of Office Design

Students were asked to illustrate and critique the office space in their respective companies. Among the findings are the following:

- *Bank #1.* This large international bank seeks to promote an eco-office through a *beehive* design or smart floors. There are neither assigned spaces nor a visible hierarchy. The distinguishing feature of this design is that workers select their workspace. Private spaces are available as are lounge areas for face-to-face interaction. There is a reliance on virtual desktop technology for coordination.
- *Bank #2*: This is also a large international bank. Similar to the beehive, this space is more like a *warehouse* where everything is open, but workers are assigned specific workspaces. Here too there is reliance on a virtual desktop for coordination.
- *Consumer Product Organization*: This large international company is a consumer product leader with more than 400 brands. Innovation in science and technology is necessary for growth and improvement. This company employs a *semi-open* office design where each department has a small area separated from other areas by low partitions. Within each area, the manager has a separate office, whereas the staff sits in an open area. Each floor has several conference rooms. See Figure 6.16.
- *Laboratory.* The authors have worked in laboratory environments and make the following observation. The office space is essentially semi-open but with a heavy cultural aspect that lends itself to K creation. The lab space is organized around a department, each of which is separated with walls. Unlike the consumer product organization discussed earlier, the departmental spaces in a lab feel more isolated from each other. The employees occupy a private office (two employees per room) with a separate office for the manager. Within each two-person office there is abundant blackboard space for design representations or as a vehicle for social communications. There are conference rooms for departmental meetings. Social interaction is supported by a combination of office design and a collegiate culture. As necessary, individuals may interact with each other in their respective offices and typically use the blackboard as one part of their interaction. In this setting, interaction is realized through what we call *socialization-on-demand.* The authors' experience with this environment was very positive—there was worker-controlled *balance* between private and social work.
- *Academia.* A brief look at our particular floor at Stevens illustrates a variation to the lab environment. Fulltime faculty have private offices around the perimeter of the floor, whereas the center of the floor contains single-person cubicles separated by high partitions. The cubicle space is for visiting faculty, administrative staff, and teaching assistants. Every office and most cubicles are private. Everyone is "equal" in that most professors have an office of about the same size; directors have corner offices, and the Dean has a large corner office. The office layout is clearly designed to support *private work.* As with the lab environment, the collegiate culture supports socialization-on-demand; on the hand, the interaction here is not as common as in the laboratory.

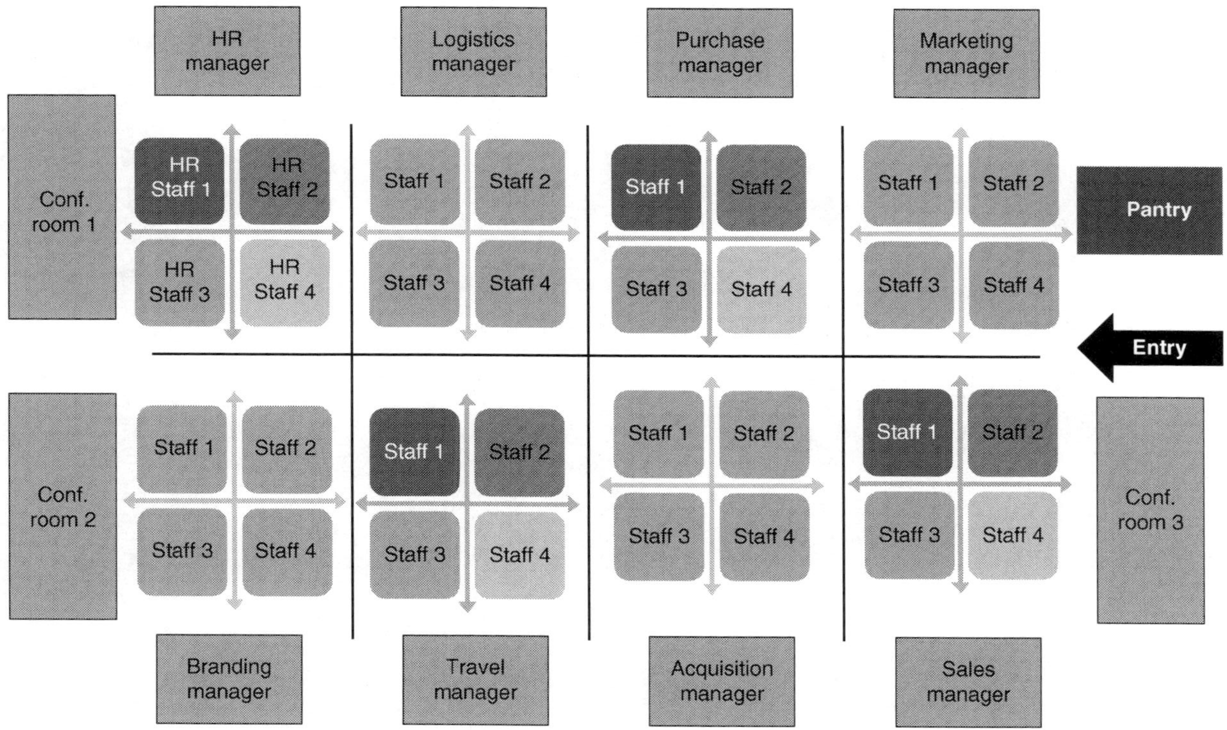

Figure 6.16 Sample office design for a consumer product organization.

- *Working from Home*. This *is* a design strategy—eliminating the office—where the worker is *physically isolated* from other workers, with periodic visits to a physical office. We know of cases where employees in large international banks were obliged to sign contracts requiring that they work from home. In this strategy, it is generally believed that technology—for example, video conferencing—will satisfactorily replace face-to-face interaction, perhaps not fully but sufficiently to justify the reduction in costs that this office arrangement provides.

6.9.3 Office Space Analysis

The office and building layout of an organization is a physical artifact of its culture. Accordingly, an organization's physical layout says something about its underlying assumptions, values, and norms, or at least is a statement of where the organization wants to evolve its culture. It also says a great deal about the nature of work and the type of interaction that is expected.

K work requires social interaction—the issue here is the form and frequency of such interaction. Interaction, for example, may center on virtual collaboration and rely primarily on technology. *On the other hand, physical space provides a unique opportunity for face-to-face socialization*. Recall too that K work is multidimensional and may require private time—to think and reflect as individuals. The issue, then, is *the right mix of social and private contexts within which work takes place*.

Figure 6.17 illustrates a range of office configurations from freeform to isolation. In the beehive design, social collisions are individually selected—workers select their space, and hence, interaction opportunities may be either serendipitous or intended. The warehouse is similar, except there is more management control over social interaction and chance collisions appear to be less likely. The semi-open space is best characterized as targeted socialization within a department. We view the semi-open space as a transition between the open design of a service organization and the more private-oriented work found in a lab or academia.

Beehive	Warehouse	Semi-open	Balanced	Private	Isolated
• Open • No assigned space • Access to private space • Subsidiary or product-specific K	• Open • Assigned space • Access to private space • Subsidiary or product-specific K	• Open within department • Partitions between departments • Product-specific expertise	• Private offices within partitioned departmental space • Socialization-on-demand • Product-specific and functional K	• Private offices • Socialization-on-demand • Functional K	• Work from home • Tech dependent

Figure 6.17 Open-private range of office configurations.

In the balanced space (lab context), the presence of *individual* offices (two workers per office) places greater emphasis on private work. On the other hand, the culture is collegiate, and hence, socialization-on-demand is quite common. Similarly, academia space is designed primarily for private work; here too there is socialization-on-demand, though in our experience it is less common than in the balanced/lab environment. There is also a cultural aspect to both the lab and academia environment: each has a history (centuries old in the case of academia) where a private office may be considered a mark of prestige.

The category of office space corresponds with the required level and depth of K as well as the frequency of social interaction. Observe in Figure 6.17 that the character of knowledge shifts from subsidiary K (beehive and warehouse) to one where there is more product-specific expertise (semi-open) to some combination of product and function (laboratory) to primarily function (academia). Note too that where product design or customer-facing activities are part of the business, there is a bias toward social interaction; where scientific or functional research is required, there is a bias toward private work. (The character of K involved with an individual working from home is somewhat open since the motivation for this design is economic and not K-specific.)

Building Design. Independent of its office design, an organization may create opportunities for accidental collisions, casual conversations, respites to unwind, and occasions to renew old acquaintances. This may be effected through (1) private areas within an open design or (2) an open floor, separate from where work takes place, that functions as a lounge with amenities such as refreshments or video streams. A third option, more common in a laboratory or academic organization, may be (3) a dedicated library space or building which facilitates both private work and where social interactions are more varied and random.

Serendipitous Design. It may be advantageous to promote serendipitous interactions in other organizational contexts, including deliberately designed informal collisions. Open office is a good start; however, the type and extent of K must be factored into the design; again, K work requires the right prescription of social and private ingredients. Other possibilities exist, such as an *offsite* where members of an organization strategize and socialize away from the distractions of the office.

Observations from Google: "Keep Them Crowded"

The underlying philosophy of Google's office design is to counter the steady state of quietude in a space consisting of offices and cubicles. In such a space, encounters are planned in conference room meetings or occasionally serendipitous at the coffee machine. Instead, Google implements a very different steady state, one that is "highly interactive, with boisterous, crowded offices brimming with hectic energy" (Schmidt and Rosenberg 2014).

6.10 Promoting Conversations and Dialogue

Recall our discussion of the *coffee metaphor* in Chapter 4: in the same way that coffee needs heat to bring forth its flavor (the coffee beans themselves representing intrinsic expertise and cognitive knowledge), it is necessary to apply heat in the form of social interaction for K to fully develop and flourish. Recall too that tacit knowledge consists of three dimensions: technical or motor skill, cognitive or mental model, and a social-group dimension that embodies practice. Thus, it is social interaction and dialogue—both formal and informal conversations—that elicit individual and group K and form the basis of all K.

Justification—rationalizing beliefs and truths—is not only an individual exercise but one that depends on discourse and justification to a wider social group. Recall that justification is the everyday work of K workers. More generally, justification through social interaction brings individual and social K into harmony.

Good conversations seem simple enough, but let us consider the possible factors that affect both the existence and the efficacy of such conversations. In fact, everything that affects knowledging is somehow related to social interaction and dialogue.

Figure 6.18 is a listing of the key factors that affect discourse in a variety of contexts. We say “key factors” in that virtually everything we have been discussing in this book affects KM and conversations; accordingly, we believe Figure 6.18 lists the most prominent factors. We have discussed several factors and continue our discussion here.

6.10.1 Dialogue Strategies: Lexus Case Study

Emi Osono (2004) studied the strategy-making process at Lexus, focusing on dialogue. Though the strategy-making process and dialogue may be viewed as two separate topics (each with a strong impact on KM), his analysis of Lexus' product development illustrates an interrelationship. Specifically, his case study revealed two distinct dialogue patterns associated with corresponding strategy-making processes.

Intended strategy is organizational strategy developed by identifiable decision-makers, such as senior managers, teams, departments, or consultants.

- Organizational dialogue strategy
- Managing conversations
- Individual personality types
- Business system personality types
- Network structures
- Office space layout
- Ba and K-spaces
- Culture and trust
- Job role and required K

Figure 6.18 Discourse factors.

It is the rational and analytical outcome of conscious choice. Among the possible issues with intended strategy are not fully understanding the organization's context (e.g., customers, competitors), the organization's capabilities, and the lead time between concept and implementation. The key risk associated with intended strategy is that it largely depends on successful communications (down the hierarchy) to all members of the organization. This is particularly true of front-line managers.

Emergent strategy occurs without organizational intent, usually as the result of administrative systems, such as the organization's structure, planning and resource allocation processes, human resource policies, and so on. There is significant environmental influence and emergent strategies may play a major role in an organization's response to crises or unexpected opportunities. Emergent strategy is close to the environment and market and is largely based on local K and action; therefore, a key issue is successful communications in the larger corporate setting (up the hierarchy).

Univocality is concerned with the degree to which communications is dominated by a single perspective. In effect, what is being communicated (intended strategy) has already been decided, with little room for other viewpoints. In contrast, *multivocality* is concerned with seeking out many perspectives. It is dynamic, heterogeneous, and the final strategy (emergent) represents a synthesis of the perspectives.

Neither combination of strategy and dialogue is inherently good or bad. Each has strengths and weakness that are relative to conditions and organizational K creation patterns. In the same way an organization needs a mix of both intended and emergent strategies to be effective, so too it needs a mix of univocal and multivocal dialogue patterns for creativity to emerge and institutionalization to take place.

Table 6.5 is a comparison of both {strategy–dialogue–K creation} constructs. Note that *univocality corresponds to early K-binding and multivocality corresponds to late K-binding,* discussed in Chapter 2. Observe too that the interdependence of both dialogue patterns may be seen as an example of middle–up–down management, *sequenced* over time as the environment changes (as opposed to occurring all at once).

Initial strategy. When Lexus first developed its entry strategy and its initial product offering, the strategy was intended and univocal, relying on only eight of its top managers. This initial strategy was codified in a constitution containing 10 guidelines to which everyone was required to adhere.

Growth strategy. The second growth phase was characterized by an intended strategy and multivocality in which Lexus expanded its product lineup and changed its product portfolio. No operating unit had full authority to make decisions and the only aspect of the Lexus line that was centrally controlled was the exterior design. The key participants, other than Lexus corporate, were its customers and dealers. Lexus formed multiple layers of

Table 6.5 Strategy making, dialogue, and K creation.

Intended strategy and univocality	Emergent strategy and multivocality
Centralized	Distributed
Unified perspective	Diverse perspectives
Original meaning	*New meaning*
Analytical	Creative
Analysis to action	Learning by doing
Rational	Redundant
Certain	Uncertain
Strategy concept	Local actions
Authoritative	
Hierarchical	

Adopted from Osono (2004).

communication networks that collectively may be seen as its K-creating ba. Among the multivocal artifacts were frequent and informal "fireside chats" among all the parties.

Repositioning strategy. Phase 3 was Lexus' response to market maturity and represents a strategic repositioning resulting from its business growth. This was characterized by a return to intended strategy and univocality. One driver of the return to intended strategy was that Lexus detected a mismatch between the strategy-making process and the changing environment, necessitating a reidentification of core values and the development of a new core concept for its line.

Among the case findings is that the emergent strategy-making process is more appropriate in the growth phase of an organization, whereas the intended strategy-making process is more appropriate in the initial entry stage and the mature stages of competition. Multivocality contains the potential for being creative and uncovering unexpected discoveries, whereas univocality helps an organization make sense of that creativity and institutionalize the findings, thus clarifying the strategy concept and building up necessary organizational capabilities. To thrive in a complex and rich world, organizations need to manage dialogue across univocality and multivocality patterns of conservation, where each interacts with the other in a dialectic manner.

6.10.2 Managing Conversations

Recall from Chapter 2 that the key process enabler for K creation is managing conversations. Recall too that Von Krogh *et al.* (2000) assert that the key principles in managing conversations are to (1) actively encourage participation, (2) establish conversational etiquette, (3) edit conversations appropriately, and (4) foster innovative language. As an example, managers should (1) first create an awareness of the role of conversations in K creation, (2) shape conversational rituals through, for example, using standard discussion formats that encourage participation, (3) promote externalization and codification, and (4) experiment with different modes of expression to generate new K, even using visual techniques in dialogue.

Also, the quality of dialogue may be raised by promoting redundancy of information, establishing a temporal continuity in the conversations, and by encouraging all parties to express their opinions freely. Finally, the dialogue should, at least initially, be multifaceted and have a temporary character that encourages revision and full participation (Nonaka and Takeuchi 1995).

6.10.3 Work-Related Individual Personality Types

Consider how our individual personality types affect our work-related social interactions. We have, for example, a Type A behavior pattern that is characterized by high levels of competitiveness, irritability, and time urgency—overall a person who appears outwardly aggressive. Research shows that Type A people tend to annoy their coworkers and are involved in more conflicts than other personality types (Greenberg and Baron 2003)—not exactly a recipe for successful networking.

Let us consider the classic "big five" personality types. Note that each personality type represents a *range* of possibilities; for example, conscientiousness covers a range of people that are dependable at one end and those who are unreliable at the other.

- *Conscientiousness.* A personality who is hardworking, organized, and dependable.
- *Extraversion.* A person who is gregarious, assertive, and sociable.
- *Agreeableness.* A warm cooperative person.
- *Emotional stability.* A calm, self-confident, and secure personality.
- *Openness to experience.* This personality is creative and curious.

It should be apparent that certain individual personality types make trust and social interaction more or less likely. We may make any number of inferences: someone who is open to experience is likely to be involved in large, diverse networks; we are likely to trust someone who is conscientious; an extrovert may be involved in large, diverse networks with either strong or weak ties and so on. This also raises questions; for example, would we trust someone who is agreeable? Based on our discussion of trust, basing our decision solely on this characteristic is not such a good idea.

6.10.4 Job Role and Individual Personality Types

It follows from earlier that a given job role and its associated K requirements will dictate the level of interaction. Also, we suspect that an individual's personality type is likely to be associated with a job role. For example, would we expect an introvert to aspire to be the head of the marketing department? It is likely that this association is self-correcting and, for example, during the hiring process a fit should emerge between individual personality type, job role, associated K, and level of expected interaction.

6.10.5 Business System Personality Types

In the same way that individual personality types have been defined, Deloitte has characterized four business personality types (CIO 2014, www2.deloitte.com). The central idea is to better understand the workings, social interactions, and K exchange among business areas. The characteristics of the four personality types are as follows:

- *Pioneer*. Adaptable, networked, and imaginative; seeks out novelty.
- *Driver*. Focused, quantitative, competitive, and experimental.
- *Integrator*. Builds relationships and consensus; open to different perspectives.
- *Guardian*. Structured; strong on architecture and method.

Presumably, there are patterns between job role, individual personality, and business personality types. For example, which business personality type is a creative artist likely to possess? Is it a guardian?—not likely; an integrator?—not likely; a driver?—possibly; or a pioneer?—most likely. What of a software engineer? And so on.

6.10.6 Interaction Effects

We hypothesize that these conversational factors interact in any number of ways: the type of K, the required level of social interaction, job role, individual personality types, business system personality types, culture and occupational subculture, and so on.

We can anticipate multidimensional interactions along and between several dimensions. Figure 6.19 shows several such interactions from which any number of questions may arise. For example, how would you characterize the individual personality traits of Steve Jobs? Are professors social—do they find it easy or difficult to collaborate? Which type of individual or business personality is best suited for each type of subculture? What are the associations between individual and business personality types? Can we predict the communication patterns and the level of K exchange between types? How do we expect the different discourse strategies to be carried through each of the business personality types?

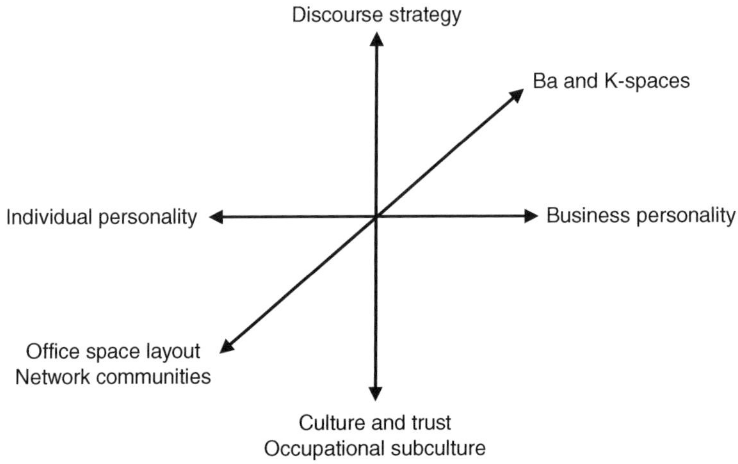

Figure 6.19 Multidimensional social interaction.

6.11 Closing Remarks

KM covers virtually every domain of human thought. Throughout this book, we have presented everything from knowledge's philosophical roots to a wide assortment of models and representations to structural arrangements. At the center of KM is the requirement for *socialization*—central to K creation at every ontological level. One key task for the organization is to design the K-spaces that facilitate socialization and its subsequent K modes and K amplification. This is basic to the Nonaka/SECI approach to KM—however, the *driving force is socialization* mediated through social interaction. This is what we have been referring to (metaphorically) as *social heat.*

The *context* for socialization includes virtually everything we have discussed in this chapter—culture, trust, the landscape of social networks and communities, the office and building layout, dialogue and conversation structures, and large-scale collaborations and collaboratories. The *functioning* of socialization is seen within each context, including an organization's cultural assumptions, network topography and K flows, the strength of network ties, relational contracting and levels of trust, the management of conversations, individual personality types, the interaction patterns of smart creatives, and so on.

Social Interaction and Socialization. How does social interaction take place? As discussed in previous chapters, the organization may provide for the transfer of K, particularly tacit K, through master–apprentice relationships, shadowing of imminent retirees, CoP, problem solving, or experimentation. Periodically, the organization can create opportunities for socialization with company offsites or retreats, and, to a lesser degree, town hall meetings.

In this chapter, we have discussed additional ways of thinking about and designing social interaction. We have considered social interaction from afar, that is, the stranger, and close up, that is, our colleagues. Each produces weak or strong ties with specific characteristics that drive the type and amount of K flows, with a corresponding impact on social innovation and development in the resident community or nation.

Collaboration. Collaboration is much more than cooperation; rather, it is a nuanced example of relational contracting combined with technology. Other considerations emerge—intellectual property, K exchange across departmental structures, data and systems, and organizations. However, everything works and is held together through trust and relational contracting.

The idea of collaboration brings social interaction closer to process and work in what is emerging as a new organizational form. The *collaboratory* is one such form with scientific research in mind; however, a collaborative structure that holds a central position in the structure and workings of the any organization may also be thought of as a business variation to a collaboratory. Therefore, we can extend our thinking in terms of K creation to include *collaborative K creation that is at once individual and organizational.*

Social Network Analysis (SNA). Analyzing the structural characteristics of network flows may provide insight into organizational performance and opportunities for improvement. We can, for example, (1) identify effective communication networks by looking at (organizationally) positive operational outcomes and (individually) at the patterns associated with high performers. Similarly, positive organizational outcomes may include successful innovations and therefore analysis may reveal the (2) locus of where in the network that innovation takes place. However, this kind of analysis may yield the opposite: ineffective networks due to (3) inefficient networking, or (4) networking gaps. Also, an analysis of cross-functional, geographical, or organizational networks may facilitate (5) the identification or necessity of bridge or information nodes to enhance large-scale networking and collaborations. Finally, SNA (6) may reveal redundant networks, which cuts two ways—inefficient cost structures (not good) but redundant info (always a good idea in classical KM). Figure 6.20 is an illustration of the organization and social network analyses that are necessary to uncover network patterns and their organizational impacts. Note too that SNA may occur at different levels of granularity and also includes (not shown) the type and amount of K and information flows between actors within and between networks.

Culture. How the organization thinks about itself and its workers is made manifest in its culture. We think of *culture as the enabler of all the other enablers.* Everything that may or may not take place in an organization is largely driven by its culture. This line of thought tells us a great deal about how K creation takes place in a given organization. We have seen, for example, an application of the combination of social ties and culture in two different models of K exchange and

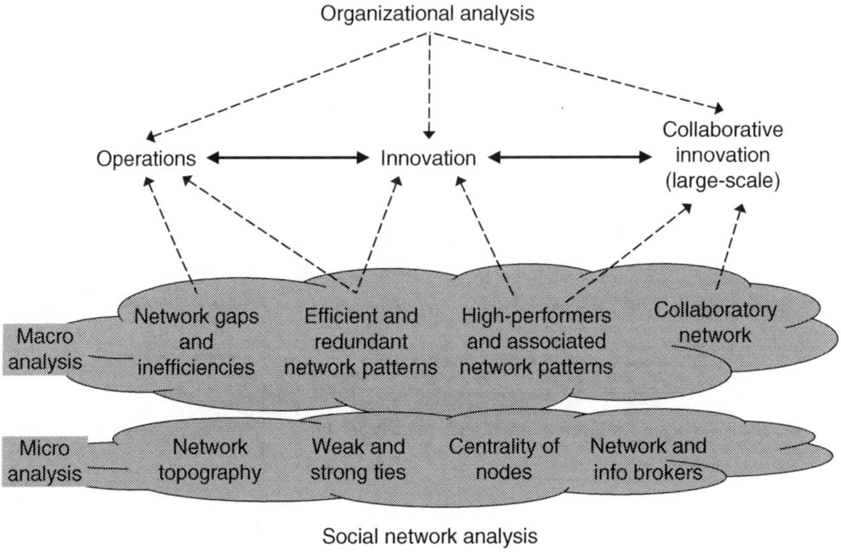

Figure 6.20 Organizational and social network analysis.

creation. Tacit organizations, embedded in tacit societies, build K-spaces on the basis of ties that are few, strong, and stable, which in turn, share and create tacit K across organizational boundaries (i.e., Toyota model). On the other hand, explicit organizations within explicit societies build spaces that promote K recombination of explicit K threads (i.e., SV model). A third possibility—an organization operating in different cultures—faces problems of resolving different assumptions, values, and expectations. As one example, we have seen how Toyota in the United States seeks to create a culturally blended set of performance policies and work-related practices.

People Value Stream. Perhaps no other organizational activity will have a greater impact on K creation and socialization than the development of its people value stream. The people value stream follows a prescribed flow, starting with attracting and recruiting people, developing, inspiring, and ending in actual practice (followed by a feedback loop). This *visible* investment in people will motivate and engage employees to create and share K and may be seen as a *cornerstone in an organization's culture.* We will revisit this topic in Chapter 7.

Trust and Social Architecture. An organization's social architecture—its network of relational contracts—may be seen as a vehicle to analyze and understand relationships within and between organizations; in fact, the organization is characterized by its social architecture. The balance between objectives and outcomes (or policies and characteristics) underpins relational contracting. Recall too that a practice may also be seen as a balance—between theoretical and practical K. The organization works through this balance, based on enriching factors such as duty and moral vision (Figures 4.9a and 5.3).

Office Space. The design of office space represents a merging of several KM ideas: social interaction, the development of ties, building trust, the transfer of tacit K, and the development of a K culture. It is unfortunate that organizations sometimes take advice from consultants who may have a limited understanding of these concepts and the culture and particulars of a given organization. We have seen, for example, how organizations design their office space—this is a manifestation of K culture. It would benefit the organization if it first did an internal analysis—driven by the requirements for specific types of K, the K flows, and the necessary interaction patterns to facilitate its development. After this assessment, an office space may be designed to the specific K requirements of the organization, using Figure 6.17 as a guide.

Creative Workers. Creatives fall into two broad categories. The bulk of smart creatives are *problem solvers* that fit in well with most organizations. As an example, Toyota drives its entire culture and organizational systems on the basis of problem solving. On the other hand, *artistic creatives* afford the organization with an opportunity for distinctiveness beyond functional features and capabilities into design and style, much as a fashion designer does more than simply stitch cloth. Artistic creatives present a different set of collaboration issues than that of problem-solving creatives. The *smart organization* would be well advised to consider both approaches to its products and services, and, accordingly, recognize that each group of creative workers has a correspondingly unique interaction pattern, work, space, and collaboration needs. See Figure 6.17.

Conversations. How do we get people to speak with each other without making it compulsory to satisfy the requirements of a performance review? We have all worked in environments where certain people appear to be more talkative than others. There are many such factors to consider—individual and business personality types, for instance. It has been our observation that several of these factors are mutually reinforcing-individual personality types, business personality types, the nature of the work environment, the required K flows, and so on tend to coalesce in unique and often unpredictable ways.

As another example, we saw in the Lexus case study that discourse strategy may drive the type of collaboration and the associated flows of information and knowledge. Furthermore, there is managerial choice in selecting one or the other—one voice (early K-binding) or many voices (late K-binding)—and this in turn may depend on the culture of the organization, its competitors, and where in the lifecycle the organization finds itself. Dialogue strategy—one or many voices—may be alternated as the organization moves through different phases of its lifecycle. In this way, it is analogous to the middle–up–down strategy and management approach that Nonaka recommends.

We have seen from Von Krogh *et al.* (2000) (separately and collaboratively) that *managing conversations* is perhaps the single most significant management competence in the K organization. In general, managing conversations requires a formalization of both the format and conduct of conversations. Further, the culture-oriented behaviors that characterize collaboration (e.g., selecting the right people) are an additional set of skills that, in combination with those suggested by Von Krogh *et al.* (2000) (e.g., establish conversational etiquette), provide the leader with guidelines for creating and managing what we have previously called the *essential* or *cognitive organization.*

Technology. Technology is a tremendous amplifier that is driving an almost insane revolution in terms of data generation, pattern development, knowledge, and social interaction. Moreover, technology shapes our understanding of the world that, in turn, affects social interaction and collaboration. We have seen, for example, that collaboration—and its new organizational form of collaboratory—is driven by technology on a bed of other KM constructs, such as relational contracting and trust.

Recall from Chapter 5 the Polanyi paradox: labor polarization and the simultaneous growth of high-education, high-wage and low-education, low-wage jobs. Technology drives the Polanyi paradox, in terms of not only employment but also societal changes. Further, technology may require *more* human work, not less work (as in the Polanyi paradox). Whereas technology appears on the surface to be explicit oriented, in fact it will require new threads of tacit understanding as to what the new technologies mean and how to use them.

What Is Organizational Structure? In traditional management thinking, lateral structures in the form of teams or lateral processes are often viewed as superimposed over the traditional organizational chart. In contrast, the collaborative organization—interconnected and thinking—is characterized by its social aspect. In this world of social networks, the classical organizational chart should be reevaluated in a broader context. We think of the organizational chart as one network in a landscape of many such networks, each with a particular purpose, structure, and characteristics. That is, the realized "organizational structure" is a super network of interconnected networks within which employees are embedded. This includes the traditional organizational chart as well as context-specific networks (see Figure 6.1). For example, collaborative innovation is an example of interorganizational, interstructure, and intersystems coordination and communication within a collaborating research community.

What Is Second Wave KM? So now we come to the question, what is the second wave in KM? Recall from earlier chapters that traditional KM—first wave KM—is viewed as an evolutionary process within traditional management thinking: management by objectives, organizational development, total quality management (TQM), business process reengineering (BPR), and finally KM. This line of thinking may in part explain why, since 2000, relatively little has taken place to change business practice (despite extensive KM literature).

In contrast, we have argued in this book that the second wave in KM is revolutionary and simultaneously more complex and nuanced that the first wave. The second wave requires that we expand our perspective to embrace and integrate many ideas from diverse domains of human thought, and includes, for example, systems theory, ethnography, law, philosophy, and so on.

Compounding. Central to second wave KM is the idea of *compounding*—what and how things come together. By compounding we mean the construction of high-level constructs from low-level or atomic KM ideas. Compounding—a series of composites with unique emergent properties—may be "mapped" to any level of abstraction. As with a pharmaceutical compound, KM compounds may be true composites (in a systems sense) or mixtures with identifiable constituents.

As an example, Toyota's DNA representation is a single *compound* that ties together thin and thick organizational elements. Similarly, the Toyota TPS House and the Toyota Way house are each *mixtures* that are intended to represent a complex K-oriented system. In this case, the compounds may take the form of a cultural artifact that guides the organization or a specific artifact or plan for a specific problem.

New Frontier in KM. We think of second divide KM as the *new frontier in KM*. The growth of the K society, including K organizations, is centered on *individual expertise and human interaction*, the compounding effect on the capabilities at the organizational level, on certain K and practice, and many other principles we have been discussing. Second wave KM is neither a collection of techniques nor a toolkit. Rather, it is a sophisticated system of *compounding* ideas, systems, structures, and policies, in which *people and their relationships play the central role.* Second wave KM is a total system of interacting knowledge constructs that is at once *imaginative* embracing a single vision and *analytical* regarding the problem at hand.

6.12 Class Exercises

1 Discuss the networks in which you participate, including organizational, professional, and interorganizational networks. Describe their size, purpose, and how you participate in each. Include in your answer the strength or weakness of your ties and the type and frequency of K flows.

2 Select a given K-oriented task in your organization and assess the role of social interaction. Identify and characterize the nature of each interaction: the type of network, frequency of interaction, and type of K exchanged, and so on.

3 Discuss the role of creative workers in your organization or school. Distinguish between artists and problem-solving creatives, identifying their job roles, interactions, and how each adds value to the organization.

4 Discuss the most prominent K-networks in your organization or school and how each came into existence. Describe how and why they were designed (intended) or how they emerged, their purpose (e.g., support of individual network members), their ongoing operations, and so on.

5 Diagnose the culture of your organization or school. Using Figure 6.9 as a guide, instantiate the Schein model for your organization. Include the interaction mechanisms between layers and how the visible artifacts reflect the lower-level assumptions and values. Identify any disconnects and how you would bring the cultural levels into alignment.

6 Identify the occupational subcultures in your organization or school. Discuss how communications and K exchange takes place between them. Include in your answer the structural aspects of the subcultures, including organizational design considerations such as departmentalization.

7 Project. (a) Discuss the patterns of trust in your organization or school. Start at the individual level (e.g., trust between individuals and their managers) and expand to the social and organizational level. Analyze trust patterns in terms of the organization's social architecture; for example, is your organization characterized by transactional or relational contracting? Is there a difference based on function or department? (b) Discuss the means used to promote *mutual trust*; include, for example, benefits, reward and promotion policies, education programs, problem-solving structures and systems, opportunities for professional and organizational growth, and so on.

8 Project. Describe and characterize K exchange and creation between your organization or school and external entities. Analyze and discuss the K exchange in terms of the following dimensions: product–process features, explicit–tacit K, culture, nature and strength of network ties, and the ba or K-spaces involved. Is your example of interorganizational K creation closer to the Toyota or Silicon Valley model? Explain. Does collaboration play a role? If so, elaborate (see next question).

9 Project. Analyze and discuss the role of collaboration in your organization. (1) In terms of a specific problem or research initiative, identify the problem at hand, the type and number of internal and external entities, the type of collaboration (e.g., virtual CoP), the flows of information and K, the roles

of the people participating in the collaboration, and the systems and structures involved in the collaboration. (2) In general, does collaboration hold a central position in the structure and workings of your organization or school? How does collaboration relate to the other structures and processes of the organization or school?

10 Discuss how conversations are managed in your organization. For example, is the importance of conversations in KM discussed? How are participants encouraged to participate considering, for example, their respective individual and business personality types? How are conversations actually conducted—the room, discussion formats, access to information, business context, temporal continuity, and so on? What are the postconversation artifacts—readouts, lessons, K codification, and so on?

11 Project. Layout the existing office space in your organization. Describe the work and patterns of K exchange, characterizing each, and what you would change to improve K flow. Include the immediate office spaces as well as the building floor layouts. *Redesign* the entire office, reconfiguring the building floors, workspaces, access to K, and patterns of social exchange. Annotate your design.

12 Evaluate the dialogue strategies in your organization. Do they change over time and in what context? Discuss the mechanisms used to implement each type of dialogue strategy.

References

Ahmadjian, C.L. (2004), "Inter-Organizational Knowledge Creation: Knowledge and Networks." In *Hitotsubashi on Knowledge Management*. Eds. Takeuchi, H., and Nonaka, I., John Wiley & Sons (Asia) Pte Ltd.

Bernstein, E. (2014), "The Transparency Trap." *Harvard Business Review*. October 2014.

Bos, N., Zimmerman, A., Olson, J.S., Yew, J., Yerkie, J., Dahl, E., Cooney, D, and Olson, G.M. (2008), "From Shared Databases to Communities of Practice: A Taxonomy of Collaboratories." In *Scientific Collaboration on the Internet*. Eds. Olson, G.M., Zimmerman, A., and Bos, N., The MIT Press.

CIO (2014), "Better Living Through 'Business Chemistry.'" May 2014. Sponsored by the Wall Street Journal.

Congdon, C., Flynn, D., and Redman, M. (2014), "Balancing 'We' and 'Me.'" *Harvard Business Review*. October 2014.

Cross, R., Gray, P., Cunningham, S., Showers, M., and Thomas, R.J. (2010), "The Collaborative Organization: How to Make Employee Networks Really Work." *MIT Sloan Management Review*. Fall 2010.

Elsbach, K.D., Brown-Saracino, B., and Flynn, F.J. (2015), “Collaborating with Creative Peers.” *Harvard Business Review*. October 2015.
Granovetter, M. (1973), “The Strength of Weak Ties.” *American Journal of Sociology*, 78(6): 1360–1380.
Granovetter, M. (1983), “The Strength of Weak Ties: A Network Theory Revisited.” *Sociological Theory*, 1: 201–233.
Greenberg, J., and Baron, R.A. (2003), *Understanding Organizations*. Eighth Edition. Prentice Hall.
Hatch, M.J. (1997), *Organization Theory: Modern, Symbolic, and Postmodern Perspectives*. Oxford University Press.
Kay, J. (1995), *Why Firms Succeed*. Oxford University Press.
Kay, J. (1996), *The Business of Economics*. Oxford University Press.
Komatsoulis, G.A. (2011), “Collaboration in Cancer Research Community: Cancer Biomedical Informatics Grib (caBIG).” In *Collaborative Computational Technologies for Biomedical Research*. Eds. Ekins, S., Hupcey, M.A.Z., and Williams, A.J., John Wiley & Sons, Inc.
Liker, J.K., and Hoseus, M. (2008), *Toyota Culture: The Heart and Soul of the Toyota Way*. McGraw Hill.
Lynch, R.P. (2011), “Collaborative Innovation: Essential Foundation of Scientific Discovery.” In *Collaborative Computational Technologies for Biomedical Research*. Eds. Ekins, S., Hupcey, M.A.Z., and Williams, A.J., John Wiley & Sons, Inc.
Mayer, R.C., Davis, J.H., and Schoorman, D. (1995), “An Integrative Model of Organizational Trust.” *The Academy of Management Review*, 20(3): 709.
Mayrand-Chung, S., Cohen-Freue, G., and Hollander, Z. (2011), “Models for Collaborations and Computational Biology.” In *Collaborative Computational Technologies for Biomedical Research*. Eds. Ekins, S., Hupcey, M.A.Z., and Williams, A.J., John Wiley & Sons, Inc.
McGregor, D.M. (1960), *The Human Side of Enterprise*. McGraw-Hill.
Nonaka, I., and Takeuchi, H. (1995), *The Knowledge-Creating Company*. Oxford University Press.
Osono, E. (2004), “The Strategy-Making Process as Dialogue.” In *Hitotsubashi on Knowledge Management*. Eds. Takeuchi, H., and Nonaka, I., John Wiley & Sons (Asia) Pte Ltd.
Pugh, K., and Prusak, L. (2013), “Designing Effective Knowledge Networks.” *MIT Sloan Management Review*, 55(1): 79–95.
Schein, E.H. (1992), *Organizational Culture and Leadership*. Second Edition. Jossey-Bass.
Schein, E.H. (1996), “Three Cultures of Management: The Key to Organizational Learning.” *Sloan Management Review*, 38(1): 9.
Schein, E.H. (1999), *The Corporate Culture Survival Guide*. Jossey-Bass.
Schmidt, E., and Rosenberg, J. (2014), *Google: How Google Works*. Grand Central Publishing.

Szalma, S. (2011), "Enabling Precompetitive Translational Research: A Case Study." In *Collaborative Computational Technologies for Biomedical Research*. Eds. Ekins, S., Hupcey, M.A.Z., and Williams, A.J., John Wiley & Sons, Inc.

Von Krogh, G., Ichijo, K., and Nonaka, I. (2000), *Enabling Knowledge Creation*. Oxford University Press.

Waber, B., Magnolfi, J., and Lindsay, G. (2014), "Workspaces That Move People." *Harvard Business Review*. October 2014.

Waller, C.L., Durvasula, R.V., and Lynch, N. (2011), "Need for Collaborative Technologies in Drug Discovery." In *Collaborative Computational Technologies for Biomedical Research*. Eds. Ekins, S., Hupcey, M.A.Z., and Williams, A.J., John Wiley & Sons, Inc.

Weston, J. (1996), *Directing Actors*. Michael Wiese Productions.

7

Strategy and Leadership for Knowledge Management

Chapter Preview

Introduction

Imagine you are a chief executive officer (CEO). You want to promote innovation and knowledge creation in your company. What are your directions for the future? What are the signposts that will help you get started and keep you on track? Where do you even start?

Knowledge management (KM) is an exercise in strategy and leadership. A key function of 21st century leadership is to develop a strategy framework that is driven by knowledge. All the attendant pieces of K formulation and implementation—business strategy, tools, people, systems, culture, and structures—are dimensions of a K strategy.

KM is neither a toolkit nor a collection of projects. Rather, it is a system of complex human interactions that embraces the larger organization, including its physical, technological, and social dimensions. Whereas specific strategies may target certain techniques or tools and have spot effects on productivity or innovation, the wider purpose of both tools and strategies is, in fact, often lost. When looked at more broadly, a K strategy and leadership (and KM generally) is about applying the principles of K to the practice of people in their organization.

Food for Thought

1) What strategy will you employ to accommodate the large-scale retirements that are anticipated in the coming years?
2) What is the best way to build a core competency in your organization?
3) How do you plan to attract and retain top talent?
4) What elements do you consider important in developing a K culture?
5) What actions would you take to promote K sharing and mutual trust in your organization?
6) What are the desirable attributes of a leader in a K organization?

Designing Knowledge Organizations: A Pathway to Innovation Leadership, First Edition.
Joseph Morabito, Ira Sack, and Anilkumar Bhate.

Topic Layout

The figure below illustrates the topic layout of Chapter 7. As shown on the left side of the figure, we discuss the structure of strategy, the role of the decision-making hierarchy, the K content of strategy, and several other factors associated with strategy. We also discuss leadership, its role in KM, and K-based or expert leadership. The left section informs the organization on the issues associated with strategy and leadership. The center of the figure represents the organization in practice. We describe how the organization should position itself and suggest several tacit-oriented strategies that are unique to the K organization. We propose the Chief K Office (CKO) as a governance structure to provide the leadership and the staging area for specific K-functions and activities, including the development of the organizational K-base. We conclude with the strategies associated with the development of the people value stream. The right side of the figure refines our perspective of strategy with a discussion of 10 strategy schools. We introduce the strategy continuum as a suite of specific K strategies for implementing both an enterprise and an integrative approach to strategy. We set the stage by revisiting intangibles, core competences, and individual competence. We also review the lessons of a DNA metaphor for integrating intangible and tangible assets.

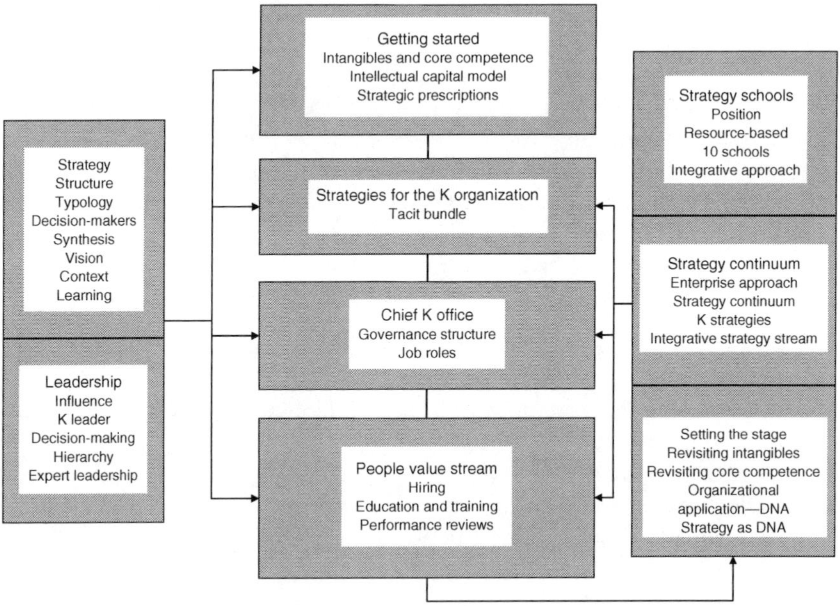

Concept roadmap to Chapter 7.

7.1 What Is Strategy?

Strategy is typically viewed as both a set of long-term goals and the course of actions necessary to carry out and attain those goals. That is to say, strategy refers to both the ends and the means to those ends (Robbins 1990). Strategy may thus be seen as a planning framework with ebbs and flows between goals and means, emphasizing one or the other as the situation demands. Further, strategy-related, decision-making processes may similarly either be planned and formal or emerge in a pattern of decisions.

Structure of Corporate Strategy

Strategy may be conceived as three levels corresponding to the decision-making hierarchy of the organization. The top level is corporate strategy and typically involves the board of directors, the CEO, and key administrative officers. In the middle of the decision-making hierarchy are the corporate and business managers who are responsible for business-level strategy, which translates the intent of the corporate strategy into concrete objectives and strategies for the individual business areas. In effect, the business strategies describe how the organization will compete in selected product and market areas. At the bottom of the decision-making hierarchy is the functional level composed chiefly of product, geographic, and functional managers. The functional level addresses annual objectives and short-term strategies. The decision makers at this level are primarily concerned with implementing and executing the strategic plan of the organization (see Pearce and Robinson 1991). At yet a lower level of granularity, K-oriented strategies may target specific technologies and systems. The decision makers at this level are typically team leaders, consultants, or domain specialists. Observe the strong link between strategies and the decision makers who formulate them.

Which Comes First, Knowledge or Strategy?

Does strategy follow K, or does K follow strategy? While it is tempting to view a given K strategy as a distilled low-level implementation of a business strategy, today it is often the reverse. Technology and K are driving business at a maddening pace; the more companies embrace and align with the exponential growth of technology and the new K, the more likely they are to thrive in the 21st century. Recall from Figure 1.2 that the source of advantage is K; hence, it follows that design domains such as strategy, process, structure should follow the development of new K.

As an example, consider Sharp's decision to develop a competence in flat-panel screen displays. In the late 1950s, Sharp was an assembler of televisions and radios—little more than a manufacturer using routine technology.

To break out of that position, Sharp's founder Tokuji Hayakawa created a corporate R&D facility. When the Japanese Ministry of International Trade and Industry blocked Sharp from designing computers, the company used its resources to produce the world's first digital calculator in 1964. To further strengthen its position, Sharp backward integrated into manufacturing its own specialized semiconductors and made a strong commitment to the new liquid-crystal display (LCD) technology. What Sharp did is to exploit a "virtuous circle" in which resources are upgraded by sequentially adding new competencies. Sharp's bet on LCD technology paid off in many ways: Sharp developed a number of new products, such as the Wizard electronic organizer and the breakthrough product, Viewcam. The superiority of its display technology gave Sharp a competitive advantage in businesses in which it previously had difficulties. Today, Sharp is the dominant player in the $7 billion LCD market and a force in consumer electronics (see Collis and Montgomery 2008).

Sharp's overall strategy has been to differentiate itself through products that are technology intensive and difficult to imitate—that is, a proprietary competence. Specifically, by developing a key component like the LCD, Sharp further developed the products that use it and then improved the components and products ahead of its competition. This is what the Japanese call "seeds and nuts."

The key lesson is that Sharp developed a business strategy based on innovation and uniqueness. By using a component-based strategy and focusing on a specific competence in a "key" component, Sharp was able to leverage that K and compete in an entire product stream. When Sharp's president encourages managers to apply LCD technology to their product lines, he is implementing a technology-push strategy. An important lesson here is that *business strategy follows technology-driven innovation and the expertise and K that accompany it.*

Observe too that Sharp sequentially developed its special competence internally, which is in keeping with a tacit-oriented culture and the notion of incremental improvement. On the other hand, recall from Chapter 6 our discussion of interorganizational K creation comparing Toyota and Silicon Valley. The key challenge for Japanese companies in the 21st century is to collaborate outside the boundaries of closely linked business groups (keiretsu). Recall too that one risk of a tacit-oriented culture is its reliance on strong ties, which tends toward a steady state with few innovations. In contrast, a culture with many weak ties is characterized by K recombination and innovation, as in Silicon Valley. Following this line of thinking, competence in a US organization is as likely to be acquired through alliances, mergers, or acquisitions as it is to be developed internally. In either case, *the structure of an organization's strategy should embody the development of proprietary competence and expertise.*

Strategic Thrusts

We see a range of strategic orientations that may be described with the typology advanced by Miles and Snow (1978) in their classic paper on strategy and structure. The typology identifies in order from the least to the most aggressive orientations: *reactors, defenders, analyzers,* and *prospectors.* The key factor in this model is the perceived level of environmental uncertainty. As the environment becomes more dynamic and uncertain, the organization shifts its strategy from efficiency and control to one that is more innovative and adaptable. Similarly, the organization shifts internally along this continuum from high to low formality and from centralized to decentralized decision making.

In this continuum, we would expect to see a similar shift in K strategy from one that promotes stability and efficiency to one that emphasizes innovation. For example, in a *defender* we are more likely to see short-term system strategies such as the development of a K directory of employees; an *analyzer* is more likely to emphasize data analytics, K reuse, and collaboration; and a *prospector* will likely adopt the same strategies as an *analyzer* and also include long-term strategies. Such long-term strategies comprise culture-oriented strategies that promote education and learning as well as process-oriented strategies such as reengineering K-oriented processes (guided, e.g., by the SECI (socialization–externalization–combination–internalization) model). Finally, a *reactor* has at best a residual business strategy, and we expect that it has no K strategy at all.

7.1.1 Strategy as K Synthesis

Nonaka and Toyama (2004) have applied the concepts of K creation to that of strategy. In their perspective, organization strategy is a *synthesis* between a strategy of position and a resource-based strategy. The former is a strategy that focuses on an organization's environment, whereas the latter focuses on its internal resources. Strategy then, is a resolution of external and internal forces and resources that produces a new level of K and understanding—the synthesis. See Figure 7.1.

The synthesis of internal and external resources is explained (in part) by structuration theory (Giddens 1984). Structuration theory views all institutional and social practices as structures. It then proceeds to study the interaction of humans and their environment and the way each produces and reproduces the other. Further, whereas tangible resources lie outside of the firm, intangible resources and firm-specific K are within an organization and are responsible for competitive advantage (Spender 1996). This too is shown in Figure 7.1.

Position. Michael Porter (1985) developed the *strategy of position* as a full-blown analytical framework. This approach includes an analysis of an organization's value chain and an assessment of environmental factors or forces. The primary environmental forces that shape strategy include an organization's *competitors,* the threat of *new entrants, supplier power, customer power,* and *substitute products* or services—this is known as the five-force model. This model has since been updated to include a sixth force, *government power.* Note that in the new Pangaea, the flow of information and K has decidedly shifted power from organizations to customers.

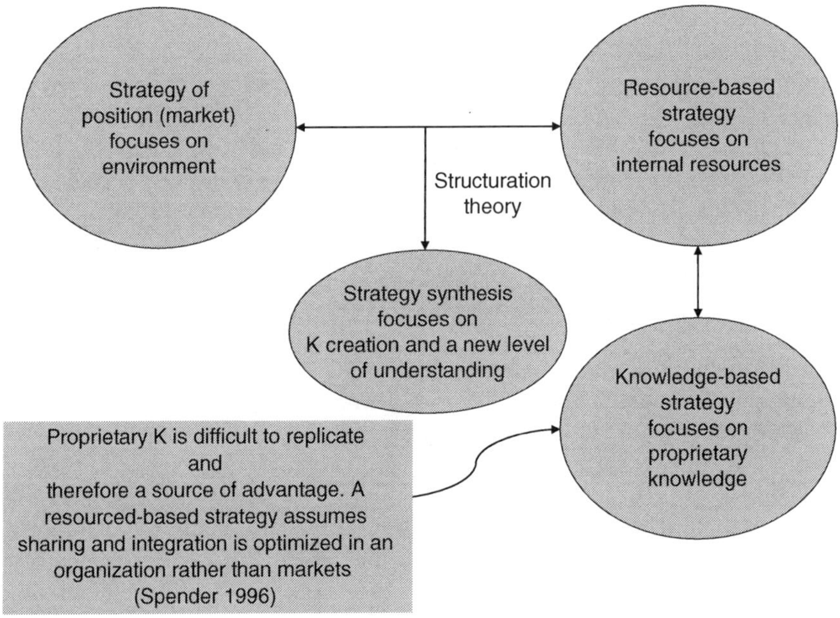

Figure 7.1 Strategy as a synthesis of position and internal resources.

Resource-based. In contrast to the strategy of position, the *resource-based strategy* is a comparatively long-term approach that develops an organization's core competencies. Recall from Chapter 1 that a core competence is a bundle of intangible resources that enables a company to provide a particular benefit to customers (Andriessen 2003; Hamel and Prahalad 1994). Among the constituents of a competence are business processes, systems, people, K, and so on (see later). For example, Peter Keen (1997) asserts, "processes are a source of 'firm-specific' special competence that makes for competitive difference far more than does strategy, products, structure or industry." Note too that a resource-based approach to strategy focuses on the development of proprietary K as a source of distinctive advantage: "capabilities are *core* only because they embody proprietary knowledge and are superior to those of competitors" (Leonard 1995). This too is shown in Figure 7.1.

Naturally, organizations should employ both approaches to strategy. From a KM perspective, the importance of the Nonaka and Toyama (2004) research is their leverage of *synthesis* to affirmatively create a higher level of strategic understanding through the interaction of internal and external approaches. Further, it should be clear that the context of competence implies *knowledge at work*; that is, the center of a core competence must include K tasks, people, and the development of K content.

7.1.2 Strategy as Unique Activities

Porter (1996) argues that the development of operational effectiveness (which includes operational efficiency), while necessary for superior performance, is insufficient for the long term. Operational effectiveness means performing certain activities better than competitors performing similar activities. The more companies use benchmarking, the more they outsource, the more their activities will look alike and become generic.

Strategic positioning. Porter reinforces the strategy of position by asserting that strategy rests on *unique activities*—deliberately choosing a different set of activities from that of competitors to deliver a unique mix of value. Strategic positioning originates from (1) *variety-based* positioning (based on product or service varieties rather than customer segments), (2) *needs-based* positioning (traditional thinking about targeting a segment of customers), and (3) *accessed-based* positioning (segmenting customers who are accessible in different ways).

Fit. In addition to choosing which activities a company will perform, strategic positioning is about *relating activities.* Whereas operational effectiveness is about achieving excellence in individual activities or functions, strategy is about combining activities. (Recall from Chapter 6 the importance of compounding ideas or activities.) First-order fit is achieved through *simple consistency* between each activity (or function) and a strategy, whereas second-order fit occurs when activities are *reinforcing,* and third-order fit is achieved through *optimizing effort.* The latter is achieved (at a low level) through coordination and information exchange and (at a high level) through product design choices and coordination between suppliers and distribution channels. Finally, it is important to note this is a systems approach where the whole is considered more important than any of the individual activities.

7.1.2.1 Strategy as Generic Positions

As part of the strategy of position, Porter (1985) introduced the notion of generic strategies. That is to say, within an industry a company may position itself as one that is a *cost leader, differentiator,* or *focused in a niche segment.* Though formulated decades ago, the generic strategies still have value. The bases for positioning mentioned earlier—*variety, needs,* and *access*—carry generic strategies to greater levels of detail and specificity (Porter 1996).

7.1.3 Schools of Strategy

Mintzberg *et al.* (1998) have further refined strategy into 10 schools that include *design, planning, positioning, entrepreneurial, cognitive, learning, power, cultural, environmental,* and *configuration* (a combination of the others). These schools are related in a variety of ways; for example, the design, planning, and positioning schools represent, in sequence, a historical progression with increasing levels of formality and analysis. From a KM perspective, the schools differ primarily in terms of both the type of K (explicit, tacit, communal) and K-binding (late, early). By type of K, we include both content and process.

Strategy as Learning. Porter (1996) asserts that Japanese companies rarely make strategy. On the other hand, Mintzberg and Lampel (1999) have challenged Porter's assertion and claim that the Japanese, in fact, do make strategy—but one based on learning rather than positioning. Whereas the design, planning, and positioning schools are prescriptive and carry an intended message of *fit, formalize,* and *analyze,* respectively, the learning school is a descriptive, emergent process with an intended message of *learning.* Among the activities that characterize the learning school/organization are *experimentation, ambiguity,* and *adaptability*—each a characteristic of the K organization. Similarly, the culture school rests on the disciplines of anthropology and ethnography—other KM characteristics we have previously discussed. In the culture school/organization, strategy is a social process focusing on *common interests* and *integration*; again, a KM characteristic of culture. Finally, note that since Japanese strategies are grounded in a tacit culture, they are not immediately visible or evident, whereas American strategies are manifestations of an explicit culture and are therefore more apparent and discernible.

In Mintzberg's assessment, *it is the learning and culture schools that constitute the resource-based approach to strategy.* The other schools do not fit neatly into a single category. For example, the cognitive school is a mental process that is based on psychology. Early forms of this school considered strategy as simple mental models or cognitive maps of reality, whereas today this school follows a subjective interpretative or constructivist view where "cognition is used to construct strategies as creative interpretations" (Mintzberg and Lampel 1999).

Multiple K Strategies. Our assessment of the strategy schools indicates that a distinguishing feature of each school is their comparative tacit and explicit K content. Figure 7.2 is our extension of the 10 schools diagramed by Mintzberg *et al.* (1998). Observe that the relative tacit/explicit content may dictate the corresponding K strategy. For example, are we seeking to establishing common assumptions (culture school); to develop ideas, methods, and tools associated with deep learning (learning school); or, say, to specify a data analysis model of customer behavior (positioning school)?

Integrative K Strategy. Whereas it is easy to presume that the resource-based school is the focus of a K strategy, in fact, strategy making itself may be seen as a single integrative process (Figure 7.3). It follows that a K strategies should be integrative and complementary. Observe too the central position of the cognitive school, shown in bold (Mintzberg's representation) thus highlighting the importance of tacit K—imagination and mental models—in strategy formation.

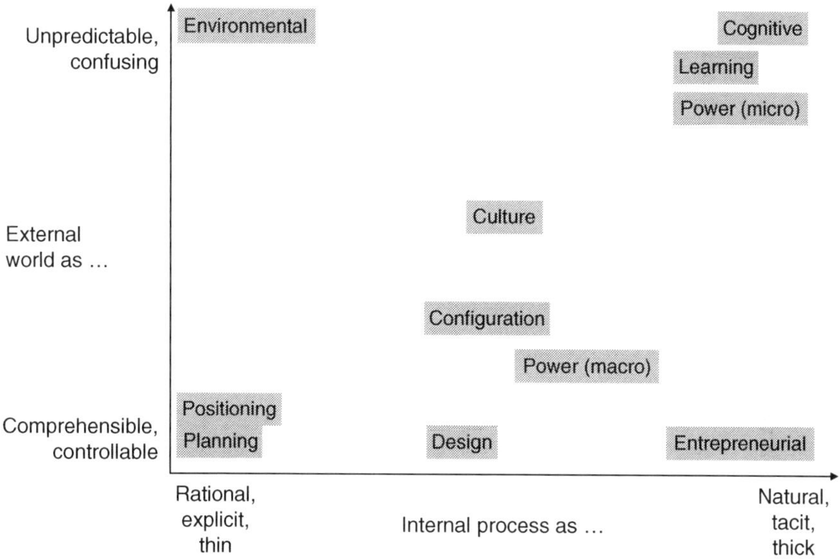

Figure 7.2 Strategy schools and thin and thick organizational states—Adapted from Mintzberg and Lampel (1999).

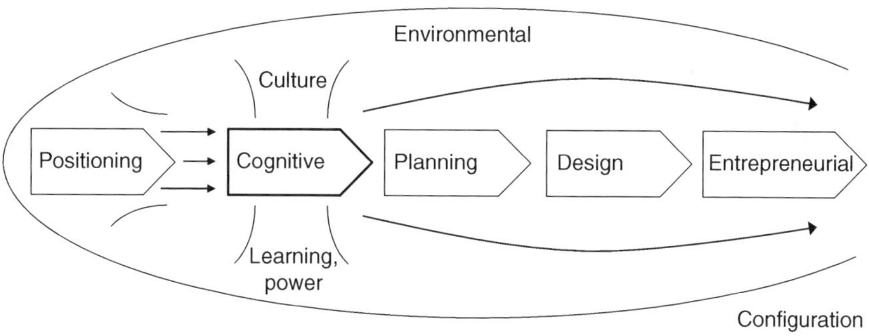

Figure 7.3 Strategy formation as a single process—Adapted from Mintzberg and Lampel (1999).

This line of thinking implies that elements of a K strategy are similarly integrative and need to be targeted to different aspects of an organization's overall strategy stream, depending on its position in Figure 7.3. Indeed, a large chunk of a K strategy may concentrate on the intangibles (i.e., culture, cognitive, and learning schools), but let us not forget, say, the development of K systems and data analytics and their important role in building customer relations and assessing competitors (i.e., positioning school). Hence, as with all strategies, K strategies should be similarly integrative and complementary.

7.1.4 The Importance of Vision in Strategy

In the same way that long-term advantage rests on proprietary K and unique activities (Leonard 1995; Porter 1985, 1996), Collins and Porras (1996) assert that long-term success arises from a vision framework that includes a core ideology and an envisioned future. By core ideology, Collins and Porras refer to both values (a stable and unchanging vision) and purpose (the organization's raison d'être—its motivation). An enduring future, in turn, refers to long-term (10–30 years) "audacious" goals. As an example, Sony (1) values include the elevation of Japanese culture and being a pioneer, (2) a corresponding purpose includes advancing technology for the benefit of the public, and (3) an audacious goal (1950) is to raise the worldwide image of Japanese product quality. Observe the K thread running from value to purpose to goals; it follows that both core competence and strategy should follow similarly, establishing a sequence of strategic contexts and embedding competencies and K. We refer to this as a *strategy hierarchy.*

{Value–purpose–goal–competency–strategy}

This sequence of interrelated beliefs and ideas are critical constituents in any strategy approach, each with some degree of tacit and explicit elements.

(Aside: There are several such hierarchies of beliefs and expectations. For example, Kaplan and Norton (2004) identify mission, values, vision, and strategy, each corresponding to "why we exist," "what's important," "what we want to be," and "our game plan" (i.e., strategy—"how the organization intends to create value for shareholders"), respectively. Such strategies may include, for instance, a productivity or growth strategy. In another hierarchy, Pearce and Robinson (1991) distinguish company mission, long-term objectives, and annual objectives, the latter two referring to a grand and an operating strategy, respectively, which, in turn, correspond to the decision-making hierarchy of the organization.)

Additional Example. We have seen the importance of values earlier in this book. Recall from Chapter 4 that Whole Foods' uniqueness rests on an essentially moral vision. The higher purpose of Whole Foods has evolved over the years and includes, briefly, (1) developing of the world's agricultural system to be both efficient and sustainable, (2) raising the public's collective awareness about principles of healthy eating, (3) helping to end poverty around the world (through their foundation), and (4) making "conscious capitalism" the dominant economic and business paradigm to promote human flourishing. This is representative of what we have previously referred to as ennobling goals and moral vision—key enablers of the K organization.

7.1.5 Organizational Context for Strategy

Kaplan and Norton (2000, 2004) have created the notions of a balanced scorecard and a corresponding strategy map, respectively. The balanced scorecard identifies the measures associated with strategy, whereas the strategy map identifies and documents goals. Both the balanced scorecard and the strategy map have the same core structure, which is shown in Figure 7.4. By employing a paired scorecard and map, the organization can simultaneously formulate and document strategies while measuring and reporting on their progress.

As illustrated in Figure 7.4, the balanced scorecard/strategy map comprises four connected perspectives. Moving from bottom to top they are learning and growth, internal (i.e., a business process), customer, and financial perspectives. The lowest level—learning and growth perspective—is where we find the intangible assets of the organization. In the Kaplan and Norton (2000, 2004) perspective, intangible assets are broken into three areas: human, organization, and information. Its context is the next highest level or the processes of the organization, similarly divided into four classes: innovation, operations, customer, and regulatory/social processes. Kaplan and Norton (2000, 2004) also argue for a set of industry standard scorecards—templates with predetermined strategic objectives.

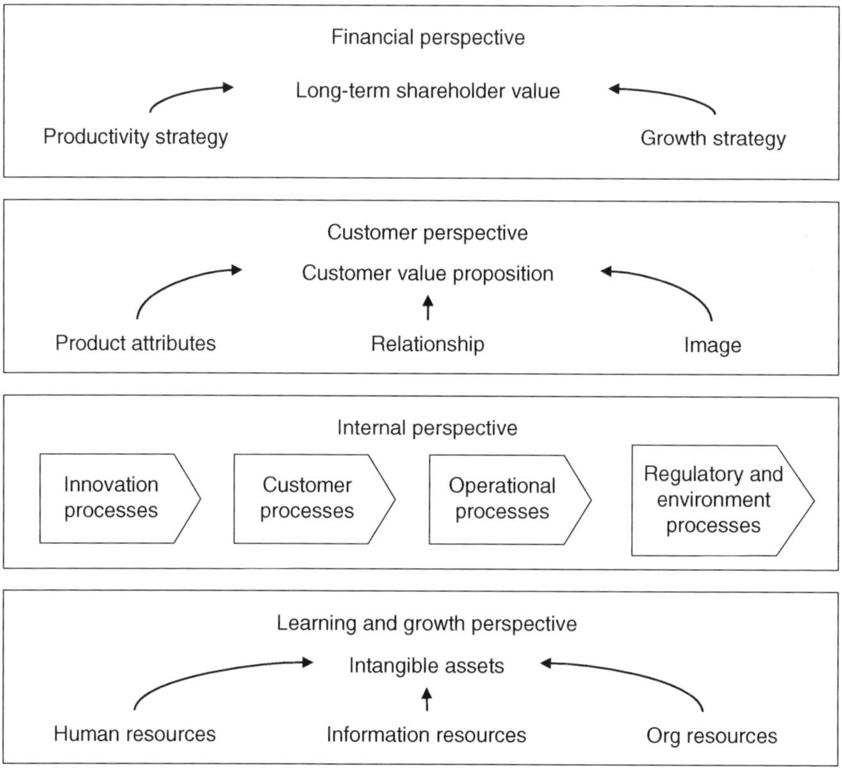

Figure 7.4 Balanced scorecard and strategy map—Adapted from Kaplan and Norton (2000, 2004).

(Aside: Observe that this is a layered model with the highest context and goal being financial performance. Kaplan and Norton (2000, 2004) assert that a not-for-profit organization has a similar map, except that fiduciary and customer perspectives (two perspectives at the same level; e.g., obligations to students and donors) replaces the financial perspective.)

The balanced scorecard/strategy map concept is useful as a guide in understanding strategic context. However, we remain cautious; as Porter has stated, strategy comes from creating something unique. We believe uniqueness implies unique activities and not merely unique instantiations of a standardized template. For example, we believe the organization should develop a unique intellectual capital (IC) model and not rely on the generic list of intangibles provided by Kaplan and Norton (2000, 2004). Our discussion in Chapter 1, revealed in the student exercises, highlights both the complexity and importance of thinking through and developing a unique IC model for your institution. Also, a K strategy should include the full breadth of K assets, which includes selected tangible resources (e.g., data analytics) as well as intangibles. Still, the strategy map concept is a valuable tool in promoting our understanding of strategic context.

7.2 Strategy Continuum and the Knowledge Organization

7.2.1 Enterprise Approach

It is instructive to examine the lessons associated with the business intelligence (BI) case studies discussed in Chapter 4. The most successful BI implementations were enterprise-wide systems interconnected with the processes of the organization. By changing the way work was done, both the culture of the organization and the work itself were changed.

Orchestra Metaphor. We view KM in a similar fashion. For the K organization to fully flower, KM is better seen as an orchestra with interacting musicians rather than a collection of isolated instruments. KM is about assembling the right musicians (K workers and their expertise) and their instruments (K tools) and then weaving them into a symphonic orchestra. In our KM orchestra, we must consider the content of K, K workers, their actions, as well as the musical score (i.e., K strategy), formed and conducted by a K leader. *A K strategy, as with a symphony, is realized through its actual performance.*

A knowledge strategy should fit with the organization's corporate or business strategy and its activities. Horizontally, a K strategy should span the wider enterprise to produce a complementary effect; vertically, a K strategy should be embedded in the organization's corporate or business strategy, work, and culture to create a compounding effect.

7.2.2 Strategy Continuum

We think of strategy as a blending dynamic where thin and thick elements are woven into organization and K strategies. Every strategy school should be accommodated (to some degree) by a K strategy. As strategic content becomes increasingly tacit, our corresponding K strategy should shift to an increasing emphasis on intangibles; similarly, as strategic content shifts to the explicit we do the reverse and emphasize the tangibles. This is shown in Figure 7.5, which associates K strategies and outcomes with corresponding strategy schools. In effect, *Figure 7.5 is a roadmap to a suite of critical K strategies for the organization.*

The list of K strategies and outcomes displayed in Figure 7.5 is not complete, but it does include most of what we have discussed in this book. Also necessary are the foundational KM principles that support the K organization and its K strategies—indicated at the bottom of Figure 7.5. This includes tacit and explicit models, the SECI model, thin and thick constructs, ba and K-spaces, K-binding, the notions of symmetrical and asymmetrical interactions, relational contracting, and so on.

7.2.3 Integrative Strategy Stream

K strategy, as with K itself, is integrative rather than modular. While we may formulate a specific strategy for, say, a K repository, the compounding effect could be lost if integration with other K elements is not considered. Note that by integrative strategy we are not referring to an "all-or-nothing" approach but to an awareness and appreciation of the complementarity associated with K strategies.

Additionally, we remain mindful of Porter's assertion that strategic actions should have a good fit (i.e., relating activities). This would include simple consistency between specific KM initiatives and our organizational strategy, reinforcing activities such as integrating access to a K-base and K workers, and optimizing activities such as the development of a K-space to include our K-base, key K-creating business processes, collaborative interaction, and its overall impact on the K culture of the organization.

Figure 7.6 illustrates an integrative strategy implementation with a good fit. An individual K worker is shown embedded in a business process performing a task or a series of tasks. The K worker has an opportunity to collaborate with peers (within or outside the organizational boundaries). There is also access to an organizational K-base. Recall from Chapter 5 that a K-base is not a single entity but a managed collection of systems, strategies, and documentation. Its content extends into cultural assumptions, K-maps, lessons learned, best practices, professional and practice communities, collaborative information, and so on. Note too that we are showing a K portal, which provides a single point of entry with semantic navigation to the contents of the K-base depending in part

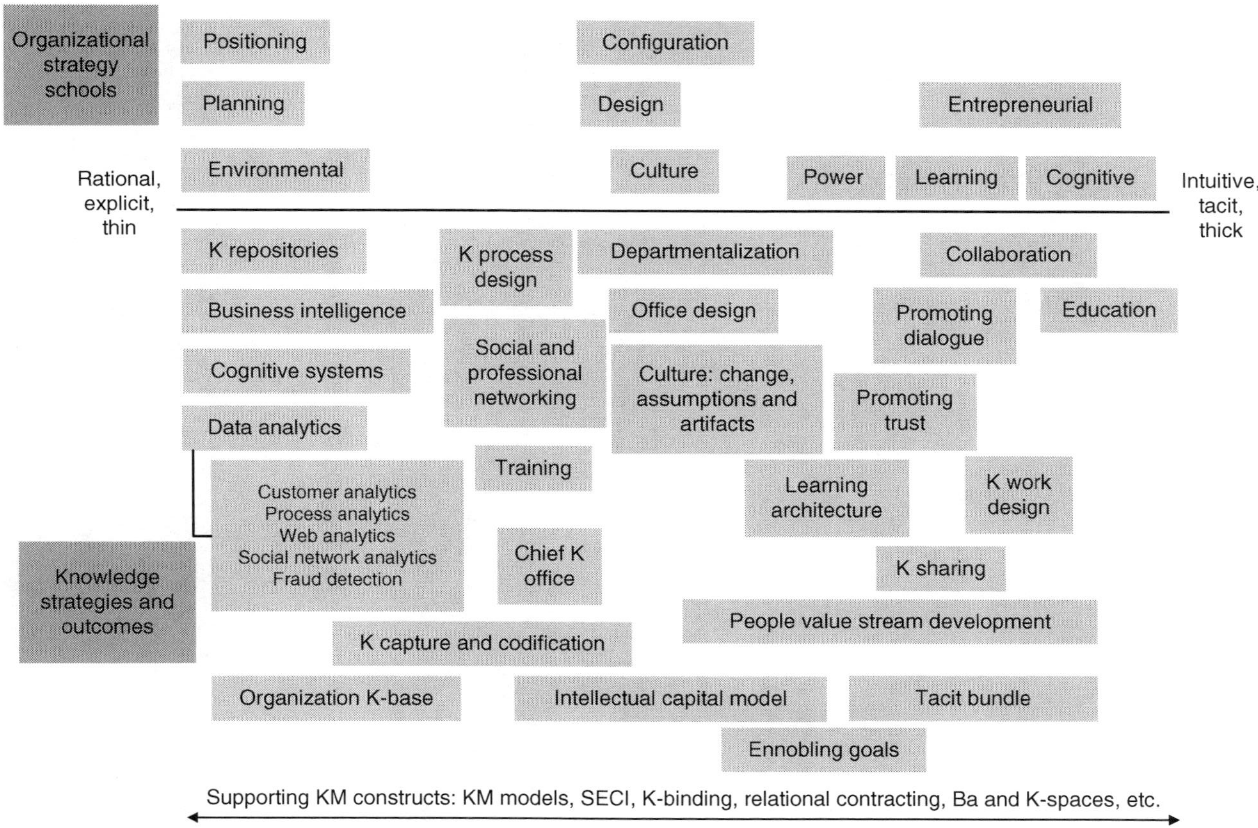

Figure 7.5 Strategy continuum—sampling of K strategies, outcomes, and strategy schools.

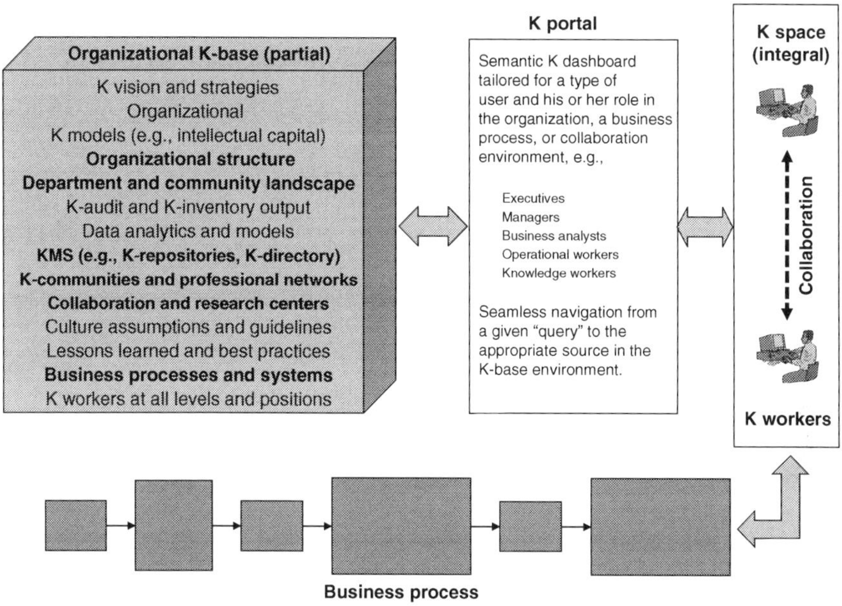

Figure 7.6 Integrated K-space and culture change strategy.

on the worker's job role and task. Thus, by integrating several implementations at once (work, worker, collaboration, and access to a K-base), there is a complementarity that affords the organization an opportunity for both learning (at all ontological levels) and culture change.

7.3 Setting the Stage: Intangibles and a Knowledge Strategy

Recall from Chapter 1 the following important characteristics of intangibles that we must keep in mind when developing a framework for a K strategy (Andriessen 2003):

- Intangible resources are nonmonetary resources without physical substance that in combination are able to produce future benefits for an organization (e.g., educated workers, brand, location).
- IC is a subset of intangible resources that includes those resources that are based on intellectual capabilities and activities.
- A core competence is a bundle of intangible resources that enables a company to provide a particular benefit to customers (sourced from Hamel and Prahalad 1994).

Additionally, we may add the following.

- The value of K is a function of its context and its interconnection to the organization's strategy, a value assessment of K threads, and their application—this is illustrated in Figure 5.26. This application or context is typically made manifest in K work and processes.
- The value of intangible assets (and IC in general) is often deferred. This is somewhat analogous to the value of education: its value is realized over time with experience and in an untold number of subsidiary cause-and-effect relationships. This is in contrast to training, where the value is often focal, requires little experience, and immediate.

7.3.1 Revisiting Core Competence

Recall from Chapter 1 (and our discussion earlier) the following key observations on competence:

1) A competence must be difficult to emulate; otherwise, it would not be core.
2) A considered mix of explicit and tacit resources is necessary; otherwise, a system or organization relying primarily on explicitness will eventually lose its distinctiveness.
3) Dorothy Leonard (1995) has stated that a *core technical capability* includes (1) physical technical systems, (2) employee skills and competence, (3) management systems, and (4) values and norms.
4) Adapting the work of Leonard (1995), Andriessen (2003) gives us an IC perspective by identifying five categories of intangible resources: (1) skills and tacit knowledge, including know-how and competencies, (2) technology and explicit knowledge, including patents, manuals, and procedures, (3) primary and management processes, including leadership, control, communications, and management information, (4) collective values and norms, including client focus, reliability, and quality, and (5) endowments, which is a category that surrounds the other four and includes customers, brand, suppliers, network of talent, and ownership of standards.
5) At a lower level of granularity, Erik Sveiby (1997) characterizes *individual competence* as (1) explicit knowledge, (2) skills, (3) experience, (4) value judgments, and (5) social networking.

7.3.1.1 Organizational Application—Integrating Intangibles and Tangibles

Recall our DNA example in Chapters 5 and 6. The Toyota DNA artifact consists of two strands: a Toyota production system (TPS) and a Toyota cultural and human strand (Toyota Way), each interconnected through policies, systems, structures, and initiatives. The TPS strand is closely associated with those activities that promote lean effectiveness and excellence and correspond to

Porter's description of operational processes. Recall too that TPS in Toyota's configuration is not a stand-alone set of (thin) activities but must be interconnected with thick activities in the Way strand. Competitive advantage thus arises from a blending of thin and thick and aligning common activities that an organization may perform exceptionally well with unique activities that distinguishes an organization. The distinguishing activities are thick and not visible and are based on culture and people.

Strategy as a DNA Framework

The Toyota DNA lesson is that creating distinctiveness rests on generating and interconnecting thick elements with operational, thin elements. In the Toyota context, *the unique activities which Porter deems essential to strategy are primarily thick* and require alignment and optimization with corresponding thin elements. Note too that we assert that unique activities should be primarily thick in order to sustain uniqueness and advantage. Toyota has stated that its advantage and success rests on that which is not visible. Thus, *the DNA molecule is a representation of a strategy framework that simultaneously identifies and builds thin and thick elements (i.e., TPS and Way strands) through specific policies, systems, structures, and initiative (i.e., cross-linkages).*

Finally, the K strategies shown in the strategy continuum (Figure 7.5) are similarly thin and thick. In the same way that the DNA molecule represents a high-level synthesis of organizational elements and policies, an effective K strategy should generate similarly complementary and reinforcing actions.

7.4 Leadership and KM

7.4.1 What Is leadership?

Yukl (2006) provides a working definition of leadership as follows: "Leadership is the process of influencing others to understand and agree about what needs to be done and how to do it, and the process of facilitating individual and collective efforts to accomplish shared objectives." The question that naturally follows is "influence what?" Here, Yukl (2006) provides a list of selected items that address this question:

- The selection of objectives and strategies
- The motivation of organizational members
- The mutual trust and cooperation of organizational members
- The development of organizational members' skills and confidence
- The learning and sharing of new K
- The design of formal structure, programs, and systems
- The shared beliefs and values of organizational members

Note that Yukl (2006) is treating leadership as both a specialized role and a social influence process. In this regard, leadership behaviors may be task oriented (e.g., organizing work activities), relations oriented (e.g., socializing with people to build relationships), or change oriented (e.g., envisioning exciting new possibilities for the organization).

7.4.2 Direct and Indirect Leadership

Direct leadership is typically seen as an organizational member influencing immediate subordinates. In contrast, indirect leadership may be seen as, first, a cascading of effects as decisions move down the hierarchy. A second form of indirect leadership involves influence over formal programs, management systems, and structural forms (Yukl 2006). One particularly important management system that is critical to the K organization is the development of the people value stream. This may be conceived at both a macro and a micro level (Figure 7.7) and is subsequently discussed. The third type of indirect leadership concerns influence over culture. For a young organization, this influence concerns the precepts of the founders and culture change will typically be evolutionary in nature. In contrast, the assumptions in a mature organization may be embedded in the systems of the organization and be resistant to change,

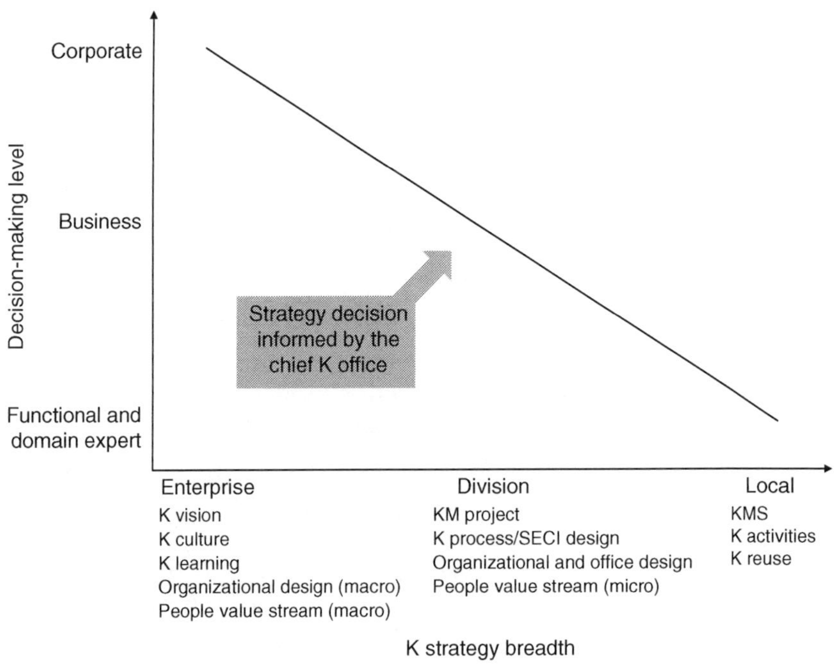

Figure 7.7 Decision-making and K strategy.

and thus culture change and influence may require transformational leadership. We discuss culture change in section 7.7.

7.4.3 Servant Leadership

Recall the idea of servant leadership shown in Figure 6.13. This figure is a leadership artifact that is a manifestation of Toyota culture. In the inverted hierarchy that is characteristic of servant leadership, team production workers are directly in the value stream and hence create value. Everyone else in the hierarchy is removed from the value stream and their purpose is to support (i.e., motivate, train, and inspire) whoever is above them in the inverted hierarchy. When a manger tells a subordinate that he or she is sorry for a mistake that the subordinate makes (not uncommon in Toyota), it is because the manager has failed in his or her responsibility to the person above in the inverted tree.

7.4.4 Expertise and Leadership

One basis of leadership is expertise. Unique K about the best course of action or the best way to solve a problem creates an opportunity for influence. Another factor is the perceived level of expertise—members of the organization must recognize the leader's expertise. Such leadership requires that expertise be maintained through continual education, practice, and the exercise of leadership itself. Also, continual teaching is important—teaching simultaneously refines K and makes expertise visible, thus building the trust necessary for influence.

For example, recall the "four power players" (Sveiby 1997) associated with K roles shown in Figure 4.5. Observe that the characteristics of a K leader include (1) high expertise, (2) high organizational competence, and (3) recognition and the according of influence by other organizational members.

Expert leadership (or power) is different than information power. Information power concerns control over the flow and distribution of information. For example, recall from Chapter 3 that the "middle line" in Mintzberg's classification (i.e., structure in fives) functions as little more than an information relay in the information processing organization. In contrast, the middle line assumes the role of a knowledge engineer in the K organization (Nonaka and Toyama 2004). Note that as the organization shifts in character from an information processing organization to a K organization, the basis of power also shifts from information (sometime, an adversarial role) to K (an enriching role).

7.4.5 Leadership and K Strategies

Recall from earlier that strategy types correspond to the decision-making level of those who formulate them. We can extend this idea and generalize K strategies and leadership, as shown in Figure 7.7. Observe that high-level decision makers are associated with long-term strategies, such as culture change initiatives. Middle-level managers focus on project- and process-related K strategies. Both high- and middle-level decision-makers are concerned with structural design and people-related strategies. Finally, low-level managers concern themselves with K systems and specific activities, such as conducting a K inventory.

Note that as we move both to the right on the horizontal axis and down the organizational hierarchy, the strategies shift in character from involving primarily intangibles to those associated with tangibles. Note too that we recommend the establishment of a CKO, which, in addition to carrying out specific activities, serves a support function in the development of K strategies. This is particularly important in the K era where K drives strategy. We discuss the CKO later.

7.5 Getting Started

7.5.1 IC Guide

The best way to initiate a K strategy is to develop your organization's IC model, assess your current state, and develop an action plan guided by your IC.

Chapter 1 contained a student exercise on the development of an organization's IC model. An IC identifies and prioritizes key leverage points in a K strategy. Recall that IC may be seen as a subset of intangible resources and includes all resources that are based on intellectual capabilities and activities (Andriessen 2003). The IC model is just that—a representation of the organization's intangible assets with intellectual merit. Yet, as we discuss in Chapter 1, IC may include selected explicit elements of an organization. This arises because (1) explicit assets may be proprietary or protected (e.g., patented) and (2) explicit assets may be interconnected with tacit-based assets (e.g., Toyota DNA). It is such a composite—for example, a mix of know-how and science—that often represents the essence of a core competence.

7.5.2 Strategic Prescriptions

Figure 7.8 shows states of knowing, states of K, and prescriptions for organizational action (Earl). We have extended Earl's prescriptions with several suggestions while retaining his classification structure. In effect, Figure 7.8 is a way to jumpstart both your strategy thinking and action plan.

From this illustration we make the following observations:

- Explicit K is knowing what you know; that is, visible K. The strategy used to identify explicit K centers on an inventory of K and information assets. This would include an assessment of existing technical and information systems

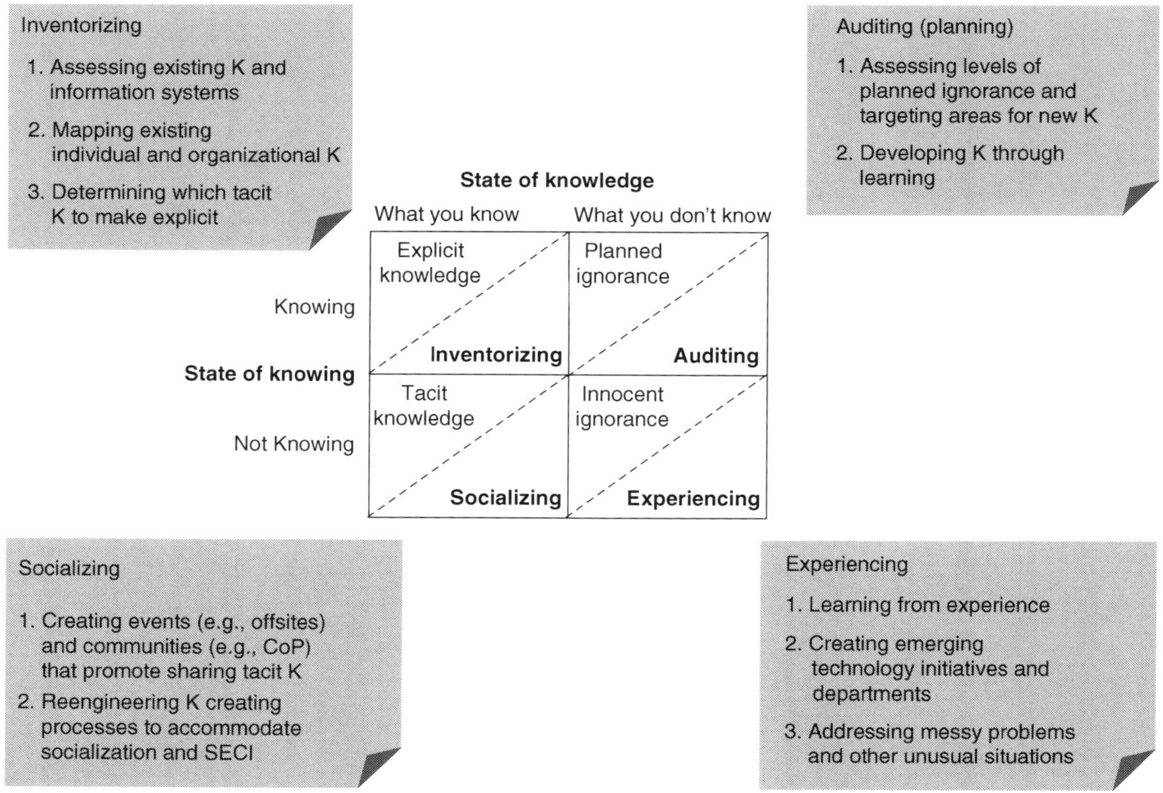

Figure 7.8 Strategies for knowing and knowledge—extension of Earl.

to uncover scripts and nuggets. Also included is a mapping to people who possess K; this possibly extends our inventory into tacit K by identifying people and the expertise they possess. As an example, expert profiling is one such system that may be used to map people and K (see Figure 4.12). Also, following a people inventory we must make decisions as to which K remains tacit (in the heads of people) and which needs to be externalized. This is illustrated in Figure 5.13. Observe that both K mapping to people and deciding which tacit K to externalize may extend our thinking into the tacit K quadrant of Figure 7.8.

- Planned ignorance is a recognized (or planned) area in which you decide to have little expertise (and presumably not to compete). The organization must decide which areas to develop expertise and which not to pursue. The action plan associated with planned ignorance and the learning necessary to reduce it should be guided by the organization's IC model and its inventory. Interestingly, it is possible to deliberately reduce expertise already possessed and not compete in a given domain; pharmaceutical giant Pfizer's decision to leave the cardiovascular research area is one such example.
- Tacit K is not knowing what you know, that is, K not visible. This is somewhat analogous to what Spender (1996) has called K "taken for granted." The suggested strategies focus on sharing tacit K. This includes the creation of events such as offsites, retreats, group activities, and so on.
- Innocent ignorance is essentially uninformed ignorance. To reduce such ignorance and promote greater awareness, the organization must deliberately pursue experiencing through external seminars and conferences, education, training, and through internal exploration. The latter may include an "emerging technologies department" or internal, exploratory initiatives. As an example, organizations became familiar with BI through proof-of-concept projects. Such projects led to a better understanding of the business opportunities that BI makes possible as well as the new technologies involved. This in turn has led to further exploration of data analytics. This sequence of exploratory projects with new technologies is an example of how one experience leads to another, and how we may sequentially build and link competences in a fashion similar to the Sharp experience mentioned earlier.

Observe that both *inventorizing* and *auditing* are periodic assessments and usually precede a KM initiative or development of a system. In contrast, *socializing* and *experiencing* are action-oriented strategies and should be ongoing activities.

Auditing has a *planning* flavor and represents an organization's possible development of a new competence, guided by the organization' IC and its overall business strategy (e.g., Sharp). This would have to be followed by specific learning and development initiatives. The Sharp experience discussed earlier is an example of this strategy.

The lower two quadrants (i.e., socializing and experiencing) give rise to the distinguishing management feature of a K organization (as opposed to an information organization)—the concentration on tacit K.

7.6 Strategies for the Knowledge Organization—Tacit Bundle

Figure 7.5 shows an assortment of K strategies, covering tacit and explicit K, structural and behavioral elements, and so on. However, keep in mind the K organization is distinguished by its emphasis on tacit elements. Accordingly, listed below is a sampling of tacit-oriented strategies that will simultaneously build the K organization and promote the development of a K culture. We call this the *tacit bundle*, meaning that it focuses primarily on social or tacit–explicit interaction. This is in contrast to those K strategies that focus primarily on explicit K, such as K codification or BI development. In practice, tacit and explicit K threads are not easily separable; hence, the tacit bundle touches explicit knowing, such as the exploitation of SECI in business process reengineering.

1) Development of the IC model for an organization. The IC serves as an umbrella in guiding the development of the organization's K strategies.
2) The development of ennobling goals that extends the purpose of an organization beyond financial goals. Consider, for example, Whole Foods' goal of eliminating hunger. In addition to their manifest effect, noble goals engage K workers beyond the immediate requirements of their work to a larger purpose. *K workers need to believe their work is socially important!*
3) Create and exploit social interaction. This includes specific activities, such as department meetings and social events; targeted activities, such as off-sites (which should also be strategic in nature); and background strategies, such as open office design. Additionally, functional design may be used to promote social interaction.
4) Promote the sharing of tacit K and experiences through focused communities, such as CoP, centers of excellence, and mentoring communities. Also, job rotation is a technique that promotes the sharing of experiences and know-how.
5) Closely related to number 4 is the creation of discourse structures. Such structures may be informal and focus on sharing expertise and experiences (number 4) or be more formal, such as project teams, committees, task forces, and lateral processes.
6) Facilitate tacit–explicit interaction. For example, consider the development of a DNA molecule as both a practical and an operational construct. Similarly, consider conceptualizing your organization as a system of

layers—thin and thick—that embody the ideas of know-why (functional expertise), know-how (thin functional integration), and know-what (thick practice). Additionally, the institution of a governance structure, such as the CKO, is one example of how we facilitate tacit–explicit interaction in the context of KM. Organizational centers of excellence are also included. The CKO should guide tacit–explicit interaction, such as the planned, formalized, and systematic capture of K and its codification, among other initiatives. Here we discuss the activities associated with the CKO.

7) Reengineer K-creating processes with a focus on exploiting SECI. This follows from earlier and was discussed in Chapter 5, including our case study on the Peace Corps.
8) Promote external interaction through professional networking. This includes the development of professional networking and collaboration initiatives, particularly academic and scientific research projects.
9) Develop competence through collaborative initiatives with external sources of K.
10) The deliberate, ongoing exploration of new technology. This may include the creation of an emerging technology department as well as the social interaction and K sharing that comes from attending conferences and seminars.
11) Developing a competence and building a learning culture through integrated design. This covers a multitude of strategies that both build and align the organization's competence, learning capability, and its K culture. This includes the management of occupational subcultures, promoting trust and dialogue, integrated work {collaboration–process–technology} design (Figure 7.6), and specific K activities, such as project readouts, K capture and codification, and building K systems (e.g., K-repositories), and so on.
12) Development of the people value stream. This includes the hiring, education and training, and career pathing of employees as well as job rotation.
13) Attempt to solve messy problems (or exploit opportunities) facing your organization or society. A good example of this was mentioned in Chapter 4 with Google and its nano-particle project, which attempts to detect early stage cancer. As with most K strategies, solving a messy problem requires interdisciplinary activities that touch on several different areas of expertise.

7.7 Culture Change

7.7.1 Culture Formation

The most important factor in forming and embedding culture is the behavior of an organization's founders and leaders. Figure 7.9 is a summary of the mechanisms for creating and developing culture in a new organization, adapted

Primary embedding mechanisms	Secondary reinforcing mechanisms
• What leaders pay attention to, measure, and control	• Organization design and structure
• How leaders react to organizational crises and incidents	• Organizational systems and procedures
• Observed criteria by which leaders allocate resources	• Organizational rites and rituals
• Role modeling, teaching, and coaching	• Design of physical space, facades, and building
• Observed criteria by which leaders allocate rewards and status	• Stories, legends, and myths about people and events
• Observed criteria by which leaders recruit, select, promote, retire, and exclude organizational members	• Formal statements of organizational philosophies, values, and creed

Figure 7.9 Primary and secondary mechanisms for culture formation—Adapted from Schein (1999).

from Schein (1999). As can readily be seen, our assessment of the tacit assumptions that are the essence of culture follows directly from the observation of leaders and their policies. The ensuing discussion is largely drawn from the works of Schein (1999).

7.7.2 Organizational Age and Culture Change

It follows from Figure 7.9 that the best way to change culture is to change the observed behavior of leaders, particularly in regard to the elements of the people value stream. However, the age of the organization is critically important—it may serve as an indicator of how rigid and resistant to change are the cultural assumptions of the organization.

7.7.2.1 Young Organizations

A relatively young organization is dominated by the assumptions of the founders. The emphasis is on differentiating the organization and making the culture manifest. Also, young companies tend to emphasize specific departments (e.g., DEC bias toward engineering and manufacturing). There is an expectation that new members will fully imbibe the organizational culture. As importantly, in a young organization, the values of the founders are often broader than purely economic ones. Again, consider the mission of Whole Foods described earlier.

Culture change in a young organization is typically *evolutionary*. Change may follow a traditional evolution pathway as the organization diversifies and differentiates itself into new forms (e.g., product or market structures). Additionally, evolution may be specific where particular parts of the organization adapt to the environment and create new subcultures. Moreover, subcultures are not exclusive to function or job role but may become stratified by organizational level. Consider, for example, the different assumptions of union members, first-line supervisors, and middle managers. Other important change strategies of the young organization include making explicit those tacit assumptions that the leaders believe should be emphasized, fostering insight as the environment changes, and empowering selected subcultures (e.g., data analysts).

The young organization may reinforce its assumptions through specific mechanisms. This includes, for example, hiring people who are consonant with the culture of the organization, rewarding behavior consistent with cultural assumptions (or, conversely, punishing behavior that violate those assumptions), and through training, socialization, and mentoring programs.

7.7.2.2 Mature Organizations

In a mature organization, professional managers have replaced the organizational founders or its entrepreneurs. The organization is characterized by a culture that is more differentiated, with cultural elements embedded in structures and processes. Culture is now "taken for granted" as culture creates leaders (as opposed to a young organization where the founders create culture). In contrast to the possibly higher goals of the founders, the mature organization (dominated by professional managers) subordinates social values to financial values.

This implies that the change strategy of the mature organization should be more *transformative* than evolutionary—culture change must be "planned and managed." That is to say, culture no longer evolves but is directed. We should not be surprised, for example, if the assumptions of the founders are no longer appropriate in today's environment.

7.7.2.3 Transformative Culture Change in Mature Organizations

Transformative culture change requires that the organization go through three stages:

1) *Unfreezing and creating a motivation for change.* Creating a motivation for change requires a certain "disconfirmation" where the organization and its members face a threat of some kind—economic, political, technological, legal, moral, or internal. For example, a merger, acquisition, or joint venture may be a source of disconfirmation.
2) *Learning new concepts and new meanings in place of old concepts.* Learning new concepts includes the identification of role models, scanning for solutions, and trial-and-error learning. For example, learning new ways of thinking may involve formal training, informal training of groups or teams, participation in practice fields where organization members may make mistakes and learn without disrupting the organization, support groups for

learning, and a consistent reward and punishment system consonant with the new concepts.

3) *Internalizing new concepts and meanings.* Internalizing new concepts may be thought of as refreezing as new behaviors become a stable part of individuals and the organization.

7.7.3 Culture and KM

From a practical perspective, the cultural assumptions of a young K organization are more likely to be consonant with the principles of KM than that found in a mature organization. It is not surprising that young people are more excited about working for Google than for a more mature traditional company. It is also important that K workers (of all ages) believe their work and organization are larger than that found in a purely mercantile organization. This is particularly impactful on the assumptions associated with K sharing, socialization, and learning. A mercenary culture is built on a foundation of largely financial values that focuses on narrow goals—it typically runs contrary to a communal culture that promotes collaboration and K sharing.

Culture change in the K organization may be targeted and involve the emergence of new subcultures, even in the mature organization. For example, recall from Chapter 4 that big data analytics is leading to specialized departments. Our suggestion for a Chief Knowledge Office governance structure (including an organizational K-base) is another specific strategy for the K organization.

Creating a *fact-based culture* is an example of combining several strategies. This involves the creation of new organizational forms and subcultures and the creation and elevation of new assumptions on the workings of the organization (e.g., fact-based decisions). This requires the support of management (e.g., recognition) and an integrated organization of systems—that is, integrated tasks and processes, technology, and collaboration that collectively are transformative to the K organization and its culture (e.g., Figure 7.6).

7.7.3.1 Long Range People Strategies—Building K Culture

It has been our experience that KM can have a manifold effect on culture and vice versa. We will discuss this in more detail, but briefly leaders should consider *long-term, K-oriented strategies that invest in people.* These strategies are cultural in that they are long term and focus on people; they are not project or task-related:

- Visible investment in people—show employees that you have a sincere interest in their development and careers!
- Continual education—actively encourage and reward the development of knowledge and expertise!
- Continual teaching and internal consulting—build expert leaders from within the organization. Teaching both refines and makes expertise visible!

- Integration of K and work—interconnect expertise, people, technology, and collaborative processes and structures as a matter of course. Work should be new and exciting!
- Inspire your organization with ennobling goals and vision—the organization and its work should be socially and professionally important!

7.8 Chief Knowledge Office

We propose the creation of a CKO as an important strategic initiative for the K organization. We choose the word "chief" because we envision a centralized governance structure that is responsible for the development and coordination of K activities and strategies on an organization-wide basis. We have previously introduced the notion of a K-base in Chapters 2 and 5. In this chapter, we extend our discussion by elaborating on the larger structure we are calling the CKO. Note that we are using the abbreviation "CKO" to refer to the Chief K Office and not the job title (i.e., chief K officer). The Chief Knowledge Office is a composite K initiative that embraces several specific K strategies identified in Figure 7.5, such as developing the organization K base and K capture and codification.

7.8.1 CKO Structure

The CKO serves as the staging area for a variety of specific K functions as well as the configuration of the organization's K-base. Figure 7.10 is a representation of its governance and functional structure.

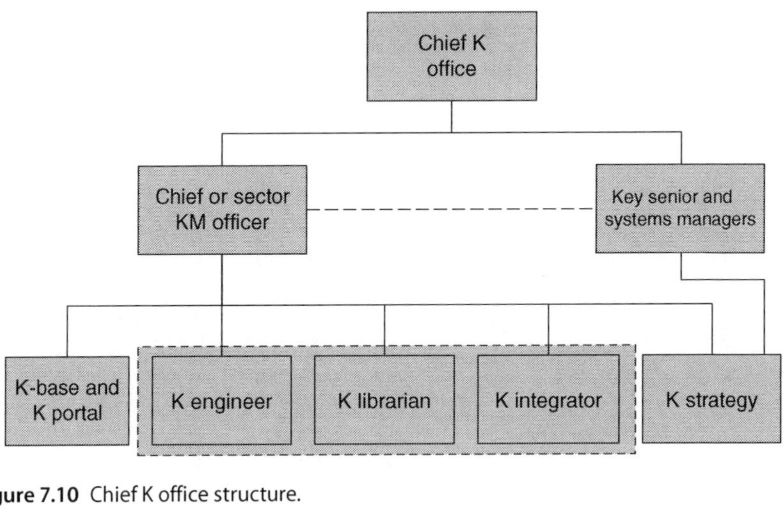

Figure 7.10 Chief K office structure.

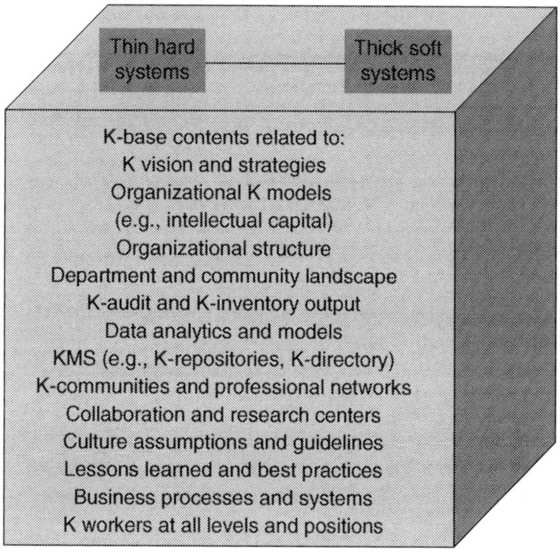

Figure 7.11 Contents of an organizational K-base.

Its primary constituents are as follows:

- The CKO is largely a governance structure with a dedicated *Knowledge Management Officer* and several affiliated *senior* and *systems managers*. The KM officer may be a chief or sector KM officer.
- The *K-base* is a collection of the organization's K threads and includes both its hard and soft systems and representations. We conceive of the K-base as an extensive repository containing the organization's critical data, representations, and K systems. The contents vary in both type and organization level and include the organization's IC model, K maps, K systems, K processes, K communities, collaboration arrangements, organizational values and mission, and so on. On the other hand, the collection should be conceived as a single repository with selected access through a *K portal* under the guidance of the CKO. See Figure 7.11.
- The CKO staff consists of specialized K workers who report to the K officer. The K workers have specific duties—*K engineer, librarian,* and *integrator*—with respect to K projects and activities. These are functional job roles and may include any number of people.
- The CKO formulates and manages the organization's *K strategies*, which serves as input to the *corporate* and *business strategies* of the organization.

7.8.2 Activities of the Chief Knowledge Office

In Chapter 4, we introduced the job roles for the each of the specialized job roles mentioned earlier—that is, K engineer, librarian, integrator, and K management officer (chief or sector). Here we further discuss their key activities in terms of our recommended functions for the CKO. In several of the following listed K activities, the CKO either is directly responsible or acts as an internal consultant:

- Leads the development and refinement of the organization's IC model, including its internal relationships and measures.
- Guides K audits and K inventories for the organization.
- Identifies, measures, and refines the organization's necessary K, based on the organization's strategy, K-audit, and K-inventory. This includes the value assessment of specific knowledge threads (e.g., flat-panel display component technology).
- Guides the extraction of K from project readouts, business sources, and external sources. The K integrator subsequently refines, aggregates, stores, and distributes corresponding knowledge threads. The captured K should lead to the development of both lessons learned and best practices documentation.
- Facilitates the acquisition of K (tacit K capture) in the development of an expert or other K system.
- Extracts, structures, refines, and maintains K in specific KM systems or the organization's K-base. This includes building K-stores and K-maps.
- Guides the development of CoP, centers of excellence, and other internal networks.
- Identifies and establishes external K-communities, such as professional communities.
- Provides guidance into (1) the reengineering of the organization's knowledging processes and (2) development of reuse strategies for K systems.
- Leads the development of collaboration arrangements and alliances with external suppliers of knowledge, such as universities, government agencies, and vendors.
- Leads the development of K-strategies, both technological and business.
- The CKO may have varying levels of participation in a middle–up–down structure. Recall from Chapter 2 (Nonaka and Toyama 2004) that the middle-level manager functions as a K engineer, and thus (1) leads new product development projects and (2) functions as the "middle" in a middle–up–down structure, that is, someone who integrates strategic intent (mostly tacit) from the top and market and resource realities (mostly explicit) from the bottom.
- Establishes, coordinates, and oversees the organization's K-base. This includes the establishment of a K portal and new subsystems, such as internal libraries and K maps. Within the K-base itself, the K librarian integrates separate knowledge threads into composite knowledge structures.
- Establishes and offers internal courses on all aspects of KM for the organization.

7.8.3 K-Base and Portal

The K-base may be configured in any number of ways. For example, Figure 7.6 illustrates an active K-base where its contents are accessed within the contexts of a business process and a collaborative structure. Also, we have highlighted those content areas that may be useful to access during performance of a business process. In contrast, Figure 7.11 is a static representation of a stand-alone K-base.

7.8.3.1 K Portal

Access to the K-base is best accomplished through a K portal. As indicated in Figure 7.6, we conceive the portal to be organized around specific semantic requirements rather than data structures or files. For example, information on a business process should be in terms of process content (e.g., process name, task names, information requirements) and not files containing that information.

There are various access strategies that may be employed, often in combination. For example, the following is a sampling (see Figure 7.12):

- Business process or task of the user. Active access as illustrated in Figure 7.6. In this case, the user role, task, and information and K requirements are the query variables.
- Business function access, such as data architecture or marketing. Closely associated with function are business systems and structures. For example, whereas accounting is a function, a corresponding business system may be an enterprise resource planning (ERP) system.
- Meta-K such as person name, date, and location. This access strategy reveals information on a specific person, including K maps, patents, seminars given or attended, and so on.

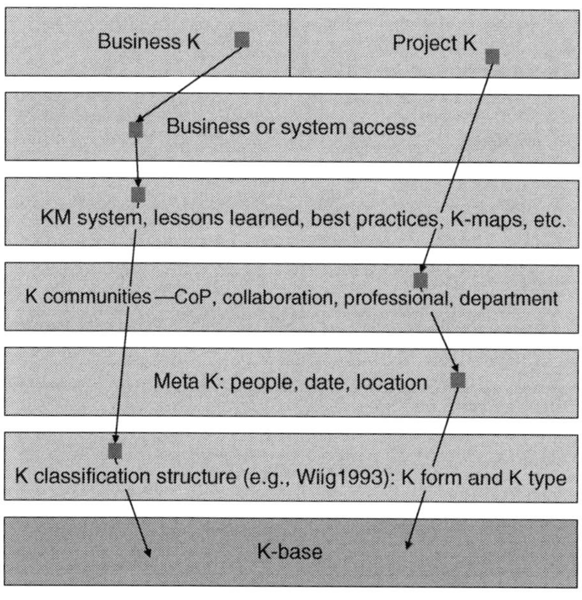

Figure 7.12 Sample K portal structure with access paths.

- Business or project K. This concept is derived from the hypertext organization; however, the categorization is probably too high level to be useful.
- Any number of K classifications may be used to access the contents of the K-base. For example, the Wiig (1993) classification is along two distinct lines. The first line is the *form* of K, and includes *public, shared,* and *personal.* The second line is the *type* of K and includes *factual, conceptual, expectational,* and *methodological.* Wiig (1993) goes on to identify the intersection of both classifications, such as shared conceptual K (e.g., "the stock market is bullish"). The combination of both K form and K type should narrow the search more quickly than either classification alone.

7.9 People Value Stream

Senge quotes performance guru Deming, "Our prevailing system of management has destroyed our people. People are born with intrinsic motivation, self-esteem, dignity, curiosity to learn, joy in learning. The forces of destruction begin with toddlers—a prize for the best Halloween costume. Grades in school, gold stars, and on up through the university. On the job, people, teams, divisions are ranked—reward for the one at the top, punishment at the bottom. MBO, quotas, incentive pay, business plans, put together separately, division by division, cause further loss, unknown and unknowable" (Senge 1990). Senge continues, "Ironically, by focusing on performing for someone else's approval, corporations create the very conditions that predestine them to mediocre performance."

7.9.1 Human Resource Policies and K Workers

Horwitz *et al.* (2003) conducted an exploratory study of human resource (HR) policies for attracting, motivating, and retaining K workers in Singapore. Structured questionnaires were sent to the CEOs of 200 knowledge-intensive firms (KIFs) in six industries. Forty-four usable responses were received. The six industries included (1) electronics, (2) IT, (3) telecommunications, (4) R&D, in particular in biotech, life sciences, and science and technology, (5) venture capital, and (6) others, including the government. Though this research is concentrated in Singapore, we believe it provides useful insight into HR practices in a K-intensive part of the world.

7.9.1.1 Literature Review

Each question in the survey was framed based on the authors' review of the literature on best HR practices for managing K workers. Several factors were considered, most of which have been previously discussed in this book. Here we highlight those elements Horwitz *et al.* (2003) list in their research:

Attributes and expectations of K workers.

- Skills are in high demand.
- Prefer autonomy.
- Resist a command and control culture.
- Value occupational advancement—commitment is more occupationally than organizationally oriented.
- Higher turnover rate with K workers than other employee groups.

Organization design considerations.

- Strategic choice of core competencies.
- Flatter, networked structures in KIFs than in traditional companies, often in boundaryless or loosely coupled organizational environments (e.g., joint ventures, strategic alliances).
- Organizational form may require that work be done independently with a measure of ambiguity.
- Decentralized, often networked decision making.

Organizational blueprint typologies of HR practices.

Whereas most research has been conducted with respect to individual attributes of K workers, the authors assert the growing importance of an "organizational blueprint." For example, is the organization commitment-based, engineering-like (e.g., "skunk-works"), a bureaucracy, an autocracy with a contractual basis for work, or a "star" organization that recruits top talent and pays high wages? Organizational blueprints extend to the culture of the organization, concern for people, job design, and challenging work, among other factors.

This leads to the idea of distinctive HR practices and K worker performance. As an example, the authors report that in one consulting firm, performance reviews (PR) included an evaluation of an individual's contribution and utilization of K assets; in another firm, rewards were based on K sharing and help given to others.

Finally, attention must be given to the career pathing of K workers. There is a shift from traditional career pathing to one that sculpts jobs to the life interests of K worker professionals—one that requires constant retooling and the development of a portfolio of careers over time. All of this is to say that organizations must rethink compensation, job design, and HR practices.

7.9.1.2 Research Results

Attracting K workers. The most effective attraction strategies, ranked in order of effectiveness, were (1) very competitive total package, (2) internal talent development, (3) reputation as employer of choice, (4) proactive recruitment initiatives, and (5) job advertisement. This is in contrast to the most popular strategies, which were (1) job advertisement, (2) internal talent development, (3) use of headhunters, (4) online recruitment, and (5) career plans for redeployment and promotion. Note too that online recruitment, general advertising, and the use of headhunters were particularly ineffective strategies for attracting experienced people.

Motivating K workers. The most effective strategies for motivating K workers, ranked in order of effectiveness, were (1) freedom to plan work, (2) challenging work, (3) access to leading-edge technology and products, (4) top management support, and (5) ensure fulfilling work. There was a high level of consistency between effective and popular strategies: the most popular was also the freedom to plan work. However, ranked number 3 in popularity was the use of incentive bonuses, which did not rank at all as effective. Note too that this research focused on employer strategies and perceptions of effectiveness rather than on K worker views on what motivates them.

Retaining K workers. The most effective strategies for motivating K workers, ranked in order of effectiveness, were (1) challenging work, (2) highly competitive pay package, (3) having performance incentives and bonuses, (4) opportunities to develop in a specialized field, and (5) top management support. Of the most popular strategies, 4 out of 5 were also found to be highly effective. The findings support the importance of both job design and a highly competitive pay package.

Need for cross-cultural research. Additional cross-cultural research was considered important to determine whether competitive pay for all three constructs (attraction, motivation, and retention) was important only for South-East Asia or more generally applicable. Also considered a subject for cross-cultural research is firm-funded further education—a benefit thought to be highly valued in Singapore and possibly in Europe and elsewhere.

Significant differences in HR strategies were found for local and foreign-owned firms; for example, multinational corporations seem to rely more strongly on internal career pathing. Such differences extended into wholly and jointly owned firms and their respective HR strategies.

Conclusions. Table 7.1 for a summary of the key research results is attracting, motivating, and retaining K workers:

- The most popular HR strategies were not necessarily the most effective in attracting, motivating, and retaining K workers.
- Most employers were able to distinguish K and non-K workers necessitating different HR strategies between the two employee groups.
- There are distinctive bundles of HR strategies for managing K workers that vary according to several factors, such as whether firm ownership is local or foreign. The bundle of HR practices for motivation and retention has more in common than those for attracting K workers—we suspect this to be the case since existing employees have at least started to imbibe the culture of the organization.

Table 7.1 Summary of Singapore research on K worker HR strategies.

Attraction strategies	Motivation strategies	Retention strategies
1) Compensation package 2) Proactive recruitment 3) Internal career pathing	1) Regular communications 2) Conducive work environment 3) Compensation 4) Opportunities for career advancement	1) Opportunities for growth 2) Conducive work environment 3) Transparent compensation

Adapted from Horwitz *et al.* (2003).

- There are three effective attraction strategies: a compensation package that may include a sign-on bonus and funding for further studies, proactive recruitment (e.g., on-campus recruitment, targeted advertisement), and internal filling of positions based on career pathing. Also, consideration is given as to whether the prospective recruit will fit in with the culture of the firm.
- Effective motivation strategies are clustered into four categories: regular communications, conducive work environment, compensation, and an opportunity for career advancement. Note that a conducive environment includes teamwork, access to technology, cash rewards, freedom to plan work, challenging work, common culture among people, and a "fun" place to work.
- There are three effective retention strategies: opportunities for growth (i.e., promotions), a conducive work environment, and transparent compensation award. All three require top management leadership and support.
- The effectiveness of strategies for attracting, motivating, and retaining K workers may vary according to industry, organizational strategy, ownership structure, and cross-cultural factors.

Limitations. The following are our reservations that prevent a generalized conclusion in developing HR strategies for K workers:

- This research represents an employer perspective. No employee input was obtained.
- A sampling of job titles may have been instructive, particularly in a concentrated geographical area. For example, were most K workers software or systems developers or was there a broader spread that included, for example, marketing analysts? This is necessary to distinguish occupational subculture effects from general characteristics of K workers.
- The results indicated cultural differences in HR strategies and hence the need for additional cross-cultural research. HR systems and strategies may vary by culture or country of organizational ownership. This research focused on firms in Singapore employing mostly Singaporeans. According to Horwitz *et al.* (2003), one such cultural effect may be the claimed tendency for Singaporeans to leave a firm primarily for higher pay.

In the same way that the K era is influencing the design of virtually every organizational domain, so too must HR strategies adapt to the expectations of K workers. The research discussed earlier gives us insight into the end-to-end management of K workers. Some of the ideas uncovered in the research are to be expected, for example, the importance of competitive packages for workers who presumably have several options in their selection of an employer. On the other hand, firm-sponsored education as an HR strategy for managing K workers may vary across firms, based on the culture of the firm. Additionally, the idea of developing a distinctive HR bundle of strategies for K workers could raise the position of the HR area into a strategic partner for the firm.

Here we continue our discussion by focusing on the organizations to which we have previously made reference, that is, student experience at their respective organizations, Google, Toyota, Whole Foods, and Starbucks.

7.9.2 Recruiting, Selecting, and Hiring People

Let us keep it simple the Google way: "Hiring is the most important thing you do." Continuing, "We learned the only way for business to consistently succeed today is to attract smart creative employees and create an environment where they can thrive at scale" (Schmidt and Rosenberg 2014).

Would you advertise for people over the Internet? Several of the large organizations in which we have taught have investigated hiring and have found that the best technical employees—in terms of performance and retention—are those acquired through internal referrals from existing employees. This does not negate the role of good recruiters, only that the organization (following the Google model) would benefit from widening the scope of its recruiters—everyone's job is to recruit great people (i.e., great people know great people).

How does Google attract smart creatives? Google has commented on the "herd effect": A workforce of great people attracts other great people, not for organizational amenities but because they want to work together. Recruiting at universities is one way to promote the herd effect, as are targeted searches. Recall too our assertion (in our discussion of culture earlier) that K workers prefer work that is exciting and important—socially and professionally. Also, as with young companies generally (see Schein 1999), culture should be documented and made manifest—*what does the organization care about, how does it do work, and how are decisions made?* The mature organization is well advised to emulate the young organization and make its culture visible and living.

Google decided it was going to follow the academic rather than the corporate model. Universities hire for the long term; hence, there are relatively few layoffs in a university. Rather than depend on hierarchical decision making, Google uses committees for both hiring and promotions. Interestingly, Google hires the best possible people, even if there is no immediate match with available job roles. (Aside: This is similar to a team in the National Football League who drafts the best available college player rather than the player who fits an existing opening.)

When assessing a candidate, Google looks for *passion*—a key characteristic of K workers we have mentioned frequently in this book. We have previously used the term *engagement* to represent passion and commitment. That is to say, K workers live their dedication. What are the indicators for passion? For Google, the markers are *persistence, grit, seriousness*, and *all-encompassing absorption*—all of which are not apparent in a resume but must be uncovered during the hiring process.

7.9.2.1 Interview Process

At Google, candidates are evaluated in four areas during an interview:

- Leadership. How has the candidate responded in various situations? Has the person asserted a leadership role at work or in an organization—perhaps through organizing a team to solve a problem?
- Role-related K. Look for people with a range of strengths and passions, not just isolated skills. This includes necessary experience and background.
- General cognitive ability. Go beyond grades and search for the way a candidate thinks. Ask questions that provide insight into how he or she will solve problems.
- Googleyness. Look for that which makes a candidate unique. Can the candidate thrive in your organization? This leads us to the LAX layover test. If you were stuck at an airport, say, Los Angeles International, for 6 h, would you be able to pass time in a good and interesting conversation with the candidate? This test—designed to discover intangibles—has been institutionalized into "Googleyness" and includes the search for "ambition and drive, team orientation, service orientation, listening and communication skills, bias to action, effectiveness, interpersonal skills, creativity, and integrity."

Here are other ideas to keep in mind when hiring and interviewing:

- Whole Foods. Whole Foods initially selects through recommendations and narrows the search through identifying those with experience in a field or department. The interview process is designed to reveal a candidate's knowledge of products (e.g., what kind of cheeses do you like?) and how he or she is likely to communicate with customers. The overall search is for cooperative people.
- Toyota. Hire for life. In Japan, Toyota is the first and usually the last employer for most employees. We understand that in the United States this may not be practical, but the intent is to select and hire people as if they were being hired for life. This is not unique to Toyota—after all, hiring for life works for universities!

7.9.3 Training and Educating People—Building the Community

The long-term, K-oriented strategies that invest in people (see Section 7.7.3.1) represent a combination of our literature review and teaching experience in several industries. There is a common theme here: one of life-long education, expertise building, and a bonding between people, organizations, and communities. The following are illustrative cases:

- Starbucks. In June 2014, Starbucks announced it would provide college tuition support for its employees who work more than 20 h/week. There are no strings attached; an employee who receives a degree is free to leave Starbucks without reimbursing the company. We make an important observation—education promotes internal trust among employees and builds communities in addition to building people and companies. Interestingly, when Starbucks announced its employee education initiative, at least one large bank announced it was scaling back on its education for employees. Starbucks has it right!
- Toyota. Training for production workers usually follows the Perrow model; that is, training is targeted to the job classifications shown in Figure 3.2. Recall too the pencil metaphor described in Chapter 2—even routine work has elements of thickness. Thus, a worker performing routine work (Perrow classification) is trained to recognize problems and respond and problem solve appropriately. Similarly, someone performing an engineering task ("technician" in Toyota) is expected to troubleshoot, intuitively problem solve, and to develop mental maps of problem situations; craft work requires intuitive problem solving and reading situations; and nonroutine work requires creative–innovative ability, intuitive problem solving, and reading situations (Liker and Meier 2007). Note too the overlap in know-how across categories of jobs.
- Toyota Again. Train all your workers as if they were surgeons—this is a good metaphor for corporate training.

What does education at companies like Starbucks and Toyota have to do with KM? Everything. As the world transforms itself into a K society, even companies not normally associated with K work may follow K-oriented strategies. K strategies are invariably people oriented and thus attract and retain good employees.

Key Lesson. Education and training is a cheap way to build a company. It simultaneously develops people, inspires them to achieve great things, promotes passion and mutual trust, and strengthens local communities.

7.9.4 PR, Recognition, and Rewards

Let us keep it simple again: the largest rewards should go to those closest to the value chain and who create the most value. Note the following:

- Toyota. Recall from Table 6.4 that Toyota uses a blended model to rewards in the United States. All production workers have the same job classification and pay (based on seniority). Everyone is expected to learn all jobs associated with a team; team members are encouraged to acquire skills, but there is no reward for doing so (see Liker and Meier 2007).
- Google. In contrast to a blended approach, Google uses disproportionate rewards for disproportionate impact. Who has the greatest impact? Similar to the Toyota representation in Figure 6.13, those who are closest to the product have the greatest impact—this means engineers and not sales or management people.

  Google argues against egalitarian pay systems. "Pay outrageously good people outrageously well, regardless of their title or tenure." While Google acknowledges that widely varying wages may negatively impact team morale and productivity, they conclude the trade-off is worth it. It has been Google's experience that getting great work out of the best performers outweighs perceived unfairness (Schmidt and Rosenberg 2014).
- Whole Foods. Similar to Toyota, Whole Foods has a compressed salary structure. Also, everyone's compensation is posted on an internal web site.

Toyota and Whole Foods do not have large numbers of K workers in the way most of us use the term (though both would argue that point—recall the pencil metaphor we have used in previous chapters). Toyota and Whole Foods share a more routine work environment with lesser skilled job requirements than does Google. This standardization of work may explain their compressed salary structure. On the other hand, what is important here is *transparency*—both the policies and actual compensations (in the case of Whole Foods) are made visible. Recall from the Singapore study earlier that transparent compensation is listed as one of three effective retention strategies for K workers.

7.9.4.1 Performance Reviews

A PR is an analysis of the match (or gap) between expected and realized goals. Included in the PR are the actions necessary to reduce the gap; this may include everything from additional employee training to better technical and management support. The criteria used in a PR often include both objective and subjective measures. The PR has evolved from simple ranking to "modern methods," such as management by objectives, behaviorally anchored rating scale (BARS), HR accounting, assessment centers, and many others.

The effectiveness of a PR is often a function of the employees' perceptions. This implies that the PR must be technically correct and that it be carried out truthfully. To be successful, the purpose of the PR must be understood, accepted, and effectively executed. The PR has become a tool for the organization to improve itself and to make decisions on employees.

Concerns with PR in the K Era

There is an assumption in a PR that objectives can be clearly defined—a characteristic of the industrial and not the K society. The PR rests on an engineering approach to work where specification and implementation are both clear and separable; that is, a thin, modular architecture (the process molecule in Figure 4.7). Thin work carries with it the luggage of industrial age—recall from our previous discussion that one source of organizational failure is to view (and evaluate) all work the same way.

The ambiguous and integral character of thick or K work, both in content and in implementation, makes evaluation as unclear as the work itself (the practice molecule in Figure 4.7). Davenport suggests peer reviews as one approach to PR with K work. However, here too there could be problems—a tendency to groupthink and reinforcement of the status quo.

There are two additional issues with PR in the K era: the evaluation itself and the way it is carried out. First, even if, say, peer reviews were acceptable, K work (as with K itself) is typically hidden and its value often deferred, making assessment difficult if not impossible. Second, almost every organization in which we teach appears to implement PR with quotas (e.g., forced ranking) that normalize performance across departments, which is neither realistic nor fair. Moreover, top-rated or ranked employees may have to wait several evaluation periods before they can receive another top ranking; again, a mismatch with reality and fairness. Additionally, extensive HR analytics are now performed on the metrics associated with PR—meaning HR decisions are being made on poorly executed appraisals based on false assumptions. Recall from Chapter 2 that this is known as "solving the wrong problem precisely."

7.9.4.2 Reinventing PR

Deloitte has redesigned their approach to performance management. One motivation for their redesign was a public survey conducted by Deloitte revealing that 58% of executives questioned believed their current performance management approach drives neither engagement nor high performance (see Buckingham and Goodall 2015).

An evaluation of the existing performance management system at Deloitte revealed the following:

- Counting. In the old system, objectives are set for each employee at the beginning of the year. After each project, the employee's manager rates the employee on how well those objectives were met. These evaluations are factored into a single end-of-year rating where groups of "counselors" discuss hundreds of people with respect to their peers. It took close to 2 million hours per year to evaluate 65,000 plus employees.
- Science of Rating. Assessing someone's skills produces inconsistent data. The process of evaluating an employee by an immediate manager is influenced by "idiosyncratic rater effects." Buckingham and Goodall (2015)

observed that the most comprehensive research on what ratings actually measure was conducted by Michael Mount *et al.*, published in the *Journal of Applied Psychology* in 2000. In a subsequent publication, the researchers concluded, "Although it is implicitly assumed that the ratings measure the performance of the ratee, most of what is being measured is the unique rating tendencies of the rater. Thus ratings reveal more about the rater than they do about the ratee."

- Self-examination. The defining characteristic of the best Deloitte teams is that they are *strengths oriented*. The three strengths that correlated best with high performance were the commitment of a given individual's coworkers to doing *quality work*, a company mission that *inspires* the individual, and the opportunity for the individual to use his or her *strength* every day. The latter was the most powerful across the organization.

In response to their research, Deloitte decided to redesign their performance management system. Their new system has no cascading objectives, no once-a-year reviews, and no 360° feedback tools. In its place, the new system is much simpler, real-time, and individualized, where the focus is on fueling future performance rather than assessing the past. In the new forward-looking system, team leaders are essentially being asked, "what they would *do* with each team member rather than what they *think* of that individual."

The new performance management system has three objectives:

- Recognize performance. Most current systems do this.
- See performance clearly. There are two issues: the idiosyncratic rater effects and the need to streamline the traditional process of evaluation, project rating, consensus meeting, and final rating. The former is addressed not by asking more people for their opinion of a team member (e.g., 360° survey), but by asking *only* the immediate team leader a very different kind of question. The team leader is not asked about the skills of each team member, but about their future actions with respect to that individual. At the end of every project (or every quarter for long-term projects) each team leader is asked to respond to four forward-looking statements on each team member. The statements cover (1) overall performance and unique value to the organization, (2) the ability to work well with others and the willingness of the leader to keep the person on the team, (3) whether the person is at risk for low performance, and (4) whether the person is ready for promotion today. Collectively, this is called the performance snapshot.
- Fueling performance. Deloitte asks: *Is performance management more about performance or about management? Isn't it better to improve performance rather than appraise it?* The new system requires that a given team leader "check in" with each team member once a week. For Deloitte, the check-ins are not in addition to the work of a team leader, but *are* the work of the team leader.

7.9.5 Rethinking Human Resources

In the following discussion, we make reference to three articles that appeared in the *Harvard Business Review* in the July–August issue that focused on rethinking human resources: (1) Cappelli (2015); (2) Charan *et al.* (2015); and (3) Boudreau and Rice (2015). Among the interesting observations are the following:

- HR managers have focused on "administrivia" and often lack vision and strategic insight. Now is the time to get past the transactional nature of benefits, compensation, recruiting, and performance management. HR managers should set the talent agenda by rethinking programs that have changed little since the 1950s.
- Other than in the United States, HR is considered one of the most powerful functions throughout the world and is growing in importance. "HR appears to be growing in influence as companies recognize the importance of organizational culture, knowledge management, and so forth. The US is the outlier."
- HR innovation is most prominent in technology giants such as Google, Apple, and Microsoft. "Human capital is practically their only major asset."
- In most companies, CEOs are typically distanced from and dissatisfied with their chief HR officers (CHROs) and the HR function generally. The time has come for the CEO to elevate HR and make the CHRO a strategic partner. In the same way that the finance function is considered a strategic business partner by raising and allocating financial resources, the HR function has a strategic role in attracting and building talent.
- In 2011, Jupiter Networks (a Silicon Valley innovator of networking technology) became one of the first global companies to abolish forced rankings. It has replaced annual reviews with frequent "conservations." As opposed to spreading people across a bell curve, it now seeks to have Best Talent throughout the company. Today, 97% of Jupiter's employees are considered Best Talent and its talent management efforts focus on putting the right capabilities in the right places to achieve its business goals.

7.10 Closing Remarks

Strategy as an Integral Construct

Michael Porter elevated strategy (of position) to a major academic and management field. When people think of strategy, they usually first think of the strategy of position and its artifacts—for example, the five-force model and generic strategies. It is formal and analytical and creates a veneer of correctness that is attractive to management, particularly in the West. Strategy making has become a key task of leadership.

The resource-based approach to strategy provides a counterpoint to the position approach. With an emphasis on core competence, the resource-based strategy introduces tacit elements into strategy, is more consonant with new K and technology, and is more long term than is the strategy of position. Finally, it is easy to see how K fits with and leads strategy (e.g., the Sharp case).

While it is tempting to visualize strategy as one based on either position or resources, Figure 7.3 illustrates a more complex system of 10 interacting schools of strategy, each with varying degrees of formality and analysis. Recall that the culture and learning schools represent the resource-based approach. In practice, companies emphasize a particular strategy approach that embodies their culture and management style; not surprisingly, US companies are biased toward the strategy of position, whereas Japanese companies emphasize the learning school. Hence, *a given strategy is not just a company business plan or scheme, but a manifestation of the culture and practice of that company*. The new work of leadership (and management in general) is crafting strategy, competence, and innovation within the framework of culture.

Strategy of Strategies

Our suprastrategy has been to create an enterprise approach with the following characteristics:

- Strategy may be characterized by any number and combination of 10 strategies, ranging from the thin to the thick. It follows that corresponding K strategies also lie on a strategy continuum. The predominant form will largely depend on the internal and external situation (Figure 7.2).
- In practice, the position school stands in opposition to the resourced-based approach (i.e., culture and learning schools) and may serve as useful analogs for thin and thick strategy schools, respectively.
- We see here Porter's idea of *fit*: K strategies and corresponding business strategies (vertically) and coordinated strategies covering the breadth of hard and soft technologies and practices (horizontally). Thus, though we typically think of K strategies as concentrating of the intangibles, it is the full range of possibilities (Figure 7.5) that must be accommodated in a 21st century organization.
- We see in Figure 7.6 how the combination (i.e., fit) of work, technology, and collaboration may impact the cultural assumptions of the organization.
- An organization's strategy is a window into its culture and practice. Hence, K and strategy are intertwined with the culture and practices of an organization.
- The collection of complementary and reinforcing strategies (thin and thick) requires some level of enterprise management to achieve fit across the organization. The CKO is one possible governance structure for supporting

and coordinating the specific K strategies, K activities and initiatives, and K technologies of the organization.

- Long-term strategies that promote a K culture are necessarily soft and center on the people value stream. This includes the tacit bundle (Section 7.6) as well as the long-range people strategies (Section 7.7.3.1).
- Specific people value stream strategies focus on attracting, motivating, and retaining K workers. Also included are the performance management strategies and policies of the organization.

Force of Culture

Culture shapes and drives human activities more than we can possibly state. Consider, for example, the case of Korean Air Cargo flight 8509. In December 1999, Korean Air flight 8509 crashed on takeoff from London. The crew banked the aircraft into the ground while multiple audible warnings were sounding. Subsequent investigation revealed that the pilot did not respond appropriately to warnings during the climb after takeoff despite prompts from the flight engineer. Though the captain's attitude director indicator (ADI) was malfunctioning, the first officer's ADI was working properly. The first officer either did not monitor the aircraft attitude during the climbing turn or, having done so, did not alert the pilot to the extreme unsafe attitude that developed. The pilot had a military background and many hours of flight time. In contrast, the first officer was relatively inexperienced and was deferential to the pilot. Investigators suggested, among other things, that Korean Air alter training materials and safety education to meet the "unique" Korean culture—a hierarchical culture that discourages input from the first officer to the pilot. British investigators called on Korean Air to revise its company culture and training, "to promote a more free atmosphere between the captain and the first officer."

Culture in Mature Organizations

We have seen that a characteristic of young organizations is the desire of leaders to make their culture manifest. In contrast, mature organizations have their culture hidden in structure and processes. Like that of Korean Airlines, the culture of a mature organization is taken for granted and sits in the background shaping and guiding behavior.

In the same way that a mature organization sometimes attempts to recreate the more innovative environment of a young company through structural or process change, we suggest that it also emulate a young organization and make its culture visible and living. This characteristic of young companies may serve as a beacon in attracting K workers. The *cultural attributes* in Figure 7.13 may be used as a guide and are elaborated here. Note that these attributes apply equally to young and mature organization.

Cultural attributes (Schein1999; several organization)	Work attributes (Deloitte strengths)	Individual attributes (Google interview)
• Values and ennobling goals	• Team effect (quality)	• Leadership
• How work is carried out	• Ennobling effect (inspiring)	• Role-related K
• How decisions are made	• Expertise effect (building knowledge)	• Cognitive ability
• Career and reward policies		• Googleyness

Figure 7.13 Integral hiring considerations and characteristics.

- What does your organization care about? How does it demonstrate its values? In mature organizations, it may be necessary for leaders to reinspire the organization through manifestation and visible statement and application of values.
- How is work carried out? Is the focus on individuals and small teams? Is the work professionally enriching? Is the work exciting and socially important?
- How are decisions made? How much control do K workers have in the selection and execution of projects? What are the communications patterns in the organization—can an individual engage a senior manager?
- How are people evaluated and rewarded? Is everyone a "best employee," or alternatively, is there individual competition for rewards and opportunities along a bell curve? Are rewards given to those closest to the value chain or is there some other pattern at work?

People Value Stream

The Deloitte research interests us on several fronts. In addition to its new performance management system, we find Deloitte's internal investigation of its top teams informative. The best Deloitte teams are characterized by "strengths" that yield high performance. Since all Deloitte team members may be considered K workers, we may use these strengths as a guide in understanding and designing K work. This is what we refer to as *work attributes* in Figure 7.13 and are elaborated here:

- The commitment of a given individual's coworkers to doing *quality work*. We think of this as the *team effect*. Everyone in the team is committed to high-quality work. That is to say, the team effect is representative of a human interaction within the immediate social environment and the work itself.
- A company mission that *inspires* the individual. We call this the *ennobling effect*. This is a function of the culture and history of the organization, its founders, and (in the case of mature companies) the willingness of executive management to reinspire the organization's culture with a higher vision. We have previously made reference to both Whole Foods and Google as examples.

- The opportunity for the individual to use his or her *strength* every day. This is the most significant strength across the organization. We consider this to be the *expertise effect*, where K is strengthened and made visible though work activities. Moreover, the expertise is concentrated and deliberately put to work. Note too that a given strength need not be technical but may be analytical, collaborative, project oriented, and so on.

Observe the synergy between the cultural effects and the Deloitte findings of its K workers, which are strengths oriented. We can now add a third set of attributes (*individual attributes*) that describe how Google evaluates candidates during the interview process: *leadership, role-related K, cognitive ability,* and *Googleyness*. See Figure 7.13.

People Value Stream 2.0

We have seen that even companies not normally associated with K work (e.g., Toyota, Whole Foods) benefit from K-oriented strategies, leadership, and HR policies. This is because K policies are people oriented and thus attract engaged and motivated people. This is in contrast to task- or methodology-oriented employee searches. Google does both—for example, assessing candidates for both role-specific K and Googleyness. It should not be surprising that most HR innovation is taking place in the Silicon Valley giants where people are their central asset.

We find the Google evaluation criteria of candidates instructive. It appears that Google is searching for something more than task-specific K, but thick (whole person) characteristics where leadership, passion, action-orientation, creativity, collaborative skills (listening and communications), tenacity, and so on are of utmost importance. These are the intangible qualities that make for a great employee.

In our classes, we have long argued against forced rankings. Deloitte asks its team leaders to use the phrase "if it were your money" when making decisions. We ask the same of companies: If it were your money, would you design a system where only, say, 10% were top performers? Is that what you really want? Wouldn't you rather have 97% top performers?

The 21st century organization is faced with an interesting situation: matching new job requirements with the skills of K workers. Organizations require employees with *deeper expertise, more independent judgment, and better problem-solving skills*. The match should be straightforward. However, if an organization jettisoned its PR; for instance, what would follow? But this is the point of design in the K organization: replacing industrial era systems with those necessary to support the K organization. Metaphorically, current performance systems count pins: counting the number of pins produced in a single day has been replaced with layers of complexity in a balanced scorecard of key performance indicators (KPIs) (see Ewenstein *et al.* 2016).

Systems such as the balanced scorecard arose as organizations grew larger and the work less routine. However, the essential character of the work has remained *thin* and *task oriented*. Organizations match task specifications with realized performance—but the underlying approach remains mechanistic and is the same as counting pins.

The new organization is a K organization, meaning its work is *thick*—integral and "whole" system—and hence assessment requires new approaches. Giants such as GE and Microsoft are doing just this as they experiment with new performance management systems. They are rethinking the link between such systems and actual performance, compensation, and so on. Interestingly, organizations are mixing and matching, and even combining several systems in the same company—for example, retaining hardwired targets for sales while applying the new approaches to other functions and business units. Whatever form organizations adopt, "change they must."

7.11 Class Exercises

1 Does new K or technology drive your company's business strategy, or does business strategy lead technology? Give examples.

2 Classify your company's strategic orientation. Is it that of a reactor, defender, analyzer, or prospector? Identify the K strategies that your company employs and discuss whether it supports the company's strategic orientation.

3 Does your organization or school emphasize a strategy of position or resources? Recall that a strategy of position is explicit oriented and involves high levels of formality and analysis. In contrast, a resource-based strategy (culture and learning) is more tacit oriented and is usually indicated through organizational experimentation, ambiguity, and adaptability.

4 Is your organization or school's strategy driven by higher values or ennobling goals? Give examples. Given it history and culture, what higher vision would you propose for your organization?

5 Describe your organization's or school's core competence. What strategies would you employ to protect and strengthen that competence?

6 Does your organization or school have a K-base? Discuss its attributes—contents and organization, and search and access methods.

7 Project. In your company or school, describe the genesis and implementation of specific K initiatives, such as performing a K inventory, acquiring K through tacit K capture, development of K maps or other K systems, and so on. Where is the new K stored and how is it accessed and shared? Who coordinates and integrates the separate K initiatives? What do you propose to improve the organization of these K efforts?

8 Project. Assess the principle assets in your organization (you may want to create or be guided by the IC model of your organization). To what extent is it people or technology oriented? Describe how this may explain your organization's HR policies and HR innovations. Consider, for example, how your organization attracts, motivates, and retains people, the criteria for interviewing and selecting new employees, the process for making decisions on hiring and promoting (e.g., committees, hierarchies), how your company evaluates performance, and so on.

9 What kinds of people are attracted to your company? Why do you think this is so?

10 Assess the relationship between your organization or school's culture and the type of people it hires.

11 Why do you remain in your company? What motivates you?

12 Project. Design a new performance management system for your organization. Use the Deloitte system as a guide, taking into account your organization's cultural attributes (e.g., age).

13 Is human capital the principal asset in your company? Discuss how it drives K strategies and performance.

14 How would you do on the LAX layover test? What of your manager?

15 Discuss career pathing in your organization. How do culture and the values of your organization affect the kinds of people and professions that move upward into the senior levels of management?

16 How would you characterize the leadership in your organization? Is it task, relations, or change oriented? Give examples.

17 Project. Assess how business and K strategies are created in your organization and how they are aligned. Include in your answer a discussion of the locus of decision making with respect to each.

References

Andriessen, D. (2003), *Making Sense of Intellectual Capital*. Routledge.

Boudreau, J., and Rice, S. (2015), "Bright, Shiny Objects and the Future of HR." *Harvard Business Review*. July–August 2015.

Buckingham, M., and Goodall, A. (2015), "Reinventing Performance Management." *Harvard Business Review*. April 2015.

Cappelli, P. (2015), "Why We Love to Hate HR … and What HR Can Do About It." *Harvard Business Review*. July–August 2015.

Charan, R., Barton, D., and Carey, D. (2015), "People Before Strategy: A New Role for the CHRO." *Harvard Business Review*. July–August 2015.

Collins, J.C., and Porras, J.I. (1996), "Building Your Company's Vision." *Harvard Business Review*. September–October 1996.

Collis, D.J., and Montgomery, C.A. (2008), "Competing on Resources." *Harvard Business Review*. July–August 2008.

Davenport, T.H. (2005), *Thinking for a Living*. Harvard Business School Press.

Earl, M. (1998), "*What on Earth Is a CKO?*" Earl and Scott, IBM Survey. London Business School. In *Knowledge Horizons: The Present and the Promise of Knowledge Management*. Eds. Despres, C., and Chauvel, D. Butterworth Heinemann.

Ewenstein, B., Hancock, B., and Komm, A. (2016). "Ahead of the Curve: The Future of Performance Management." *McKinsey Quarterly*. 2016 Number 2.

Giddens, A. (1984), *The Constitution of Society*. University of California Press.

Hamel G., and Prahalad, C.K. (1994), *Competing for the Future*. Harvard Business School Press.

Horwitz, F.M., Heng, C.T., and Quazi, H.A. (2003), "Finders, Keepers? Attracting, Motivating and Retaining Knowledge Workers." *Human Resource Management Journal*, 13(4): 23–44.

Kaplan, R.S., and Norton, D.P. (2000), "Having Trouble with Your Strategy? Then Map It." *Harvard Business Review*. September–October 2000.

Kaplan, R.S., and Norton, D.P. (2004), *Strategy Maps: Convertible Intangible Assets into Tangible Outcomes*. Harvard Business School Publishing.

Keen, P. (1997), *The Process Edge: Creating Value Where It Counts*. Harvard Business School Press.

Leonard, D. (1995), *Wellsprings of Knowledge: Building and Sustaining the Sources of Innovation*. Harvard Business Review Press.

Liker, J.K. and Meier, D.P. (2007), *Toyota Talent: Developing Your People the Toyota Way*. McGraw-Hill.

Miles, R.E., and Snow, C.C. (1978), "Organizational Strategy, Structure, and Process." *Academy of Management Review*. July 1978.

Mintzberg, H., Ahlstrand, B., and Lampel, J. (1998), *Strategy Safari*. The Free Press.

Mintzberg, H., and Lampel, J. (1999), "Reflecting on the Strategy Process." *Sloan Management Review*, 40(3): 21–30.

Nonaka, I., and Toyama, R. (2004), "Knowledge Creation as a Synthesizing Process." In *Hitotsubashi on Knowledge Management*. Eds. Takeuchi, H. and Nonaka, I. John Wiley & Sons (Asia) Pte Ltd.

Pearce, J.A., and Robinson, R.B. (1991), *Strategic Management: Formulation, Implementation, and Control*. Fourth Edition. Irwin.

Porter, M.E. (1985), *Competitive Strategy*. The Free Press.

Porter, M.E. (1996), "What Is Strategy?" *Harvard Business Review*. November–December 1996.

Robbins, S.P. (1990), *Organization Theory: Structure, Design, and Applications*. Prentice Hall.

Schein, E.H. (1999), *The Corporate Culture Survival Guide*. Jossey-Bass.

Schmidt, E., and Rosenberg, J. (2014), *Google: How Google Works*. Grand Central Publishing.

Senge, P.M. (1990), "The Leader's New Work: Building Learning Organizations." *Sloan Management Review*. Fall 1990.

Spender, J.C. (1996), "Making Knowledge the Basis of a Dynamic Theory of the Firm." *Strategic Management Journal*, 17(Winter Special Issue): 45–62.

Sveiby, K.E. (1997), *The New Organizational Wealth*. Berret-Koehler Publishers, Inc.

Wiig, K. (1993), *Knowledge Management Foundations*. Schema Press.

Yukl, G. (2006), *Leadership in Organizations*. Sixth Edition. Person Prentice Hall.

8

Knowledge Horizons

8.1 Knowledge Arises

Knowledge arises through cognitive effort—knowledge and imagination—the two aspects of intellectual experience. Unlike a machine or the manual labor that characterizes the industrial era, K is neither manufactured nor compelled. K creation cannot be forced, and therefore must be inspired. This does not mean that K creation is haphazard—the primary purpose of this book is to discuss how we can contrive the environment to create a suitable context for knowledge and knowledging. "Innovation can't be owned or ordained, it needs to be *allowed*. You can't tell innovative people to be innovative, but you can *let* them" (see Schmidt and Rosenberg of Google 2014).

8.1.1 Opportunities to Promote Knowledge

There is a recurring theme in the creation of K—the interplay between contextual dimensions. *Cognitivity* arises through the interplay of {awareness–activity}; *knowing* through {tacit knowing–explicit knowing}; *knowledge* through {truth–belief}; and *organization knowledge* through {tacit K–explicit K}. We may add {knowledge–imagination}; that is, conscious activity and K content. Finally, wisdom arises through the interaction of theoretical and social knowledge, as scientific and social truths collide and are brought into harmony. No doubt, new K may be characterized through many other such interactions.

There is an opportunity for intervention and design throughout this cycle of give and take. In effect, K design involves scheming interaction possibilities to allow K to flourish. Recall, for example, that in Plato's justified true belief (Figure 1.11) justification solidifies {truth–belief} into objective knowledge and not superstition. Justification in turn gives rise to various methods of inquiry and science as design tools in the pursuit of knowledge.

In an organization, we may create an environment that simply "allows" thinking to take place through, for example, some combination of human interaction

Designing Knowledge Organizations: A Pathway to Innovation Leadership, First Edition.
Joseph Morabito, Ira Sack, and Anilkumar Bhate.

(in an appropriate form) and private time (in a quiet space or a library, for example). As yet another example, we have commented that thick design (e.g., K work) requires the design of a *surrounding* context or frame, that is, a ba or set of K-spaces that facilitates and allows thick work to shape itself. (This is in contrast to the thin design of the industrial era where *internal* task structure and behavior are specified in an engineering sense.).

As K becomes more visible, the design opportunities become more varied and explicit. For example, in the Nonaka approach to design, we reference rich frameworks such as SECI (socialization–externalization–combination–internalization) as well as both organizational and process enablers. This too may be extended and refined and may include, for example, an organizational K-base. Refinement extends into the nitty-gritty—designing office space, for example.

On the other hand, opportunities to create K are sometime fortuitous. The combination of expertise, social interaction, motivation, context, and endless other variables, is sometimes beyond forethought. Hence, the Google observation from earlier is instructive—allow innovation to take place and follow up with implementation. Not all opportunities repeat—"one time one chance."

8.1.2 Three Triangles of Knowledge

Knowledge is a complex system, meaning no abstraction level exists in isolation of another. That is to say, when formulating K, we must account for the entire system. K arises through simultaneous interactions within a given context and between successive levels of human experience.

We have created three levels of understanding: tacit, practice, and culture. Recall the K-triangles—the essential tacit K-triangle (Figure 4.8), the practice of K (Figure 4.9b), and the cultural integration of K (Figure 1.10). These triangles are not independent but represent abstractions of *knowledge as such*. This is illustrated in Figure 8.1, where (in order from inner to outer triangles) the tacit, practice, and culture contexts are embedded in each other.

The three triangles of K suggest that K cannot be isolated (except for reasons of analysis) at a given level of understanding. Every level has an impact on every other level. For example, our worldview, our understandings, and our interpretations—all depend on our culture and where we live. Science is unique in that it seeks truth as such, independent of social or situational contexts. However, even with scientific facts, our interpretations remain qualified, dependent on the situation, our social group and its facts, and our wider culture. We may encapsulate several observations that we have been making throughout this book.

- K arises through the interaction of cognitive, practice, and cultural contexts. Each is a separate and complex interaction, yet connected and intertwined through common themes that underscore the essential meaning and application of knowledge. For example, observe in Figure 8.1 that the "social" is characterized, at once, by tacit embedding in a group, the structure of the

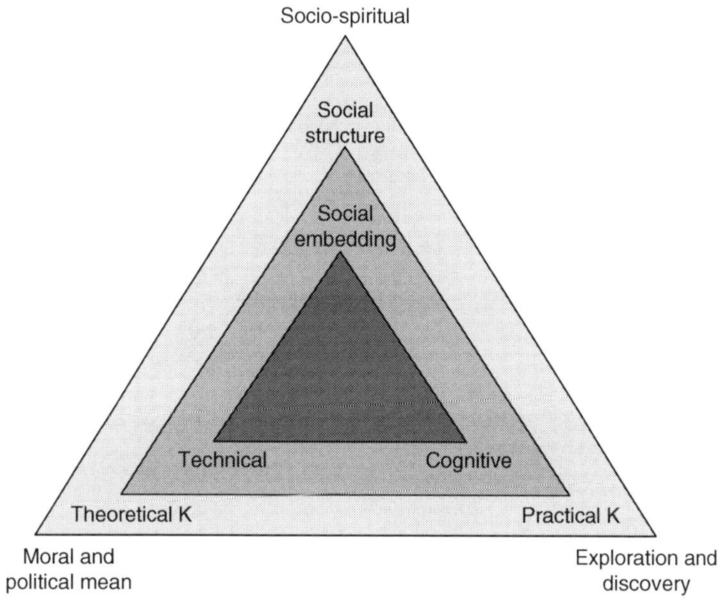

Figure 8.1 Three triangles of knowledge—tacit-practice-culture.

practice or community group, and the spirituality of the wider culture. Each is inextricable from the other.

- K is inherently noble and human. Design is a manifestation of our ideals. Hence, we have emphasized ennobling goals for organizations. Recall, for example, one goal of Whole Foods is to eliminate world hunger. Furthermore, we witness young organizations doing more than making money, but doing exciting and good things—going to the moon, for instance.
- K is ethical. The organization must decide if it will use technology to automate or to augment work. Do we automate because we can or because it is part of a broader strategy involving people and technology? One strategy adopted by Toyota, for example, is to use automation to free workers to do more value-added work rather than replacing them. Recall too that the ethical behavior associated with a K practice (and K work generally) is developed and sometimes enforced through the structural component of a K practice (see Figure 4.9b). Furthermore, ethics extends to all areas and levels of design—for example, a performance review should be ethical and not just a position on a bell curve.
- K is social. Design is a manifestation of our social groups, meaning it should account for social practices and facts (not just scientific facts) in addition to the responsibilities that the possessor of K (individual or organization) has to the community. As shown in Figure 8.1, {tacit–practice–culture} implies that moral

and ethical behavior is a constituent of the individual mind made manifest in his or her K work and practice. For example, recall Jonas Salk's rejection of profit in favor of rapid and widespread distribution of his polio vaccine. Recall the Chinese mandate: *moral preparation* and *social action* underlie K.

Designing social space is perhaps the most varied K design activity and includes, for example, developing a practice around an area of expertise, creating a CoP, social networking opportunities with both internal and external parties, collaboration agreements with partner organizations and groups, and office space design.

- K is action. The combination of K and action has been a cornerstone in our study of K and knowledge management (KM). Recall from Chapter 1 that the relationship between knowledge and conduct has been a persistent question in Chinese philosophy. It has long been argued which came first and which is the more difficult to achieve. Generally, it has been concluded that the two cannot be separated—"knowledge cannot be realized without action, and unless coupled with true knowledge, no action can be intelligent or correct." The ancients had it right!

  Recall from Chapter 4 that justification—judgment and action—is at the heart of building new knowledge, whether theoretical or practical. Justification is thus the principal work of K workers. Justification extends to the balance between theoretical and practical K (Figure 4.9b) and the emergence of wisdom (Figure 5.3). Action and K cannot be separated.

  Recall too from Chapter 4 that a K activity is embodied in a practice that "houses the semantic spaces"—(1) a person's state of *being* (e.g., feelings, desires), (2) the *situation*; that is, the people, events, and things with which a person interacts, and (3) the *behavior* the person engages in response to the situation. Action usually involves some measure of social interaction—recall our coffee metaphor where coffee beans represent intrinsic K and heat represents the social interaction necessary to elicit and refine K. Therefore, as we have stated, K in action arises through mental effort and social heat.
- K is scientific. Design must always search for truth and elucidate physical reality.

8.1.3 Evolution of Knowledge and the Knowledge Organization

Organizations must be flexible and embrace a variety of lenses to define knowledge and (re) design themselves to successfully achieve competitive advantage in the 21st century. The two principle types of knowledge are cognitive-based (theoretical K) and practice-based (social K). These must not merely be scene as dichotomous forms, but as essential to employ in a selective and harmonious fashion. Organizations must not fall into the trap of thinking that there is one best way of treating knowledge. Each type has its own strengths and weaknesses. For example, the cognitive-base type lends itself to generalization, synoptic form (rationalism), prediction, and explicitness, whereas the practice-based type takes into account history, culture, incrementalism, social interaction, and tacitness within unique situational circumstances. Each approach has the potential to serve as a corrective to the other. Further research (e.g., cognitive science, design theory, social networking, etc.) and practical reports (e.g., case studies) should be pursued to help increase organizational understanding and promote the advancement and evolution of today's knowledge organizations.

8.2 Digital Economy

As knowledge shapes the new Pangaea, there is yet another revolution materializing before our eyes. The K economy is being augmented with bits and bytes, even more than currently is the case. A new pillar is emerging—really a subtype of KM—that of the *digital economy.*

What are its characteristics? We have seen glimpses throughout this book. The Internet and the Internet of Things are foundational to the new digital world. *Central to the digital society is the tidal wave of data*! And robots! Organizations will be compelled into real-time decisions—speed is the new metric. Customer interaction will be an everyday, every second occurrence. How will people and organizations adapt?

8.2.1 K Society Paradox

The Polanyi paradox will not go away but will become more visible. Recall from Figure 4.6 that work is not entirely thin or thick but exists as a range. The evolution of work means eliminating routine or thin tasks and replacing them with thick tasks. Exponential growth in technology is driving a corresponding transformation in our lives. Two centuries ago most people found themselves working on farms, today only 2–3% of the US population is agriculture-based—we do not have 97% unemployment. In the 19th century, people found new employment in factories. In the 20th century people found employment in offices. There has been a hand-in-hand evolution of individuals and society. In the near term, a social upheaval exists as more jobs are replaced than created. On the other hand, we believe that in the long term K work will become the norm as education and training take hold on a society-wide basis.

8.2.2 Data

The *central theme of the digital economy will be more data and more data streaming at an ever-increasing speed.* New capabilities will be required to exploit big data. Recall from Chapter 4 that the scale of big data implies that the common organizational practice of separating data processing (in the IT department) from data analytics (in the business units) is no longer feasible (see Belanger 2014). This leads to the requirement for new organizational configurations, such as discussed here.

A second characteristic of the digital economy in that data is "on" 24×7. This implies that organizations will have to adopt data streaming practices in addition to database management. This extends to data streaming analytics as well. Big data is primarily real-time data.

8.2.3 Organizational Upheaval

The digital age is creating yet another upheaval in our society. In many ways, the issues are not very different than those associated with KM. For example, recall from Chapter 1 that there has been a shift in the source of wealth in organizations—from land to capital to knowledge. The digital perspective makes the K perspective more visible.

Bughin *et al.* (2016) have suggested five questions for organizations to consider in the digital economy, briefly described here.

1) Do you understand the competitive landscape? Digital platforms allow companies of any size to quickly roll out products and deliver them to global markets.
2) Do you have the right assets and capabilities to compete? The issues normally associated with large companies now apply to companies of all sizes. This includes marketing data and customer relationships, for example. We would add that companies must hire and develop talent with deep analytical skills (McKinsey Global Institute 2011). Companies will be driven to explore new ways of making money.
3) Can we simplify our product strategy? While digitalization can simplify the tailoring of products, there is a parallel trend toward more streamlined global product portfolios. For example, Apple offers a limited number of products, all with consistent design and branding throughout the world.
4) Should we retool our organization and supply chain? Digital tools promote collaboration and instant communication, making it possible to centralize selected global functions (e.g., R&D, back-office operations) and to create virtual global teams. Similarly, supply chains are being simplified and made shorter to lessen time to market. For example, Bughin *et al.* (2016) report that in a UPS survey, approximately one-third of high-tech companies are moving their manufacturing or assembly closer to end-user markets.
5) What are the risks? Maintaining data security will continue to be a top priority as hackers become more sophisticated.

8.2.4 Organizational Design

Increased clock speed in decision making is an essential objective for the digital organization. Speed in decision making is tied to all aspects of the organization, including its customer interactions, product development (strategies and manufacturing), and supply chains.

Recall from Chapter 4 that Galbraith (2014) suggests new organizational design strategies for managing the tidal wave of data. In particular, departmentalization and job design will have to be revisited. Recall that managing big data requires a *chief digital officer* (CDO) role. This responsibility may be shared with the chief information officer (as in P&G) or may be a separate CDO job (as in Starbucks).

Galbraith (2014) suggests two organization design strategies for increasing clock speed in decision making. The central issue is one of governance. Do we distribute data and analytical talent into the business units or do we colocate in a semi-independent department? Is there a role for the traditional IT department? Answers are beginning to emerge as organizations experiment with all the issues of big data—managing the data itself, developing analytics, hiring talent, and reconfiguring the organization.

8.3 End of Course Questions for Discussion and Research

We believe that the learning in our course neither begins nor ends with the presentation of the coursework in this book. We carry our knowledge and patterns of thinking with us into other classes and into other organizations. A tacit person understands asking the question "why" is so essential to our learning. We hope that students will continue their explorations and deepen their understandings of the issues presented in this book.

The following questions are both new and sourced from existing "food for thought" questions. For each question, use logical argumentation, case studies, examples from your organization or school, or further research to develop your answer.

1) Can knowledge arise solely from the faculty of the mind?
2) Do you believe there are cultural or social tendencies in our thinking patterns that may lead to different perspectives on the meaning of knowledge?
3) What is the difference between knowledge and experience?
4) Compare theoretical and practical knowledge, elaborating on how they interact and give rise to each other.
5) Discuss how social knowledge and social facts affect decision making in your organization or school.

6) Discuss the role of noble ideas in designing a knowledge system or organization. Is your organization or school motivated by higher ideals? What would you change?
7) Would you automate every activity that could be automated? Extend your discussion beyond economic considerations and include, for example, the impact of automation on your community.
8) What elements do you consider important in developing a K culture? How would you encourage learning in your organization?
9) What actions would you take to promote K sharing and mutual trust in your organization?
10) What 21st century aspects of knowledge can be incorporated in your organization's culture to increase organizational effectiveness? How would you encourage all K workers in your organization to consider them in their daily activities?
11) How would you promote collaboration between your company or school and external institutions from around the world? Discuss the types of problems (including messy problems) you would address.
12) Design a system—organizational structure and departmentalization, distribution of analytical talent and data (database and data streams), and distribution of key managers—to increase the clock speed of decision making in your company.
13) How would you go about creating a global community of practice around your area of expertise? Discuss the critical elements such as culture, technology, the nature of your expertise, organizational structure and departmental considerations, the locus of decision making in your organizational, and the type of problems (including messy problems) in need of analysis and decisions.
14) What are the desirable technical and leadership attributes of a chief digital officer? How do they compare with those of the chief K officer and the chief information officer?
15) Thought problem. Take a complex idea or pattern that you have been thinking about and trace its roots to the social communities in which you live and work. Include communities such as your family, religion, school, company, occupational subcultures, city and country, and global culture.
16) How can emotions (help) create or facilitate individual or organizational K?
17) What metaphors can you create to promote K creation, transfer, or application in your organization or school?
18) Describe how policies can promote or hinder K sharing in your organization or school.
19) Discuss you organization's strategy for using big data. How will it affect product or service offerings?
20) Discuss how your organization or school will be configured and will operate in the coming years. Consider the factors you consider important—for

example, K creation and sharing, departmentalization, politics, big data, culture and globalization, product strategies, key competencies and K domains, the availability and hiring of K worker, and so on.

21) Project. Examine recent research literature on motivation and retention of knowledge workers. Based on what you learn, what recommendations would you make to your organization to increase retention cognition of its knowledge workers?
22) Does your organization provide sufficient creative tension to challenge its knowledge workers to innovate and perform at increasingly higher levels of effectiveness?
23) How would you describe those aspects of the today's psychological contract as it pertains to knowledge workers (as opposed to industrial workers)? In light of the accelerating development of technology (e.g., robotics, machine learning), how do you anticipate the psychological contract will evolve?
24) What K strategies does your organization employ? What does this imply about your organization's thin and thick K policies, how it views K, and the locus of its core competencies?
25) Discuss your organization or school's social network landscape. How are individual, organizational, and interorganizational networks driving K generation? Characterize the knowledge that is being transferred or developed and the nature of the social ties.
26) Project. Design a new performance management system for your company. Identify the goals and key elements or characteristics of your system and how it differs from the current system. Discuss how you would socialize and implement the new system.

References

Belanger, D. (2014), *Seminar Notes*. Stevens Institute of Technology.

Bughin, J., Lund, S., and Manyika, J. (2016), "Five Priorities for Competing in an Era of Digital Globalization." *McKinsey Quarterly*. 2016 Number 2.

Galbraith, J. (2014), "Organization Design Challenges Resulting from Big Data." *Journal of Organization Design. JOD,* 3(1): 2–13.

McKinsey Global Institute (2011), *Big Data: The Next Frontier for Innovation, Competition, and Productivity*. McKinsey Global Institute.

Schmidt, E., and Rosenberg, J. (2014), *Google: How Google Works*. Grand Central Publishing.

Appendix

Table A Key topics from Nonaka and colleagues'/coauthors' research publications together with text references from *Designing Knowledge Organizations* (DKO).

Topics	DKO References	Research Publications
Definition of Knowledge (Nonaka)	Chapter 1 (p.56)	Nonaka and Takeuchi (1995); Takeuchi and Nonaka (Eds.). (2004)
Justified True Belief (Plato's definition of Knowledge)	Chapter 1 (pp.19, 38–40, 234)	Nonaka and Takeuchi (1995); Takeuchi and Nonaka (Eds.). (2004)
Knowledge Creation	Chapter 2 (pp. 74–92)	Nonaka and Takeuchi (1995); Takeuchi and Nonaka (Eds.). (2004); Von Krogh, Ichijo, Nonaka (2000); Nonaka and Toyama (2004); Nonaka, von Krogh, and Voelpel (2006)
Tacit Knowledge	Chapters 2, 4, 5, passim	Nonaka and Takeuchi (1995); Takeuchi and Nonaka (Eds.) (2004); Nonaka and von Krogh (2009)
Cognitive Dimension (Nonaka)	Chapters 1,4 (pp. 183–184)	Nonaka and Takeuchi (1995); Takeuchi and Nonaka (Eds.). (2004)
Dialectic and Synthesis	Chapter 2 (pp. 64, 74–77)	Nonaka and Takeuchi (1995); Takeuchi and Nonaka (Eds.). (2004)
Organization and Process Enablers	Chapter 2 (pp. 90–92)	Von Krogh, Ichijo, Nonaka (2000)
Ba	Chapter 2 (pp. 81–84)	Nonaka and Takeuchi (1995)

(*Continued*)

Designing Knowledge Organizations: A Pathway to Innovation Leadership, First Edition.
Joseph Morabito, Ira Sack, and Anilkumar Bhate.

Table A (Continued)

Topics	DKO References	Research Publications
Middle-Up-Down Management	Chapter 2 (pp. 74–87)	Nonaka (1988); Nonaka and Takeuchi (1995)
SECI model for Knowledge Conversion/SECI spiral	Chapter 1	Nonaka and Takeuchi (1995); Takeuchi and Nonaka (Eds.). (2004); Nonaka, Toyama, and Hirata (2008)
Response to Criticisms of SECI Model	Chapter 5 (pp. 250–252)	Nonaka and von Krogh (2009)
Promoting Conversations and Dialogue	Chapter 6 (pp. 339–343)	Nonaka and Takeuchi (1995)
Organizational Culture	Chapter 6 (p.315), passim	Nonaka and Takeuchi (1995)
Strategy and Knowledge Creation	Chapter 7, passim	Von Krogh, Ichijo, and Nonaka; Osono (2004)
Leadership and Knowledge Management	Chapter 7, passim	Nonaka, Toyama, and Konno (2000); Nonaka, Toyama, and Hirata (2008); Nonaka. Von Krogh, Rechsteiner (2011)

The table above presents some of the principal topics and publications created by or focused on by the eminent scholar Ikujiro Nonaka and several of his coauthors and colleagues. It also shows references to these topics in *Designing Knowledge Organizations.*

More than any other management scholar, Nonaka has had a significant impact on both management theory and practice in the field of knowledge management. The research publications provided much of the source material and inspiration for the writing of our text as well as for our teaching activities. It is our sincere hope that students who use this book will also consult at least some of the listed publications.

Index

Designing Knowledge Organizations: A Pathway to Innovation Leadership, First Edition.
Joseph Morabito, Ira Sack, and Anilkumar Bhate.

w

y

Made in the USA
Columbia, SC
17 September 2026

2adc7eb8-9b0b-4e18-8cfa-db763c6d7f8eR02